T0211198

This graduate/research-level text describes in a unified fashion the statistical mechanics of random walks, random surfaces and random higher-dimensional manifolds with an emphasis on the geometrical aspects of the theory and applications to the quantization of strings, gravity and topological field theory.

With chapters on random walks, random surfaces, two- and higher-dimensional quantum gravity, topological quantum field theories and Monte Carlo simulations of random geometries, the text provides a self-contained account of quantum geometry from a statistical field theory point of view. The approach uses discrete approximations and develops analytical and numerical tools. Continuum physics is recovered through scaling limits at phase transition points and the relation to conformal quantum field theories coupled to quantum gravity is described. The most important numerical work is covered, although the main aim is to develop mathematically precise results that have wide applications. Many diagrams and references are included.

This book will be of interest to graduate students and researchers in theoretical and statistical physics, and in mathematics.

CAMBRIDGE MONOGRAPHS ON MATHEMATICAL PHYSICS

J. Ambjørn, B. Durhuus and T. Jonsson *Quantum Geometry: A Statistical Field Theory Approach*

A. M. Anile *Relative Fluids and Magneto-Fluids*

J. A. de Azcárraga and J. M. Izquierdo *Lie Groups, Lie Algebras, Cohomology and Some Applications in Physics*

J. Bernstein *Kinetic Theory in the Early Universe*

G. F. Bertsch and R. A. Broglia *Oscillations in Finite Quantum Systems*

N. D. Birrell and P. C. W. Davies *Quantum Fields in Curved Space*[†]

D. M. Brink *Semiclassical Methods in Nucleus-Nucleus Scattering*

J. C. Collins *Renormalization*[†]

P. D. B. Collins *An Introduction to Regge Theory and High Energy Physics*

M. Creutz *Quarks, Gluons and Lattices*[†]

F. de Felice and C. J. S. Clarke *Relativity on Curved Manifolds*[†]

B. De Witt *Supermanifolds, 2nd edition*[†]

P. G. O. Freund *Introduction to Supersymmetry*[†]

F. G. Friedlander *The Wave Equation on a Curved Space-Time*

J. Fuchs *Affine Lie Algebras and Quantum Groups*[†]

J. A. H. Futterman, F. A. Handler and R. A. Matzner *Scattering from Black Holes*

M. Göckeler and T. Schücker *Differential Geometry, Gauge Theories and Gravity*[†]

C. Gómez, M. Ruiz Altaba and G. Sierra *Quantum Groups in Two-dimensional Physics*

M. B. Green, J. H. Schwarz and E. Witten *Superstring Theory, volume 1:* INTRODUCTION[†]

M. B. Green, J. H. Schwarz and E. Witten *Superstring Theory, volume 2:* LOOP AMPLITUDES, ANOMALIES AND PHENOMENOLOGY[†]

S. W. Hawking and G. F. R. Ellis *The Large-Scale Structure of Space-Time*[†]

F. Iachello and A. Arima *The Interacting Boson Model*[†]

F. Iachello and P. van Isacker *The Interacting Boson–Fermion Model*

C. Itzykson and J.-M. Drouffe *Statistical Field Theory, volume 1:* FROM BROWNIAN MOTION TO RENORMALIZATION AND LATTICE GAUGE THEORY[†]

C. Itzykson and J.-M. Drouffe *Statistical Field Theory, volume 2:* STRONG COUPLING, MONTE CARLO METHODS, CONFORMAL FIELD THEORY, AND RANDOM SYSTEMS[†]

J. I. Kapusta *Finite-Temperature Field Theory*[†]

V. E. Korepin, A. G. Izergin and N. M. Boguliubov *The Quantum Inverse Scattering Method and Correlation Functions*

D. Kramer, H. Stephani, M. A. H. MacCallum and E. Herlt *Exact Solutions of Einstein's Field Equations*

N. H. March *Liquid Metals: Concepts and Theory*

I. M. Montvay and G. Münster *Quantum Fields on a Lattice*

L. O'Raifeartaigh *Group Structure of Gauge Theories*[†]

A. Ozorio de Almeida *Hamiltonian Systems: Chaos and Quantization*[†]

R. Penrose and W. Rindler *Spinors and Space-time, volume 1:* TWO-SPINOR CALCULUS AND RELATIVISTIC FIELDS[†]

R. Penrose and W. Rindler *Spinors and Space-time, volume 2:* SPINOR AND TWISTOR METHODS IN SPACE-TIME GEOMETRY[†]

S. Pokorski *Gauge Field Theories*[†]

V. N. Popov *Functional Integrals and Collective Excitations*[†]

R. Rivers *Path Integral Methods in Quantum Field Theory*[†]

R. G. Roberts *The Structure of the Proton*[†]

J. M. Stewart *Advanced General Relativity*

A. Vilenkin and E. P. S. Shellard *Cosmic Strings and other Topological Defects*

R. S. Ward, R. O. Wells Jr *Twister Geometry and Field Theories*[†]

[†] Issued as paperback

CAMBRIDGE MONOGRAPHS ON MATHEMATICAL PHYSICS

General editors: P. V. Landshoff, D. R. Nelson, D. W. Sciama, S. Weinberg

QUANTUM GEOMETRY

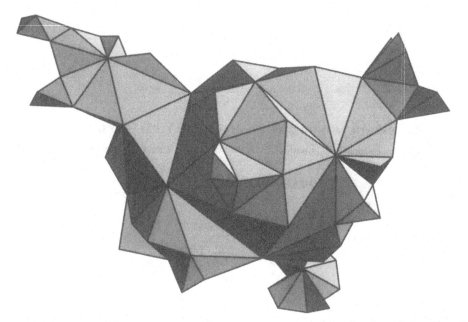

A generic universe in two-dimensional
gravity, generated by the computer
algorithms described in Chapter 5,
and embedded in three-dimensional
space for illustrative purposes.

Quantum Geometry

A statistical field theory approach

JAN AMBJØRN
BERGFINNUR DURHUUS
University of Copenhagen

THORDUR JONSSON
University of Iceland

CAMBRIDGE
UNIVERSITY PRESS

CAMBRIDGE UNIVERSITY PRESS
Cambridge, New York, Melbourne, Madrid, Cape Town, Singapore,
São Paulo, Delhi, Dubai, Tokyo, Mexico City

Cambridge University Press
The Edinburgh Building, Cambridge CB2 8RU, UK

Published in the United States of America by Cambridge University Press, New York

www.cambridge.org
Information on this title: www.cambridge.org/9780521461672

© Cambridge University Press 1997

This publication is in copyright. Subject to statutory exception
and to the provisions of relevant collective licensing agreements,
no reproduction of any part may take place without the written
permission of Cambridge University Press.

First published 1997

A catalogue record for this publication is available from the British Library

Library of Congress Cataloguing in Publication Data

Ambjørn, Jan.
Quantum geometry: a statistical field theory / Jan Ambjørn
and Bergfinnur Durhuus, Thordur Jonsson.
p.   cm. – (Cambridge monographs on mathematical physics)
Includes bibliographical references and index.
ISBN 0 521 46167 7 (hc)
1. Quantum gravity. 2. Quantum field theory.
I. Durhuus, Bergfinnur J. II. Thordur Jonsson. III. Title. IV. Series
QC174.17.G46A43   1997
530.1'43–dc20   956-9418   CIP

ISBN 978-0-521-46167-2 Hardback
ISBN 978-0-521-01736-7 Paperback

Cambridge University Press has no responsibility for the persistence or
accuracy of URLs for external or third-party internet websites referred to in
this publication, and does not guarantee that any content on such websites is,
or will remain, accurate or appropriate. Information regarding prices, travel
timetables, and other factual information given in this work are correct at
the time of first printing but Cambridge University Press does not guarantee
the accuracy of such information thereafter.

Contents

Preface

The main topic of this book is the work that has been carried out during the last 15 years under the general heading of *random surfaces*. The original motivation for the study of random surfaces came from lattice gauge theory, where one can represent various quantities of interest as weighted sums over surfaces embedded in a hypercubic lattice. A few years later, with the resurrection of string theory, random surfaces were used as regularization of that theory and, most recently, random surface models have been applied to two-dimensional quantum gravity. There is also an impressive body of work on random surfaces that has been carried out by membrane physicists, as well as by condensed matter physicists, so one often finds mathematically identical problems being studied in different branches of physics. Random surfaces are therefore not a physical theory but, rather, a theoretical tool and a methodology that can be applied to various physical problems in the same way as random walks find applications in many branches of science. The formalism that has been developed to deal with random surfaces carries over to the study of higher-dimensional manifolds, which are important for quantizing gravity in higher dimensions.

We address this book primarily to advanced graduate students in theoretical physics but we hope that more experienced researchers in the field, as well as mathematicians, may find it useful. As far as applications are concerned, our point of view is very much biased towards string theory and quantum gravity, where the geometric point of view is most important and powerful. This perhaps justifies the book's title.

The purpose of writing this book was not to provide an encyclopaedic account of all the work on random geometry; rather, we have chosen to focus on mathematically precise results that have wide applications, as well as on key results from numerical simulations which have played an important role in guiding the development of the theory. We do not provide complete proofs of the statements in the text where their content is not illuminating or where they are long and tedious, but try to warn the reader when the discussion becomes conjectural.

At the end of most chapters are notes that are intended to guide the reader to the original literature, as well as review articles and extensions of the discussion in the main text. There is no doubt that we have left out many important papers and we apologize in advance for all those omissions.

Important parts of the work reported in this book have been carried out in collaboration with our friends and colleagues A. Beliakova, D. Boulatov, J. Burda, M. Carfora, L. Chekhov, J. Fröhlich, J. Greensite, H. P. Jakobsen, J. Jurkiewicz, V. A. Kazakov, C. F. Kristjansen, Y. M. Makeenko, A. Marzuoli, R. Nest, P. Orland, B. Petersson, G. Thorleifsson, S. Varsted, Y. Watabiki and J. Wheater. We are indebted to J. Jurkiewicz, C. F. Kristjansen and Y. M. Makeenko for reading parts of the first draft of the book and pointing out many things that could be improved. M. Lund and G. Xander gave very valuable assistance with the diagrams, the references and the index.

Jan Ambjørn
Bergfinnur Durhuus
Thordur Jonsson

Notation

The purpose of this note is to explain a few notational conventions that we have used throughout the text, some of which may not be completely standard.

If f is a function of a positive real variable x, then the equation

$$f(x) = O(x^n)$$

means that there is a constant $C > 0$ such that

$$f(x) \leq Cx^n$$

for all values of x or for x in an asymptotic region, i.e. for either large or small x, depending on the context.

If f is as above, then the equation

$$f(x) = o(x^n)$$

means that

$$\lim_{x \to 0} \frac{f(x)}{x^n} = 0.$$

We use $O(x, y)$ as shorthand notation for $O(x) + O(y)$, and similarly for $o(x, y)$.

Finally, let f and g be functions of a real or complex variable x. We employ the notation

$$f(x) \sim g(x)$$

to mean one of three things: which one we have in mind is either explained explicitly in the text or it is supposed to be clear from the context. The first meaning is that the functions f and g are asymptotic as x approaches some limiting value x_0, i.e.

$$\lim_{x \to x_0} \frac{f(x)}{g(x)} = C,$$

where C is a non-zero constant. It should always be clear from the context what x_0 is. In particular, x_0 can be infinite. The second meaning is that

$$\lim_{x \to x_0} \frac{\log f(x)}{\log g(x)} = C.$$

This case arises in particular when we wish to say that two functions have the same exponential decay. The third meaning is that f and g have a singularity of the same kind at the limiting point x_0. For example, if f and g are real analytic for $x > x_0$ and the nth derivative of f, $f^{(n)}$, is the lowest derivative of f which does not have a finite limit as $x \downarrow x_0$, then $f(x) \sim g(x)$ means that the limit

$$\lim_{x \to x_0} \frac{f^{(n)}(x)}{g^{(n)}(x)}$$

exists and is non-zero.

1
Introduction

The idea of describing the physical world entirely in terms of geometry has a history dating back to Einstein and Klein in the early decades of the century. This approach to physics had early success in general relativity but the appearance of quantum mechanics guided the development of theoretical physics in a different direction for a long time. During the past quarter of a century the programme of Einstein and Klein has seen a renaissance embodied in gauge theories and, more recently, superstring theory. During this time we have also witnessed the happy marriage of statistical mechanics and quantum field theory in the subject of Euclidean quantum field theory, a development which could hardly have taken place without Feynman's path integral formulation of quantization. In this book we shall work almost exclusively in the Euclidean framework.

The unifying theme of the present work is the study of quantum field theories which have a natural representation as functional integrals or, if one prefers, statistical sums, over geometric objects: paths, surfaces and higher-dimensional manifolds. Our philosophy is to work directly with the geometry as far as possible and avoid parametrization and discretizations that break the natural invariances. In this introductory chapter we give an overview of the subject, put it in perspective and discuss its main ideas briefly.

Lagrangian field theories whose action can be expressed entirely in terms of geometrical quantities such as volume and curvature have a special beauty and simplicity. The simplest example is the theory of a free relativistic particle moving in $\mathbb{R}^d$. If the particle moves from a point x to a point y in $\mathbb{R}^d$ along a path ω, we associate with this movement the action

$$S(\omega) = m \int_\omega dl, \qquad (1.1)$$

where dl is the length element along the path and m is the mass of the

1

particle. The action is proportional to the length of the path ω and the equation of motion of the particle is given by the variational equation for the action given by Eq. (1.1). In order to use the action for analytic calculations it is necessary to introduce a parametrization of the path ω, and once this is done one can analyse the equation of motion. The extremum of $S(\omega)$ on the set of paths from x to y is of course given by the straight line from x to y. Although the equations of motion have straight line solutions they do not appear in a particularly simple form, the reason being the *reparametrization invariance* of the action (1.1), i.e. $S(\omega)$ is independent of the parametrization of ω and only depends on its diffeomorphism class, which we shall refer to as the *geometric path* from x to y. The equation of motion reflects this invariance and any parametrization of the straight line from x to y yields a solution. A virtue, the extreme simplicity of the action in terms of geometry, turns into a slight nuisance when it comes to doing calculations.

Quantum theory presents a similar picture. The first quantization of the action (1.1) is in principle straightforward if we follow Feynman's path integral prescription. In order to construct the propagator from x to y we integrate over all paths from x to y with the appropriate weight. By "all paths" here we mean all geometric paths. The formal expression for the propagator is usually written as

$$G(x, y) = \int_{(x,y)} \mathcal{D}\omega \, e^{-S(\omega)} \tag{1.2}$$

in Euclidean space, where the suffix (x, y) indicates integration over paths from x to y. We are faced with complications when we try to make sense of this integral over paths. If a parametrization is chosen one must ensure that each geometric path is counted only once. If it is technically inconvenient to choose a parametrization and if the integration is performed over all parametrized paths from x to y, one must divide out the over-counting. This means that one should formally divide by the "volume of the diffeomorphism group".

The geometric significance of the action (1.1) allows immediate generalizations to objects of higher dimension than the one-dimensional paths. In particular, we can define the action of a surface σ embedded in $\mathbb{R}^d$, with a boundary consisting of two loops γ_1 and γ_2, by

$$S(\sigma) = \int_\sigma dA, \tag{1.3}$$

where dA is the area element of the surface induced by the metric in $\mathbb{R}^d$. In the same way as ω was viewed the (Euclidean) world line of a particle travelling from a space-time point x to another space-time point y, the surface σ may be viewed as the world-sheet swept out by a closed string which moves from an initial configuration γ_1 to a final configuration

γ_2. The action (1.3) is the simplest geometric action for the motion of such a string and it is obviously independent of any parametrization of the surface σ. First quantization by a path integral is again simple in principle. We obtain the propagator $G(\gamma_1, \gamma_2)$ by integrating over all *geometric surfaces* (i.e. diffeomorphism classes of surfaces) with boundary $\gamma_1 \cup \gamma_2$:

$$G(\gamma_1, \gamma_2) = \int_{(\gamma_1, \gamma_2)} \mathscr{D}\sigma \, e^{-S(\sigma)}. \qquad (1.4)$$

We encounter technical problems in trying to give a meaning to this integral that are considerably worse than in the case of Eq. (1.2) since the invariance under diffeomorphisms imposes much stronger constraints in the two-dimensional case.

The action functionals given by Eqs. (1.1) and (1.3) are defined on geometric objects (paths and surfaces) embedded in $\mathbb{R}^d$. In general relativity the action only refers to the *intrinsic geometry* of space-time since space-time is not embedded in an ambient space. The Einstein–Hilbert action for the gravitational field (the Riemannian metric) g_{ab} on a d-dimensional space-time manifold M is given by

$$S([g_{ab}]) = \frac{1}{16\pi G} \int_M dV \, (2\Lambda - R), \qquad (1.5)$$

where G is Newton's gravitational constant, Λ is the cosmological constant, R is the scalar curvature of the metric g_{ab} and $dV = d^d\xi \sqrt{g(\xi)}$ is the volume element on M. Here g denotes the determinant of the metric tensor g_{ab}. The above action is diffeomorphism invariant, i.e. it is not a function of the metric g_{ab}, but only depends on the equivalence class of g_{ab} under diffeomorphisms which we denote by $[g_{ab}]$. Two metrics $g_{ab}(\xi)$ and $g'_{a'b'}(\xi')$ are equivalent if there exists a diffeomorphism $\xi \to \xi'$ of M onto itself which transforms $g_{ab}(\xi)$ to $g'_{a'b'}(\xi')$. In Eq. (1.5) we view the equivalence classes of metrics as the natural variable. One can find other actions, expressed in terms of other variables, which are equivalent to Eq. (1.5) at the classical level in the sense that they yield the same classical equations of motion as one obtains from the Einstein–Hilbert action. For example, one can choose the so-called viel-bein e^α_a and the spin connection ω^α_{ab} as the independent variables. The invariance group is even larger in this case and, again, the dynamical variables are equivalence classes under the action of the invariance group. Such an enlargement of the class of dynamical variables is often useful and offers alternative approaches to quantization.

An important example of this, relating to the action (1.3), is the introduction of an intrinsic geometry on the surface σ, independent of its embedding in $\mathbb{R}^d$. This geometric structure is described by a Riemannian metric g_{ab} on a fixed parameter surface M. Letting

$$X : M \mapsto \mathbb{R}^d \tag{1.6}$$

denote the embedding of σ in $\mathbb{R}^d$ we define the action of the pair (X, g_{ab}) by

$$S([X, g_{ab}]) = \int_M dV \left(\Lambda + \frac{1}{2} g^{ab}(\xi) \frac{\partial x_\mu}{\partial \xi^a} \frac{\partial x_\mu}{\partial \xi^b} \right), \tag{1.7}$$

where the x_μ are the coordinate functions of X, we sum over repeated indices and the volume element $dV = d^2\xi \sqrt{g(\xi)}$ is given by the intrinsic metric. In varying the action (1.7) we keep the boundary loops γ_1 and γ_2 fixed in $\mathbb{R}^d$. Treating both the embedding X and the intrinsic metric g_{ab} as dynamical variables, the equations of motion derived from Eq. (1.7) will coincide with the variational equation derived from Eq. (1.3) if we use the equation derived by varying with respect to g_{ab} to express g_{ab} in terms of X.

The action (1.7) also has an interpretation as the action of two-dimensional quantum gravity coupled to d matter fields, since the integral of the scalar curvature R in Eq. (1.5) is a topological invariant in two dimensions and plays no dynamical role. We therefore see that an enlargement of the set of dynamical variables can make intrinsic geometry relevant for the description of strings and particles. Similar remarks apply to paths which can be endowed with an intrinsic metric.

The identification of Euclidean quantum field theory with classical statistical mechanics has allowed us to use results from the theory of critical phenomena to improve our understanding of renormalization. The concepts of universality classes, critical points, marginal operators, etc., have made many aspects of the renormalization procedure and the renormalization group equations much more transparent, while the machinery of Feynman graphs and functional integrals has influenced the theory of critical phenomena. In two dimensions a fascinating class of conformally invariant theories has been discovered. This development has important repercussions for our discussion.

In connecting statistical mechanical systems and Euclidean quantum field theory, the discretization of space is often a key ingredient. By discretizing ordinary space and restricting the volume to be finite we approximate the field theoretical problem by a finite-dimensional problem, which can often be viewed as a generalized lattice spin problem. In the infinite volume limit we can look for phase transitions of the spin system, and if these are characterized by divergent correlation lengths, it is possible to construct scaling or continuum limits. In such a limit the ratio between the lattice spacing and the correlation length goes to zero and one often recovers a Euclidean invariant quantum field theory. The renormalized masses and coupling constants are defined not at the critical point but by the approach to it in the scaling limit.

In the process described above, space only plays a spectator role. If we do not demand any local invariance, space can usually be discretized in a straightforward way. Lattice observables which break Euclidean invariance explicitly will normally be suppressed in the scaling limit. On the other hand, local invariance usually mixes high and low frequencies and is much more difficult to discretize in a simple way. This is often used as an argument against attempts to discretize geometric theories. It is presumably correct that it is in general rather fruitless to discretize a fixed parametrization of geometrical objects. It is more natural to discretize the geometric objects directly and we shall see that this works beautifully in general. As an example consider the action (1.1) which is defined on all piecewise smooth paths. Let us denote the infinite-dimensional space of all continuous paths from x to y by $\tilde{\Omega}(x, y)$. The class of piecewise linear paths from x to y, where each linear part has a fixed length a, provides us with an approximation to $\tilde{\Omega}(x, y)$ which is a countable union of finite-dimensional spaces, i.e. the spaces of paths with a fixed number of steps. On these finite-dimensional spaces we can define the path integral in Eq. (1.2) rigorously and sum over the number of steps. It is thus natural to view a as a cutoff in $\tilde{\Omega}(x, y)$ and it is obviously reparametrization invariant. This allows us to define the measure $\mathcal{D}\omega\, e^{-S(\omega)}$ on $\tilde{\Omega}(x, y)$ as a limit of well-defined measures, and show that the path integral (1.2) indeed gives the propagator for a free relativistic particle of mass m.

In Chapter 2 we consider the construction of the path integral along the lines described here in considerable detail. Much of this theory is classical and has its origin in the work of Wiener in the 1930s but we present it from our point of view and set the stage for analogous constructions in later chapters where a host of new complications arises and where we are forced to relax the mathematical rigour at times. Random paths have certainly inspired the study of higher-dimensional random objects and they also serve as an ideal laboratory to test methods and ideas that one would like to apply in the higher-dimensional cases. We shall see that the statistical theory of critical phenomena fits perfectly in this context, even though we deal with a reparametrization invariant theory. The continuum limit is constructed at a critical point; the mass (inverse correlation length) of the theory is determined by the approach to the critical point; critical exponents are calculable and have either a simple geometric meaning or correspond, in field theoretical terms, to wave function renormalization. Furthermore, we observe *universality*: it is not important which class of piecewise linear paths we work with, and a large class of discretized actions yield the same continuum theory when the cutoff is removed. The mathematical reason for universality is particularly transparent in the study of random paths: it boils down to the central limit theorem and its generalizations.

It is natural to try to define reparametrization invariant measures on surfaces and higher-dimensional manifolds in analogy with the measures we find for random paths. The actions are well defined for piecewise smooth surfaces and manifolds and, therefore, in particular on piecewise linear surfaces and manifolds. This it obvious for the action (1.3), and the Einstein–Hilbert action has a natural definition on piecewise linear manifolds, often referred to as *Regge calculus*. The strategy is to introduce a cutoff a directly on the piecewise linear manifolds, as for the paths. In this way one obtains an approximation to the space of continuous surfaces or manifolds on which we can define the functional integral in a rigorous way such that the cutoff is, by definition, reparametrization invariant. There are many different ways to implement this programme. The approach we use here is to introduce a *fundamental building block* with side lengths a from which the piecewise linear manifolds are constructed. Random walks are constructed step by step in this way and the construction is simple because the steps are well ordered and come one after the other. In higher dimensions one can imagine a process which adds or deletes fundamental building blocks at random in such a way that the manifolds stay in the class under study (i.e. have, for example, a fixed boundary or topology). In computer simulations this is precisely the way the manifolds are simulated, as we shall describe in Chapter 5. In the case of surfaces embedded in a hyper-cubic lattice the building blocks are usually taken to be elementary plaquettes (minimal lattice squares). In d-dimensional gravity the natural building blocks are d-dimensional simplexes which one glues together to obtain d-dimensional piecewise linear manifolds. In this formalism the functional integral over equivalence classes of metrics will, for a finite volume, be replaced by a finite sum over piecewise linear manifolds which can be constructed by gluing together d-dimensional equilateral simplexes of volume proportional to a^d. However, since we study an ensemble of manifolds and ask questions of a statistical nature, we need not worry about whether the manifolds come about by a stochastic process.

In general it is not yet possible to control the scaling limit $a \to 0$ in the same way as for paths, but in the case of two-dimensional gravity, i.e. for the action (1.5) in two dimensions, we construct the limit $a \to 0$ and argue that the limiting Green functions are identical to Green functions calculated by formal continuum manipulations. This will be discussed in Chapter 4, where we also see that two-dimensional gravity fits into the framework of critical phenomena. We verify the universality of the continuum limit and, in particular, of the critical exponents which either determine the fractal geometry of the ensemble of manifolds (e.g. the Hausdorff dimension) or the scaling properties of the continuum Green functions. The solution extends to a class of theories which can be regarded as two-dimensional quantum gravity interacting with matter fields.

Lattice theories have been used successfully as regularizations of continuum field theories, but often formal continuum calculations have proved more effective in producing new exact results. In the case of two-dimensional gravity the situation is the opposite. The discretized models allow us to evaluate observables in two-dimensional gravity which at present seem out of reach using continuum methods. We believe that the combinatorial nature of the solution of two-dimensional gravity is responsible for the power of the discretized approach, and our exposition of the topic will stress this point.

In theories defined by a geometric action a natural set of quantities to study are functional integrals over classes of manifolds with fixed boundaries and perhaps fixed fields like metrics on the boundaries. Simple examples of this kind are the propagators given by Eqs. (1.2) and (1.4). In quantum gravity one defines in general such Green functions (often called Hartle–Hawking wave functionals) as

$$G(B, [h_{cd}]) = \int_{(B)} \mathscr{D}[g_{ab}] \, e^{-S([g_{ab}])}, \qquad (1.8)$$

where we integrate over the equivalence classes of metrics on a manifold M with boundary B (which may have several components) and the metrics are required to equal a fixed metric h_{cd} on the boundary, up to equivalence, i.e. $[g_{ab}|_B] = [h_{cd}]$. The action $S([g_{ab}])$ is given by Eq. (1.5) plus boundary terms which we do not write down here. Since we integrate over all equivalence classes of metrics the only link of the metrics to $G(B, [h_{cd}])$ is the boundary condition that h_{cd} should be equivalent to the metric induced by g_{ab} on B. In two dimensions this link is so weak that the major parts of the Green functions $G(B, [h_{cd}])$ have an equivalent description in terms of a *topological theory*, i.e. a theory which only depends on the diffeomorphism class of the manifold M and is independent of any metric structure. We show this by an explicit calculation in Chapter 4.

Three-dimensional quantum gravity might also be solvable by purely combinatorial methods. In order to see this it is convenient to use a different starting point from the Einstein–Hilbert action and replace it by a different action where the dynamical variables are the drei-bein and the spin connection. In terms of these variables the action of three-dimensional gravity can be written as a Chern–Simons gauge theory action. This theory is topological in nature and is related to the so-called Turaev–Viro theory, which is a topological field theory defined on simplicial manifolds and can in a certain sense be regarded as a regularization of three-dimensional gravity. These topics are discussed in Chapter 7.

In Chapter 6 we consider the discretized versions of the Einstein–

Hilbert theory in dimensions higher than two. Almost no analytic tools are presently available to study these theories, but it is possible to explore the phase diagram using Monte Carlo simulations. We outline the present understanding of these theories, which are still in their infancy, and discuss some problems that arise in numerical simulations due to the fact that there exist four- and higher-dimensional manifolds which are not algorithmically recognizable. We explain the principles behind Monte Carlo simulations of geometric theories in Chapter 5.

As mentioned, the string or *random surface* theory with the action (1.7) can be viewed as d Gaussian fields coupled to two-dimensional quantum gravity. It is well known that d Gaussian fields in flat two-dimensional space represent the simplest example of a conformal field theory with central charge d. Unfortunately, the solvability of two-dimensional matter coupled to quantum gravity is restricted to theories with central charge $c \leq 1$. In Chapter 3 we study discretizations of Eqs. (1.3) and (1.7) and present some of the results which can be rigorously established. One can show that the theory has a critical point where one can hope to be able to construct a scaling limit. Unfortunately, the *string tension* does not scale in the expected way, and we find strong indications that the scaling limit degenerates into a theory of so-called *branched polymers*, i.e. the typical surfaces are tree-like thin structures. Branched polymers will play an important role in the following. The reason is that branched polymer-like structures seem to be generic in most higher-dimensional geometric quantum theories for entropical reasons: they are numerous, and if we want to suppress them we have to choose our action and coupling constants with care. This issue will be discussed in detail in Chapter 3, where we prove that surfaces embedded in a hyper-cubic lattice are dominated by branched polymers. Branched polymers are studied in their own right at the end of Chapter 2 as a generalization of the theory of random paths.

The fact that we do not find any non-trivial limit of the simplest discretization of a bosonic string theory is consistent with continuum results stating that the bosonic string has a tachyon in the spectrum for $d > 1$. Since the statistical model of random surfaces is perfectly well defined and reflection positive there cannot be a tachyon in its spectrum, so one may perhaps view the branched polymers as a healthy theory's version of the tachyon.

Our discussion has so far been focused on the basic geometric actions, given by Eqs. (1.1), (1.3), (1.5) and (1.7), and how one might go about defining in a constructive way the associated reparametrization invariant measures on the corresponding infinite-dimensional spaces of continuous paths, surfaces, etc. It is of course of interest to study more elaborate geometric theories which are typically obtained by adding new terms to the basic geometric action functionals. The actions for such theories usually contain higher-derivative curvature terms which also appear naturally in effective models of polymers, membranes, interfaces and other phenomena studied in condensed matter physics and biophysics. These models often exhibit phase transitions (experimentally or numerically) as one varies the coupling constants associated with the new terms in the action. Often we see a phase dominated by wildly fluctuating geometry and others with flat or at least smoother geometry. The smoothing is described by saying that the generic paths, surfaces, etc., become *rigid*.

From a purely field theoretical point of view models of this type are also of interest. Any model of strings, such as the Nielsen–Olesen string or effective *QCD* strings, will be models with an action depending on the *extrinsic curvature*,[*] and their properties are likely to be quite different from strings described by the action (1.3). But even for string theories of elementary particles the extrinsic curvature actions are relevant. Superstrings have both bosonic and fermionic degrees of freedom, and if one integrates out the fermions the effective bosonic string theory has an action with extrinsic curvature terms.

For these reasons it is important to study the geometric theories with extrinsic curvature and other higher-derivative terms of a geometric nature. The simplest such theory describes the propagation of a point particle where an extrinsic curvature term has been added to the action (1.1), i.e.

$$S(\omega) = m \int_\omega dl + \lambda \int_\omega dl \, k, \qquad (1.9)$$

where k denotes the extrinsic curvature of the path ω. In principle we can add higher powers of the curvature as well as torsion terms to the action, but the term in Eq. (1.9) is singled out as being scale invariant, i.e. the coupling constant λ is dimensionless. In the language of the theory of critical phenomena the term is a *marginal* perturbation, which might or might not be important in the scaling limit, while one would naively expect the higher-derivative terms to be irrelevant. We shall study this theory in detail in Chapter 2 since it is solvable: there is a non-trivial scaling limit of the theory, different from the one found for the free particle. This scaling limit corresponds to a different universality class with paths of fractal dimension one, contrary to the ordinary random walk paths, which have fractal dimension two.

In the case of random surfaces the generalization of Eq. (1.3) analogous

---

[*] For surfaces the extrinsic curvature is the average of the inverse principal radii of curvature, while the intrinsic curvature is their product.

to Eq. (1.9) is given by

$$S(\sigma) = \mu \int_\sigma dA + \lambda \int_\sigma dA\, H^2, \qquad (1.10)$$

where H is the extrinsic curvature of the surface embedded in $\mathbb{R}^d$. The second term on the right-hand side of Eq. (1.10) is again singled out as being scale invariant, and it might or might not be important in the scaling limit, presumably depending on the scaling of λ. Unfortunately, this model cannot be solved analytically. Numerical simulations of the discretized version of Eq. (1.10) are consistent with a transition at a finite value for λ to a new phase of rigid, flat surfaces. If there is a transition, we have the possibility of constructing a scaling limit of the discretized version of Eq. (1.10) at the transition point. The corresponding continuum theory is seemingly characterized by a non-vanishing finite string tension and a non-vanishing mass. It suggests that here we may have a new kind of string theory, but so far the theoretical underpinnings are missing. We discuss this scenario in Chapter 3.

There are many topics in the theory of random manifolds which we have completely left out of this book. Among the most important ones are self-avoiding manifolds, which cannot be described in terms of a local action on the manifolds and therefore fall out of the framework we work in. In this area there has been considerable progress in recent years which is briefly discussed in the notes to Chapter 3. Another subject missing from the present work is the theory of quantized black holes which has bloomed in the past few years but has so far not found its proper place in the discretized theories. Also, we have not systematically explored the connections between the theory of quantum gravity described here and the many other promising approaches to the subject. The many conceptual problems in quantum gravity are not touched upon at all.

In the preceding description we have focused on the geometrical aspects of the theory of random surfaces and in particular on the application to string theory and quantum gravity. We would like to conclude this introduction by emphasizing the importance of random surfaces as a universal theoretical tool with potential applications in almost any branch of physics where fluctuating two-dimensional objects arise either directly or in mathematical representations. The systematic study of random walks began less than a century ago and this theory has by now become a huge mathematical subject with applications not only in physics but in almost every branch of science. Considering the variety of subjects that already use random surfaces in constructing mathematical models of the objects under investigations there is every reason to believe that the study of random surfaces and random higher-dimensional manifolds will become equally important in the future.

2
Random walks

The role of random walks as a theoretical tool in physics dates back at least to the explanation of the origin of Brownian motion at the turn of the century. The universal features of the large-scale phenomena associated with random walks are already transparent in Einstein's derivation of the diffusion equation in his famous 1905 paper on Brownian motion. Ever since, theories of random walks have played an increasingly important role in virtually every branch of physics and now form the basis of statistical theory in general in the subject of stochastic processes. The appearance of random walks in elementary particle physics was mediated by Feynman's path integral formulation of quantum theory, and a mathematically rigorous approach to the subject was made possible by the introduction of Euclidean quantum field theory in the 1960s building on Wieners's earlier work on random walks and the diffusion equation.

We start in the next three sections by introducing various discrete random walk models describing the propagation of scalar particles in space-time. We pay particular attention to critical behaviour in these models and their universality properties: many different discrete models give rise to the same continuum limit. As explained in the Introduction we are primarily interested in viewing random paths as geometric objects. Hence, we focus on such aspects as reparametrization invariance which are usually not stressed in the more standard treatments. In Section 4 we consider from this point of view a random walk model for the propagator of spin-$\frac{1}{2}$ particles. In the final section of this chapter we consider models of branched polymers which are the simplest natural generalizations of random paths and turn out to be important for the study of random surfaces and higher-dimensional random manifolds in subsequent chapters. Branched polymer-like structures appear generically in these more complicated models.

2.1 Parametrized random walks

2.1.1 *The Wiener measure*

We begin by reviewing the construction of the most basic measure on the space of paths. Let Δ denote the Laplace operator in $\mathbb{R}^d$. The solution to the diffusion (or heat) equation in $\mathbb{R}^d$

$$\frac{\partial \varphi}{\partial t} = \frac{1}{2}\Delta\varphi, \tag{2.1}$$

with the initial condition $\varphi(x, 0) = \varphi_0(x)$ is given by

$$\varphi(y, t) = \frac{1}{(2\pi t)^{d/2}} \int_{\mathbb{R}^d} dx \, e^{-\frac{|x-y|^2}{2t}} \varphi_0(x).$$

The function $\varphi_0(x)$ is interpreted as the initial distribution of particles at time $t = 0$, and $|x - y|$ denotes the Euclidean distance between x and y in $\mathbb{R}^d$.

Thus the kernel $K_t(x, y)$ of the operator $e^{\frac{t}{2}\Delta}$, called the *heat kernel*, is given by

$$K_t(x, y) = \frac{1}{(2\pi t)^{d/2}} e^{-\frac{|x-y|^2}{2t}}, \tag{2.2}$$

and represents the probability density of finding a particle at y at time t given its location x at time 0. From the semigroup property

$$e^{(t+s)\Delta} = e^{t\Delta} e^{s\Delta},$$

for $s, t \geq 0$, we have

$$K_t(x, y) = \int dx_1 \ldots dx_{N-1} K_{t/N}(x_N, x_{N-1}) \ldots K_{t/N}(x_1, x_0) \tag{2.3}$$

for each $N \geq 1$, where we have set $x_0 = x$ and $x_N = y$.

There is an obvious one-to-one correspondence between configurations $(x_1, \ldots, x_{N-1})$ and parametrized piecewise linear paths $\omega : [0, t] \to \mathbb{R}^d$ from x to y consisting of the line segments $[x_0, x_1], [x_1, x_2], \ldots, [x_{N-1}, x_N]$, such that the segment $[x_{i-1}, x_i]$ is parametrized linearly by $s \in \left[\frac{i-1}{N}t, \frac{i}{N}t\right]$ (see Fig. 2.1). We denote the collection of all such paths by $\Omega_{N,t}(x, y)$. Hence, we may consider

$$D_t^N \omega = \left(2\pi \frac{t}{N}\right)^{-\frac{d}{2}N} dx_1 \ldots dx_{N-1}$$

as a measure on the (finite-dimensional) space $\Omega_{N,t}(x, y)$.

Noting that

$$\sum_{i=1}^{N} \frac{|x_i - x_{i-1}|^2}{t/N} = \sum_{i=1}^{N} \frac{t}{N} \left(\frac{|x_i - x_{i-1}|}{t/N}\right)^2 = \int_0^t |\dot{\omega}(s)|^2 ds,$$

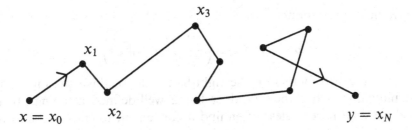

Fig. 2.1. A piecewise linear path from x to y.

where $\dot{\omega}$ is the (piecewise constant) velocity of the trajectory ω, we may rewrite Eq. (2.3) as

$$K_t(x, y) = \int_{(x,y)} D_t^N \omega \, \exp\left(-\frac{1}{2}\int_0^t |\dot{\omega}(s)|^2 ds\right), \qquad (2.4)$$

where the suffix (x, y) indicates that paths are restricted to go from x to y. *We refer to Eq. (2.4) as a random walk representation of $K_t(x, y)$ on $\Omega_{N,t}(x, y)$.*

More generally, given an action functional S on piecewise linear parametrized paths, we call the equation

$$H_t^N(x, y) = \int_{(x,y)} D_t^N \omega \, e^{-S(\omega)} \qquad (2.5)$$

a random walk representation of the kernel $H_t^N(x, y)$ on $\Omega_{N,t}(x, y)$.

The fact that the kernel $K_t(x, y)$ associated to the quadratic action

$$S_2(\omega) = \frac{1}{2}\int_0^t ds \, |\dot{\omega}(s)|^2 \qquad (2.6)$$

is independent of N implies (by Eq. (2.4)), see e.g. [85, 348], the existence of a measure $D_t\omega$ on the space $\Omega_t^0(x, y)$ of all paths $\omega : [0, t] \mapsto \mathbb{R}^d$ from x to y, such that for arbitrary $N \geq 1$ the equation

$$\int_{(x,y)} D_t^N \omega \, \exp\left(-\frac{1}{2}\int_0^t |\dot{\omega}(s)|^2 ds\right) f\left(\omega\left(\tfrac{1}{N}t\right), \ldots, \omega\left(\tfrac{N-1}{N}t\right)\right)$$

$$= \int_{(x,y)} D_t\omega \, f\left(\omega\left(\tfrac{1}{N}t\right), \ldots, \omega\left(\tfrac{N-1}{N}t\right)\right) \qquad (2.7)$$

holds for all functions f of $(N-1)d$ variables which are, say, bounded and continuous.

It turns out that the measure $D_t\omega$ is supported on the subset $\Omega_t(x, y) \subseteq \Omega_t^0(x, y)$ consisting of continuous paths $\omega : [0, t] \mapsto \mathbb{R}^d$ from x to y, whereas the subset of continuously differentiable paths from x to y has measure zero, see e.g. [348]. The measure $D_t\omega$ restricted to $\Omega_t(x, y)$ is called the *Wiener measure* on $\Omega_t(x, y)$. Formally, the Wiener measure can be written as the product

$$D_t \omega = D_t^\infty \omega \ \exp\left(-\frac{1}{2}\int_0^t |\dot\omega(s)|^2 ds\right). \tag{2.8}$$

Note that the two factors on the right-hand side of this equation do not exist individually but their product has a well-defined meaning. In the sequel we shall nevertheless often find it convenient to employ the formal symbol $\mathscr{D}\omega$ to denote an infinite product of Lebesgue measures which becomes meaningful in combination with an appropriate Boltzmann factor $e^{-S(\omega)}$.

We see that the measure $dy D_t\omega$ on the space

$$\Omega_t(x) = \bigcup_{y\in\mathbf{R}^d} \Omega_t(x,y) \tag{2.9}$$

of all continuous paths emerging from x and parametrized on $[0,t]$ is a probability measure. It is called the Wiener measure on $\Omega_t(x)$ and by abuse of notation we shall also denote this measure by $D_t\omega$. The initial point will be indicated by suffix (x) on the integral sign. Similarly, streamlining the notation and also denoting the measure $dy\, D_t^N\omega$ on the space

$$\Omega_{N,t}(x) = \bigcup_{y\in\mathbf{R}^d} \Omega_{N,t}(x,y)$$

by $D_t^N\omega$, it follows that

$$\int_{(x)} D_t^N\omega \ \exp\left(-\frac{1}{2}\int_0^t |\dot\omega(s)|^2 ds\right) f\left(\omega\left(\tfrac{1}{N}t\right),\dots,\omega\left(\tfrac{N-1}{N}t\right),\omega(t)\right)$$
$$= \int_{(x)} D_t\omega \, f\left(\omega\left(\tfrac{1}{N}t\right),\dots,\omega\left(\tfrac{N-1}{N}t\right),\omega(t)\right) \tag{2.10}$$

for all continuous bounded functions f of Nd variables.

Of course, the construction of $D_t\omega$ can be generalised to the case where a diffusion constant $c \neq 1$ is present in Eq. (2.1), in which case K_t is everywhere replaced by K_{ct}. Since K_{ct} has random walk representation

$$K_{ct}(x,y) = \int_{(x,y)} D_{ct}^N\omega \ e^{-S_2(\omega)},$$

one obtains a measure $D_t^c\omega$ on $\Omega_t(x,y)$ characterized by

$$\int_{(x,y)} D_{ct}^N\omega \ \exp\left(-\frac{1}{2}\int_0^{ct} |\dot\omega(s)|^2 ds\right) f\left(\omega\left(\tfrac{1}{N}ct\right),\dots,\omega\left(\tfrac{N-1}{N}t\right),\omega(t)\right)$$
$$= \int_{(x,y)} D_t^c\omega \, f\left(\omega\left(\tfrac{1}{N}t\right),\dots,\omega\left(\tfrac{N-1}{N}t\right),\omega(t)\right) \tag{2.11}$$

for bounded continuous functions of $(N-1)d$ variables. Naturally, we call $D_t^c\omega$ the Wiener measure on $\Omega_t(x,y)$ with diffusion constant c. As above, we likewise denote the measure $dy D_t^c \omega$ by $D_t^c \omega$ and call it the Wiener measure on $\Omega_t(x)$ with diffusion constant c.

The measures $D_t^c \omega$ can be viewed differently by considering the mapping $\varphi_{t_1,t_2} : \Omega_{t_1}(x,y) \mapsto \Omega_{t_2}(x,y)$ obtained by rescaling the variable t, i.e.

$$\varphi_{t_1,t_2}(\omega_{t_1})(s) = \omega_{t_1}\left(\frac{t_1}{t_2}s\right), \quad s \in [0,t_2] \tag{2.12}$$

for $\omega_{t_1} \in \Omega_{t_1}(x,y)$, where the $t_1, t_2 > 0$ are fixed. It then follows from the definition of $D_t^c \omega$ (or from Eq. (2.11)) that $D_t^c \omega$ is the measure obtained by transporting $D_{ct}\omega$ by the mapping $\varphi_{ct,t}$, i.e. we have

$$\int_{(x,y)} D_t^c \omega \, f(\omega) = \int_{(x,y)} D_{ct}\omega \, f(\varphi_{ct,t}(\omega)) \tag{2.13}$$

for integrable functions f on $\Omega_t(x,y)$. A corresponding statement holds where the endpoint y is not fixed.

Without entering into technicalities we mention that henceforth we consider the spaces $\Omega_t(x,y)$ and $\Omega_t(x)$ as metric spaces equipped with the uniform norm, i.e. the distance $d(\omega_1, \omega_2)$ is given by

$$d(\omega_1, \omega_2) = \sup_{s \in [0,t]} |\omega_1(s) - \omega_2(s)|. \tag{2.14}$$

Clearly, the mappings φ_{t_1,t_2} are then homeomorphisms. Moreover, the measures we consider are finite Borel measures, i.e. their domain of definition contains the open sets and they are completely determined by their values on these, or, alternatively, by the integrals of bounded continuous functions.

2.1.2 Universality of the Wiener measure

Equation (2.7), and more generally Eq. (2.11), represents our first example of a discrete approximation to the Wiener measure $D_t^c \omega$. Indeed, for large N the finite-dimensional affine subspace $\Omega_{N,t}(x,y) \subseteq \Omega_t(x,y)$ may be viewed as an approximation to $\Omega_t(x,y)$ as follows. Given $\omega \in \Omega_t(x,y)$ we let $\omega_N \in \Omega_{N,t}(x,y)$ be defined by

$$\omega_N\left(\frac{i}{N}t\right) = \omega\left(\frac{i}{N}t\right)$$

for $i = 0, 1, \ldots, N$ (see Fig. 2.2). Then $\omega_N \to \omega$ uniformly on $[0,t]$ as $N \to \infty$, and in view of Eq. (2.11) we may regard the measures

$$c^{-\frac{d}{2}} D_t^N \omega \, \exp\left(-\frac{1}{2c}\int_0^t |\dot{\omega}(s)|^2 ds\right)$$

as approximations to $D_t^c \omega$ for N large. In fact they yield identical integrals for functions $\omega \mapsto f(\omega(\frac{1}{N}t), \ldots, \omega(\frac{N-1}{N}t))$ depending only on

Fig. 2.2.   A path ω in $\Omega_t(x,y)$ and its approximation ω_N in $\Omega_{N,t}(x,y)$.

the "discretized" variable ω_N. On the other hand, the values of those integrals for arbitrary N determine the measure.

In general, given an action functional S and a measure $D\omega$ on $\Omega_t(x,y)$, we shall use the notation

$$D_t^N\omega\, e^{-S(\omega)} \to D\omega$$

if

$$\int D_t^N\omega\, e^{-S(\omega)}f(\omega) \to \int_{(x,y)} D\omega\, f(\omega) \qquad (2.15)$$

for all continuous bounded functions f on $\Omega_t(x,y)$, as $N \to \infty$. Similarly, the notion of convergence of $D_t^N\omega\, e^{-S(\omega)}$ to a measure $D\omega$ on $\Omega_t(x)$ is defined.

With this notation we may now formulate a version of a theorem by Donsker, see e.g. [348], which establishes that many different finite-dimensional measures $D_t^N\omega\, e^{-S(\omega)}$ converge to the Wiener measure $D_t^c\omega$. The phenomenon when many different discrete approximations yield the same continuum limit is often referred to as *universality* in physics.

Theorem 2.1 *Let φ be a real valued function of a real variable such that*

$$e^\rho \equiv \int_{\mathbf{R}^d} \frac{dx}{(2\pi)^{d/2}}\, e^{-\varphi(|x|)} < \infty \qquad (2.16)$$

and

$$\sigma^2 \equiv \frac{1}{e^\rho d}\int_{\mathbf{R}^d} \frac{dx}{(2\pi)^{d/2}}\, |x|^2 e^{-\varphi(|x|)} < \infty \qquad (2.17)$$

and let the action S be given by

$$S(\omega) = \sum_{i=1}^{N} \varphi\Big(\sqrt{\tfrac{N}{t}}\,\Big|\,\omega\Big(\tfrac{i}{N}t\Big) - \omega\Big(\tfrac{i-1}{N}t\Big)\Big|\Big). \qquad (2.18)$$

Then

$$D_t^N\omega\, e^{-S(\omega)-\rho N} \to D_t^{\sigma^2}\omega \qquad (2.19)$$

as $N \to \infty$.

We do not give a detailed proof of this result here: this may be found in [85]. For later reference we shall, however, indicate how the result follows from the central limit theorem. Replacing φ by $\varphi - \rho$ one may assume that $\rho = 0$. The proof then consists of reducing the statement (2.19) to the standard form of the central limit theorem as follows. One first establishes that it is sufficient to verify Eq. (2.15), with the measure $D\omega$ equal to the Wiener measure $D_t^{\sigma^2}\omega$, for functions f of the form

$$f(\omega) = F\left(\omega\left(\tfrac{1}{n}t\right),\ldots,\omega\left(\tfrac{n-1}{n}t\right),\omega(t)\right),$$

where F is a continuous bounded function of nd variables. For fixed n and F one next verifies that the N in Eq. (2.15) may be restricted to be of the form $N = kn$, i.e. one has to show that

$$\int_{(x)} D_t^{kn}\omega\, e^{-S(\omega)} f(\omega) \;\rightarrow\; \int_{(x)} D_t^{\sigma^2}\omega\, f(\omega) \qquad (2.20)$$

as $k \to \infty$. First we perform the integrations on the left-hand side of Eq. (2.20) over the variables $\omega\left(\tfrac{i}{kn}t\right)$, where $i \notin \{k, 2k, \ldots, (n-1)k, nk\}$, i.e. the arguments which do not occur in the function F, and recall that the right-hand side equals

$$\int dx_1 \ldots dx_{n-1}dy\, K_{\sigma^2 t/n}(y, x_{n-1})\ldots K_{\sigma^2 t/n}(x_1, x)f(x_1, \ldots, x_{n-1}, y)$$

where $y = \omega(t)$. Then it is sufficient to show that

$$\int_{(x)} D_t^k\omega\, e^{-S(\omega)}g(y) \;\rightarrow\; \int dy\, K_{\sigma^2 t}(x, y)g(y) \qquad (2.21)$$

for an arbitrary continuous bounded function g on $\mathbb{R}^d$, as $k \to \infty$. The convergence in Eq. (2.21) is a consequence of the central limit theorem, which states that

$$\int_{(x,y)} D_t^N\omega\, e^{-S(\omega)} \rightarrow K_{\sigma^2 t}(x, y) \qquad (2.22)$$

as $N \to \infty$ if σ^2 is defined by Eq. (2.17).

We note that the analogue of Theorem 2.1 holds for the measures $D_t^c\omega$ on $\Omega_t(x, y)$ by formally inserting for f a function of the form

$$g(\omega)\delta(\omega(t) - y),$$

where g is a function on $\Omega_t(x, y)$.

From a quantum field theoretical point of view one may regard

$$S(\omega) = \sum_{i=1}^{N} \varphi\left(Z\left|\omega\left(\tfrac{it}{N}\right) - \omega\left(\tfrac{i-1}{N}t\right)\right|\right) \qquad (2.23)$$

as the action of a lattice quantum field theory on the interval $[0, t]$ with ultraviolet cutoff $a = t/N$. We have inserted a wave-function renormalization Z, and from Eq. (2.18) and Theorem 2.1 it follows that a continuum limit exists if we choose $Z = \sqrt{1/a}$. Phrased in these terms Theorem 2.1 asserts that the quadratic action S_2 is a fixed point under the action of the renormalization group corresponding to subdivision of the interval $[0, t]$.

The importance of Eq. (2.19) is well known in non-relativistic quantum mechanics and statistical mechanics. For example, Eq. (2.19) leads in a straightforward way to the Feynman–Kac formula for transition amplitudes or correlation functions. In such applications t is a physical parameter such as a time or distance. However, in applications to relativistic quantum field theory (of scalar fields) t is an unphysical parameter which is integrated over. Thus, in particular, the relativistic propagator $G(x, y)$ of a massive scalar field in dimension d, i.e. the kernel of the operator

$$(-\Delta + m^2)^{-1} = \frac{1}{2} \int_0^\infty dt \, e^{-\frac{1}{2}t(-\Delta + m^2)}$$

is obtained as

$$
\begin{aligned}
G(x, y) &= \frac{1}{2} \int_0^\infty dt \, e^{-\frac{1}{2}m^2 t} K_t(x, y) \\
&= \frac{1}{2} \int_0^\infty dt \, e^{-\frac{1}{2}m^2 t} \int_{(x,y)} D_t \omega.
\end{aligned}
\tag{2.24}
$$

Now let

$$\Omega(x, y) = \bigcup_{t > 0} \Omega_t(x, y) \tag{2.25}$$

be the space of all continuous parametrized paths from x to y. We remark that since the spaces $\Omega_t(x, y)$ are all canonically homeomorphic (for fixed x and y) by the homeomorphisms φ_{t_1, t_2} given by Eq. (2.12), we may define a topology on $\Omega(x, y)$ by requiring the map $\Phi : \Omega(x, y) \mapsto \Omega_1(x, y) \times \mathbb{R}_+$ defined by

$$\Phi(\omega_t) = (\varphi_{t,1}(\omega_t), t), \quad \omega_t \in \Omega_t(x, y) \tag{2.26}$$

to be a homeomorphism. It is then straightforward to define a measure $D^{(m)}\omega$ on $\Omega(x, y)$ such that

$$
\begin{aligned}
\int f(\omega) D^{(m)}\omega &= \frac{1}{2} \int_0^\infty dt \, e^{-\frac{1}{2}m^2 t} \int_{(x,y)} D_t \omega \, f(\omega) \\
&= \frac{1}{2} \int_0^\infty dt \, e^{-\frac{1}{2}m^2 t} \int_{(x,y)} D_1^t \omega \, f(\varphi_{1,t}(\omega))
\end{aligned}
\tag{2.27}
$$

for any bounded continuous function f on $\Omega(x, y)$. In particular, it follows from Eq. (2.24) and Eq. (2.27) that the volume of the measure $D^{(m)}\omega$ is

$G(x, y)$, or in the language of statistical mechanics: $G(x, y)$ is the partition function for paths with fixed endpoints.

2.2 Geometric random walks

By factoring out parametrizations in the measure $D^{(m)}\omega$ we shall now see how to define a measure on geometric paths, i.e. on paths regarded as one-dimensional manifolds in $\mathbb{R}^d$, and corresponding reparametrization invariant discrete approximations. We shall sometimes refer to such a measure as a *reparametrization invariant measure* on paths.

There are several different ways to proceed. We shall discuss two below which are quite different in spirit. In the first approach the geometric paths are just one-dimensional manifolds without any additional structure whereas in the second approach the parametrization degrees of freedom are converted into intrinsic metric degrees of freedom such that the geometric paths are regarded as one-dimensional Riemannian manifolds. It may be useful to think about the former as a discretization of functional integrals of the form

$$\int \mathscr{D}[X] \, \exp\left(-m \int_I |\dot{X}| dt\right) f([X]), \qquad (2.28)$$

where the integration is over equivalence classes of paths $X : I \mapsto \mathbb{R}^d$, where two paths are declared to be equivalent if they are related by an element of the diffeomorphism group of the interval I. The latter approach corresponds to continuum functional integrals of the form

$$\int \mathscr{D}[X, e] \, \exp\left(-\int_I \left(\frac{\dot{X}^2}{2e} + \mu e\right) dt\right) f([X, e]), \qquad (2.29)$$

where the integration is over equivalence classes of paths X in $\mathbb{R}^d$, together with an independent metric e, and the equivalence is given by the action of the diffeomorphism group of I. Here X can equally well be thought of as a set of d scalar fields on I and the action in Eq. (2.29) as representing the standard coupling of these to one-dimensional gravity with μ as a cosmological constant.

2.2.1 Embedded random walks

Let us start by defining more precisely a geometric path in the first case. We say that two paths $\omega^1 \in \Omega_{t_1}(x, y)$ and $\omega^2 \in \Omega_{t_2}(x, y)$ are *equivalent* if one is a reparametrization of the other, i.e. if there exists an increasing homeomorphism $\varphi : [0, t_1] \mapsto [0, t_2]$ such that

$$\omega^1 = \omega^2 \circ \varphi.$$

This defines an equivalence relation on $\Omega(x, y)$. We denote by $[\omega]$ the equivalence class containing ω and by $\tilde{\Omega}(x, y)$ the set of all such equivalence classes. The elements of $\tilde{\Omega}(x, y)$ are called *geometric paths* from x to y.

We can now transport the measure $D^{(m)}\omega$ to a measure $D^{(m)}[\omega]$ on $\tilde{\Omega}(x, y)$ by using the quotient mapping $\omega \mapsto [\omega]$. Then $D^{(m)}[\omega]$ is the reparametrization invariant measure announced and it satisfies

$$\int_{\tilde{\Omega}(x,y)} D^{(m)}[\omega]\, f([\omega]) = \int_{\Omega(x,y)} D^{(m)}\omega\, f([\omega]) \qquad (2.30)$$

for integrable functions f on $\tilde{\Omega}(x, y)$. In other words, $D^{(m)}[\omega]$ is obtained by integrating $D^{(m)}\omega$ over equivalence classes.

Since the measure $D^{(m)}[\omega]$ is constructed from the Wiener measures $D_t\omega$, $t > 0$, it follows that discrete approximations to those measures, discussed in the previous section, also indirectly provide discrete approximations to $D^{(m)}[\omega]$. However, from our point of view, those approximations suffer from the drawback of being parametrization dependent, as is particularly clear from the manifestly parametrization dependent cutoff $\frac{t}{N}$ used.

In order to exhibit a parametrization independent approximation to $D^{(m)}[\omega]$ we shall need a substitute for the spaces $\Omega_{N,t}(x, y)$ of the previous section in the form of an approximation $\tilde{\Omega}^a(x, y)$ to $\tilde{\Omega}(x, y)$ and also a reparametrization invariant action S defined on $\tilde{\Omega}^a(x, y)$.

It is convenient to choose $\tilde{\Omega}^a(x, y)$ to be the set of paths from x to y on the hyper-cubic lattice $a\mathbb{Z}^d$, where a denotes a lattice spacing. Thus the elements of $\tilde{\Omega}^a(x, y)$ are sequences $x_0, x_1, \ldots, x_n \in a\mathbb{Z}^d$ such that each step $x_i - x_{i-1}$ is a vector of length a along one of the coordinate axes for $i = 1, 2, \ldots, n$, and $x_0 = x$, $x_n = y$. Each such lattice path can be identified in an obvious way with an element in $\tilde{\Omega}^a(x, y)$ and different lattice paths give rise to different elements in $\tilde{\Omega}^a(x, y)$. In other words, we may view $\tilde{\Omega}^a(x, y)$ as a subset of $\tilde{\Omega}(x, y)$, and it is easily verified that any geometric path $[\omega] \in \tilde{\Omega}(x, y)$ can be approximated arbitrarily well by paths in $\tilde{\Omega}^a(x, y)$ for a sufficiently small, in the quotient topology of $\tilde{\Omega}(x, y)$.

Another equally natural choice of approximation to $\tilde{\Omega}(x, y)$ is the space of all piecewise linear paths in $\mathbb{R}^d$ with all linear pieces of length a. This approximation is a limiting case of the theory considered in the next section. For the moment we stick to the lattice paths.

Next, we choose the action $S([\omega])$, for $[\omega] \in \tilde{\Omega}^a(x, y)$, to be proportional to the length $|\omega|$ of ω, which is the simplest geometric quantity associated to $[\omega]$, i.e.

$$S([\omega]) = \tilde{\beta}|\omega| = \tilde{\beta}an, \qquad (2.31)$$

where $\tilde{\beta}$ is a coupling constant and n denotes the number of steps in $[\omega]$. Of course, if ω is an arbitrary piecewise smooth representative of $[\omega]$ we have

$$S([\omega]) = \tilde{\beta} \int_0^t ds \, |\dot{\omega}(s)|. \qquad (2.32)$$

Note that if the representative ω is chosen such that each step $x_i - x_{i-1}$ is parametrized by $s \in \left[\frac{i-1}{n}t, \frac{i}{n}t\right]$, which can be considered as a gauge choice for the parametrization, then

$$S([\omega]) = \tilde{\beta} \sum_{i=1}^n |x_i - x_{i-1}| = \tilde{\beta} \sum_{i=1}^n \left| \omega\left(\frac{i}{n}t\right) - \omega\left(\frac{i-1}{n}t\right) \right|.$$

We see that this gauge fixed version of the action (2.32) is of the form (2.23), with $N = n$ depending on $[\omega]$.

Thus, for a continuous bounded function f on $\tilde{\Omega}(x, y)$, we propose

$$\langle f \rangle_{(x,y)}^a = \sum_{[\omega] \in \tilde{\Omega}^a(x,y)} w([\omega], a) f([\omega]) e^{-\tilde{\beta}|\omega|} \qquad (2.33)$$

as an appropriate discrete approximation to

$$\int_{\tilde{\Omega}(x,y)} D^{(m)}[\omega] \, f([\omega]), \qquad (2.34)$$

for $a \to 0$. We have inserted a weight function $w([\omega], a)$ multiplying the counting measure on $\tilde{\Omega}^a(x, y)$, which is to be chosen together with $\tilde{\beta} = \tilde{\beta}(a)$ so as to produce the desired result Eq. (2.34) in the limit $a \to 0$.

In order to fix $w([\omega], a)$ it should first be noted that for $x \in a\mathbb{Z}^d$ the number of geometric paths in

$$\tilde{\Omega}^a(x) = \bigcup_{y \in aZ^d} \tilde{\Omega}^a(x, y)$$

of length $\ell = an$ equals

$$N(n) \equiv (2d)^n = e^{\frac{\beta_0}{a}\ell}, \qquad \beta_0 = \log 2d, \qquad (2.35)$$

by a simple counting argument. Even if the second endpoint $y \in a\mathbb{Z}^d$ is fixed the leading behaviour of the number $N_{(x,y)}(n)$ of paths in $\tilde{\Omega}^a(x, y)$ of length an is still given by Eq. (2.35), i.e.

$$\lim_{n \to \infty} \frac{\log N_{(x,y)}(n)}{n} = \beta_0. \qquad (2.36)$$

The divergent entropy factor given by Eq. (2.35) has to be cancelled by $w([\omega], a)$. This can be achieved by shifting $\tilde{\beta}$ in Eq. (2.33) by β_0/a. We shall see that this represents the $[\omega]$-dependence of w. In addition we only need to insert a power of the cutoff, which is needed for dimensional reasons. It follows that Eq. (2.33) reduces to the simple form

$$\langle f \rangle^a_{(x,y)} = \sum_{[\omega] \in \tilde{\Omega}^a(x,y)} a^\lambda f[\omega] e^{-\tilde{\beta}(a)|\omega|}. \tag{2.37}$$

In order to fix λ we observe that for $f = 1$ the expression

$$G^a(x,y) = \sum_{[\omega] \in \tilde{\Omega}^a(x,y)} a^\lambda e^{-\tilde{\beta}(a)|\omega|} \tag{2.38}$$

should, by Eqs. (2.24) and (2.27), converge to $G(x,y)$ (up to a constant factor). The function $G(x,y)$ has dimension (length)$^{2-d}$ so we expect

$$\lambda = 2 - d. \tag{2.39}$$

We now prove that Eq. (2.38) with λ given by Eq. (2.39) does indeed converge to $G(x,y)$ if $\tilde{\beta}(a)$ is suitably chosen. It is convenient to rewrite Eq. (2.38) as

$$G^a(x,y) = a^{2-d} G_{\beta(a)}(i,j), \tag{2.40}$$

with

$$i = a^{-1}x, \quad j = a^{-1}y, \quad \beta(a) = a\tilde{\beta}(a), \tag{2.41}$$

and

$$G_\beta(i,j) = \sum_{[\omega] \in \tilde{\Omega}^1(i,j)} e^{-\beta|\omega|}, \tag{2.42}$$

where $i, j \in \mathbb{Z}^d$. According to Eq. (2.36) $G_\beta(i,j)$ is well defined and finite for $\beta > \beta_0$, and by the preceding discussion we have to choose $\beta(a)$ such that $\beta(a) \to \beta_0$ as $a \to 0$. In order to determine $\beta(a)$ we first identify $G_\beta(i,j)$ as the massive lattice propagator, i.e. $(-\Delta_L + m^2(\beta))^{-1}_{ij}$, where Δ_L denotes the lattice Laplacian given by

$$(\Delta_L f)(i) = \sum_{v=1}^{d} (f(i + e_v) + f(i - e_v) - 2f(i)). \tag{2.43}$$

In Eq. (2.43) f denotes a function $f : \mathbb{Z}^d \mapsto \mathbb{R}$, and $e_1, \ldots, e_d$ are unit vectors along the positive coordinate axes in $\mathbb{Z}^d$. Considered as an infinite matrix whose entries are labelled by pairs of points in $\mathbb{Z}^d$ we may write $-\Delta_L + m_0^2$, where m_0 is a bare mass parameter, in the form

$$(-\Delta_L + m_0^2)_{i,j} = (2d + m_0^2)\delta_{ij} - Q_{ij} \tag{2.44}$$

where

$$Q_{ij} = \begin{cases} 1 & \text{if } i \text{ and } j \text{ are nearest neighbours,} \\ 0 & \text{otherwise.} \end{cases} \tag{2.45}$$

It follows that $-\Delta_L + m_0^2$ is invertible for $m_0^2 > 0$ and $(-\Delta_L + m_0^2)^{-1}$ is given by the Neumann series

$$(-\Delta_L + m_0^2)_{ij}^{-1} = (2d + m_0^2)^{-1} \sum_{n=0}^{\infty} \left(\frac{Q}{2d + m_0^2} \right)^n_{ij}. \qquad (2.46)$$

Substituting Eq. (2.45) in Eq. (2.46) we see that

$$(-\Delta_L + m_0^2)_{ij}^{-1} = \sum_{[\omega] \in \tilde{\Omega}^1(i,j)} (2d + m_0^2)^{-|\omega|}. \qquad (2.47)$$

Comparing Eq. (2.47) with Eq. (2.42), and using the definition (2.35) of β_0, we thus have

$$G_\beta(i,j) = (-\Delta_L + m^2(\beta))_{ij}^{-1}, \qquad (2.48)$$

where

$$m^2(\beta) = e^\beta - e^{\beta_0}, \qquad (2.49)$$

which proves our claim, with the lattice mass $m(\beta)$ given by Eq. (2.49).

As $\beta \to \beta_0$ the lattice mass scales to zero as

$$m(\beta) \sim (\beta - \beta_0)^\nu, \qquad (2.50)$$

where

$$\nu = \frac{1}{2}$$

is called the *critical exponent of the mass*.

The Fourier transform of Eq. (2.48) is given by

$$\hat{G}_\beta(p) = \frac{1}{4 \sum_{\mu=1}^{d} \sin^2(p_\mu/2) + m(\beta)^2}. \qquad (2.51)$$

From the analyticity properties of $\hat{G}_\beta(p)$ it follows that $G_\beta(i,j)$ decays exponentially as

$$G_\beta(0, ne_1) \sim e^{-m(\beta)n} \qquad (2.52)$$

as $n \to \infty$. Thus, in order to obtain a finite limit of Eq. (2.40) as $a \to 0$ with $x = ai$ and $y = aj$ fixed we must choose $\beta(a)$ such that

$$\frac{m(\beta(a))}{a} \to m, \qquad (2.53)$$

where m on the right-hand side is a positive constant which, in fact, equals the mass m in Eq. (2.34). In other words, by Eq. (2.50) we have to choose

$$\beta(a) = \beta_0 + m^2 a^2 + o(a^2). \qquad (2.54)$$

By analysing the behaviour of the lattice propagator close to the critical point β_0, we have now seen how to fix the scaling of the coupling $\beta(a)$ and the "wave-function" renormalization a^λ. With these choices one obtains

$$G^a(x, y) \to G(x, y), \tag{2.55}$$

as $a \to 0$, i.e.

$$\sum_{[\omega] \in \widetilde{\Omega}(x,y)} a^{2-d} f([\omega]) \, e^{-\tilde{\beta}(a)|\omega|} \to \int_{\widetilde{\Omega}(x,y)} D^{(m)}[\omega] \, f([\omega]) D^{(m)}[\omega] \tag{2.56}$$

as $a \to 0$ in the special case of constant functions f on $\widetilde{\Omega}(x, y)$. To our knowledge this fact has not been proved for general continuous bounded functions f on $\widetilde{\Omega}(x, y)$. What we have shown above can be viewed as an analogue of the central limit theorem for the measure $D^{(m)}[\omega]$. It is reasonable to expect that a proof of Eq. (2.56) should be reducible to this special case by analogy with the proof of Donsker's theorem described in the previous section. The problem in carrying out detailed arguments is that reparametrization invariant analogues of functions of the form $\omega \mapsto f(\omega(\frac{1}{N}t), \ldots, \omega_t(\frac{N-1}{N}t))$, which played a major role in the proof of that theorem, are missing. We shall, however, here and in the following, be content with the validity of Eq. (2.55).

2.2.2 Riemannian random walks

Let us next turn to the second type of reparametrization invariant discretization mentioned at the beginning of this section. It is based on a simple reinterpretation of the action S_2 given by Eq. (2.6) as well as of the measure $D^{(m)}\omega$, which we first explain.

Let $\varphi_t = \varphi_{t,1} : \Omega_t(x, y) \mapsto \Omega_1(x, y)$ be the homeomorphism given by Eq. (2.12), i.e.

$$\varphi_t(\omega)(s) = \omega(ts), \quad s \in [0, 1],$$

for $\omega \in \Omega_t(x, y)$. Setting

$$X = \varphi_t(\omega)$$

we have

$$S_2(\omega) = \frac{1}{2} \int_0^t ds \, |\dot{\omega}(s)|^2 = \frac{1}{2} \int_0^1 ds \, \frac{|\dot{X}(s)|^2}{t} \tag{2.57}$$

for any piecewise smooth path $\omega \in \Omega_t(x, y)$.

Next, let us consider the *Gaussian action functional* $S(X, e)$ defined by

$$S(X, e) = \frac{1}{2} \int_0^1 ds \, \frac{|\dot{X}(s)|^2}{e(s)} \tag{2.58}$$

for pairs (X, e), where $X : [0, 1] \mapsto \mathbb{R}^d$ is a collection of d piecewise smooth scalar fields on $[0, 1]$ with boundary conditions $X(0) = x$ and $X(1) = y$, and $e(s)$, $s \in [0,1]$, is a positive function. We interpret

$$g(s) = e(s)^2, \quad s \in [0,1], \tag{2.59}$$

as a metric on the manifold $[0,1]$. It is easily verified that $S(X,e)$ is invariant under reparametrizations of $[0,1]$, i.e. increasing diffeomorphisms $\xi : [0,1] \mapsto [0,1]$, which transform (X,e) according to

$$(X(s), e(s)) \mapsto (X^\xi(s), e^\xi(s)) = (X \circ \xi(s), \dot\xi(s) e \circ \xi(s)). \tag{2.60}$$

Moreover, given (X,e) we may choose $\xi = \xi_e$ such that the transformed metric is constant and equal to

$$t = \int_0^1 ds\, e(s) \tag{2.61}$$

by setting

$$\xi_e^{-1}(s) = t^{-1} \int_0^s ds\, e(s). \tag{2.62}$$

We recognize the action $S_2(\omega)$ given by Eq. (2.57) as a gauge fixed form of the action $S(X,e)$ with respect to the group $Diff_+[0,1]$ of increasing diffeomorphisms of $[0,1]$, and the parameter t in Eq. (2.57) labels the orbits of metrics under the action of $Diff_+[0,1]$, i.e. t is a modular parameter.

This observation suggests that it should be possible to realize the Wiener measure $D^{(m)}\omega$ as a $Diff_+[0,1]$-invariant measure, i.e. as a measure defined on a suitable orbit space under the action of $Diff_+[0,1]$, and that reparametrization invariant approximations should be obtainable by exhibiting appropriate discrete approximations to the action $S(X,e)$.

In fact, considering the action (2.60) of $Diff_+[0,1]$ on the space

$$\Gamma(x,y) = \Omega_1(x,y) \times C_+[0,1]$$

of pairs (X,e), where $X(0) = x$, $X(1) = y$ and e is a positive and continuous function on the unit interval, it follows from the remarks given above that the corresponding orbit space $\tilde\Gamma(x,y)$ can be identified with $\Omega_1(x,y) \times \mathbb{R}_+$, where the variable in $\mathbb{R}_+$ represents the volume of the metric e. On the other hand, we have previously observed that $\Omega(x,y)$ can be canonically identified with $\Omega_1(x,y) \times \mathbb{R}_+$ simply by rescaling the path parameter to the interval $[0,1]$; see Eq. (2.26). We shall hence in the following identify the three spaces

$$\tilde\Gamma(x,y) = \Omega(x,y) = \Omega_1(x,y) \times \mathbb{R}_+ \tag{2.63}$$

and denote the elements interchangeably by $[X,e]$, ω or (ω,t), depending on which realization we have in mind. In particular, we may consider the Wiener measure $D^{(m)}\omega$ as a measure on $\tilde\Gamma(x,y)$ which we shall denote by $D^{(m)}[X,e]$. Similarly, we can of course realize the measure $D^{(m)}\omega$ on the orbit space $\tilde{\Gamma}(x) = \Gamma(x)/Diff_+[0,1]$, where $\Gamma(x)$ is defined in the same manner as $\Gamma(x,y)$ except that the boundary condition $X(1) = y$ is dropped.

It follows from Eq. (2.57) and Eq. (2.58) that under the identification Eq. (2.63) the action S is equal to S_2. Hence, *discrete approximations to S_2 and $D^{(m)}\omega$ can be viewed as manifestly reparametrization invariant discrete approximations to the action S and the measure $D^{(m)}[X,e]$, or, equivalently, as approximations on the orbit space $\tilde{\Gamma}(x,y)$.* It is therefore straightforward to construct such an approximation, since we know from the previous section how to discretize $\Omega_1(x,y)$. We introduce an ultraviolet cutoff $a > 0$ and replace $\mathbb{R}_+$ by $a^2\mathbb{N}$, i.e. we constrain the metric $g(s) = e(s)^2$ such that its (intrinsic) total length assumes a discrete set of values given by

$$\int_0^1 ds\, e(s) = a^2 n, \tag{2.64}$$

where $n \in \mathbb{N}$. Thus, for fixed a, the number n labels a discrete set of equivalence classes of metrics under the action of $Diff_+[0,1]$. We furthermore replace $\Omega_1(x,y)$ by the approximating space $\Omega_{1,n}(x,y)$ consisting of continuous mappings $X : [0,1] \mapsto \mathbb{R}^d$ that are linear on each interval $\left[\frac{i-1}{n}, \frac{i}{n}\right]$ and satisfy the boundary conditions $X(0) = x$ and $X(1) = y$. Then the subset

$$\bigcup_{n\in\mathbb{N}} \Omega_{1,n}(x,y) \times \{a^2 n\} \tag{2.65}$$

is an approximation to $\Omega_1(x,y) \times \mathbb{R}_+$ for the parameter a small. Since $\omega \in \Omega_{1,n}(x,y)$ is uniquely specified by

$$\underline{x} = (x_0, x_1, \ldots, x_{n-1}, x_n), \tag{2.66}$$

where

$$x_i = \omega\left(\frac{i}{n}\right), \quad i = 0, \ldots, n, \tag{2.67}$$

we shall in the following write $(\underline{x}, na^2)$ instead of (ω, na^2). For such paths the action S assumes the simple form

$$S(\underline{x}, na^2) = \frac{1}{2} \sum_{i=1}^n \frac{|x_i - x_{i-1}|^2}{a^2}. \tag{2.68}$$

From Eq. (2.68) it follows that a has the dimension of a length, since $S(X,e)$ is dimensionless, but note that the dimension of the intrinsic length of the paths, as defined in Eq. (2.64), is *two*. This is a reflection of the fact that the average number of steps in a random walk between two points separated by a distance L is proportional to L^2. A natural dimensionless measure on $\Omega_{1,n}(x,y)$ is therefore

$$\prod_{i=1}^{n-1} \frac{dx_i}{a^d}.$$

Given a bounded continuous function F on $\tilde{\Gamma}(x,y)$ we are thus led to the expression

$$\langle F \rangle_{(x,y)}^a = \sum_{n=1}^{\infty} w(n,a) \int \prod_{i=1}^{n-1} \frac{dx_i}{a^d} F(\underline{x}, na^2)$$

$$\times \exp\left(-\frac{1}{2a^2} \sum_{i=1}^{n} |x_i - x_{i-1}|^2 - \frac{1}{2}m^2 a^2 n\right) \quad (2.69)$$

as our candidate approximation to $\int D^{(m)}[X,e]F[X,e]$ for an appropriate weight function $w(n,a)$ which we shall choose so as to cancel the entropy divergences in the measure.

Now note that

$$\int \prod_{i=1}^{n-1} \frac{dx_i}{a^d} \exp\left(-\frac{1}{2a^2} \sum_{i=1}^{n} |x_i - x_{i-1}|^2\right)$$

$$= \frac{(2\pi)^{nd/2}}{(2\pi n)^{d/2}} \exp\left(-\frac{1}{2a^2 n}|x_n - x_0|^2\right). \quad (2.70)$$

As in the case of hypercubic paths we shall assume the simplest possible form of $w(n,a)$, i.e. that it is given as a product of the inverse of the factor $(2\pi)^{nd/2}$ in Eq. (2.70) and the factor a^{2-d}, where the latter is required for dimensional reasons. Thus, setting

$$e^{\mu_0} = (2\pi)^{d/2}, \qquad w(n,a) = \frac{1}{2}a^{2-d} e^{-\mu_0 n}, \quad (2.71)$$

Eq. (2.69) reduces to

$$\langle F \rangle_{(x,y)}^a = \sum_{n=1}^{\infty} \frac{e^{-\mu_0 n}}{2a^{d-2}} \int \prod_{i=1}^{n-1} \frac{dx_i}{a^d} F(\underline{x}, na^2)$$

$$\times \exp\left(-\frac{1}{2a^2} \sum_{i=1}^{n} |x_i - x_{i-1}|^2 - \frac{1}{2}m^2 a^2 n\right). \quad (2.72)$$

We now claim that

$$\langle F \rangle_{(x,y)}^a \rightarrow \int_{(x,y)} D^{(m)}[X,e]F([X,e]) \quad (2.73)$$

as $a \rightarrow 0$ and a similar statement for $\langle F \rangle_x^a$. Considering first the case $F = 1$ and defining the regularized propagator

$$G^a(x,y) = \langle 1 \rangle_{(x,y)}^a \quad (2.74)$$

we may use Eq. (2.3) to evaluate the left-hand side of Eq. (2.73) and obtain

$$G^a(x,y) = \frac{1}{2} \sum_{n=1}^{\infty} a^2 \frac{1}{(2\pi a^2 n)^{d/2}} \exp\left(-\frac{|x-y|^2}{2na^2} - \frac{1}{2}m^2 a^2 n\right), \qquad (2.75)$$

which we recognize as a Riemannian sum approximating the integral

$$\frac{1}{2} \int_0^{\infty} \frac{dt}{(2\pi t)^{d/2}} \exp\left(-\frac{|x-y|^2}{2t} - \frac{1}{2}m^2 t\right) = G(x,y)$$

for $a \to 0$.

The validity of Eq. (2.73) for arbitrary F follows by a slight extension of the arguments leading to Donsker's theorem. In fact, from Eq. (2.27) it follows that Eq. (2.73) is equivalent to

$$\langle F \rangle_{(x,y)}^a = \sum_{n=1}^{\infty} \frac{e^{-\mu_0 n}}{2a^{d-2}} \int \prod_{i=1}^{n-1} \frac{dx_i}{a^d} F(\underline{x}, na^2)$$

$$\times \exp\left(-\frac{1}{2a^2} \sum_{i=1}^{n} |x_i - x_{i-1}|^2 - \frac{1}{2}m^2 a^2 n\right)$$

$$\to \frac{1}{2} \int_0^{\infty} dt \, e^{-\frac{1}{2}m^2 t} \int_{(x,y)} D_1^t \omega F(\omega, t) \qquad (2.76)$$

as $a \to 0$. Donsker's theorem (for y fixed) implies the convergence

$$\frac{e^{-\mu_0 n}}{a^d} \int \prod_{i=1}^{n-1} \frac{dx_i}{a^d} F(\underline{x}, na^2) \exp\left(-\frac{1}{2a^2} \sum_{i=1}^{n} |x_i - x_{i-1}|^2\right)$$

$$\to \int_{(x,y)} D_1^t \omega \, F(\omega, t) \qquad (2.77)$$

for fixed $t = na^2$ and $n \to \infty$, $a \to 0$. Converting the sum over n into an integral over t we see that Eq. (2.76) follows from Eq. (2.77). We omit the details.

We have now seen that Eq. (2.72) represents a proper regularization of the Wiener measure on $\widetilde{\Gamma}(x,y)$ and a similar statement holds for the Wiener measure on $\widetilde{\Gamma}(x)$. As the preceding arguments indicate, it is possible to generalize this result to a class of $Diff_+[0,1]$-invariant action functionals with the property that the central limit theorem is applicable to their gauge fixed forms. In particular, let us consider an action of the form

$$S(X,e) = \int_0^1 ds \, e(s) \, \varphi_0\left(\frac{|\dot{X}(s)|}{e(s)}\right), \qquad (2.78)$$

where φ_0 is a suitable real function. This action is $Diff_+[0,1]$-invariant by Eq. (2.60). Recall that e has the anomalous scaling dimension 2, so

$\dot{X}/e$ has dimension $(\text{length})^{-1}$. It follows that the function φ_0 contains dimensional coupling constants. For $\omega \in \Omega_{1,n}(x, y)$ we have, using the notation Eq. (2.66) and Eq. (2.67),

$$S(\underline{x}, na^2) = \sum_{i=1}^{n} \varphi_0 \left(\frac{|x_i - x_{i-1}|}{a^2} \right) a^2. \qquad (2.79)$$

Adjusting the coupling constants in φ_0 we can write Eq. (2.79) as

$$S(\underline{x}, na^2) = \sum_{i=1}^{n} \varphi \left(\frac{|x_i - x_{i-1}|}{a} \right), \qquad (2.80)$$

where φ is independent of a. The arguments leading to Eq. (2.73) now suggest the more general definition

$$
\begin{aligned}
\langle F \rangle_{(x,y)}^a &= \sum_{n=1}^{\infty} \frac{\sigma^2 e^{-\mu_0 n}}{2a^{d-2}} \int \prod_{i=1}^{n-1} \frac{dx_i}{a^d} F(\underline{x}, na^2) \\
&\quad \times \exp \left(-\sum_{i=1}^{n} \varphi \left(\frac{|x_i - x_{i-1}|}{a} \right) - \frac{1}{2} \sigma^2 m^2 a^2 n \right)
\end{aligned} \qquad (2.81)
$$

for continuous bounded functions F on $\tilde{\Gamma}(x, y)$, where μ_0 and σ are defined by

$$e^{\mu_0} = \int dx \, e^{-\varphi(|x|)} \qquad \sigma^2 = \frac{1}{de^{\mu_0}} \int dx \, |x|^2 \, e^{-\varphi(|x|)}. \qquad (2.82)$$

In particular, the regularized propagator $G^a(x, y)$ is obtained by setting $F = 1$. Applying Donsker's theorem we obtain, in analogy with Eq. (2.75), that

$$G^a(x, y) \to G(x, y) \qquad (2.83)$$

for $a \to 0$ and $n \to \infty$ with $na^2 = t$ fixed. More generally, one finds

$$\langle F \rangle_{(x,y)}^a \to \int_{(x,y)} D^{(m)}[X, e] F([X, e/\sigma^2]). \qquad (2.84)$$

Of course, a similar statement holds for paths with one free endpoint. We have thus established the important universal validity of Eq. (2.81) as an appropriate regularization of the Wiener measure for arbitrary functions φ with finite μ_0 and σ^2.

For the purpose of comparison with the hyper-cubic regularization discussed previously it is useful to re-express $\langle F \rangle_{(x,y)}^a$ in terms of its dimensionless analogue corresponding to $a = 1$ and defined by

$$\langle F \rangle_{(x,y)}(\mu) = \sum_{n=1}^{\infty} \int \prod_{i=1}^{n-1} dx_i \, F(\underline{x}, n) \exp \left(-\sum_{i=1}^{n} \varphi(|x_i - x_{i-1}|) - \mu n \right). \qquad (2.85)$$

We note that the right-hand side of Eq. (2.85) is finite for $\mu > \mu_0$, whereas for $\mu < \mu_0$ it is in general divergent. Moreover, if F is a continuous bounded function on $\widetilde{\Gamma}(x, y)$, we have

$$\langle F \rangle^a_{(x,y)} = a^{2-d} \langle F^a \rangle_{(\frac{x}{a}, \frac{y}{a})}(\mu(a)), \tag{2.86}$$

where

$$\mu(a) = \mu_0 + \frac{1}{2}\sigma^2 m^2 a^2, \tag{2.87}$$

and the function F^a is given by

$$F^a(\underline{x}, n) = F(a\underline{x}, na^2). \tag{2.88}$$

Setting $F = 1$ in Eq. (2.86) we obtain the relation

$$G^a(x, y) = \frac{1}{2}\sigma^2 a^{2-d} G_{\mu(a)}\left(\frac{x}{a}, \frac{y}{a}\right) \tag{2.89}$$

between the regularized propagator $G^a(x, y)$ and the dimensionless propagator

$$G_\mu(x, y) = \sum_{n=1}^{\infty} \int \prod_{i=1}^{n-1} dx_i \, \exp\left(-\sum_{i=1}^{n} \varphi(|x_i - x_{i-1}|) - \mu n\right). \tag{2.90}$$

Eq. (2.89) is seen to be analogous to Eq. (2.40) and we recognize Eq. (2.87) as the analogue of Eq. (2.54).

In contrast to the lattice progagator, $G_\mu(x, y)$ is invariant under rotations and translations in $\mathbb{R}^d$. We therefore define the dimensionless mass in analogy with Eq. (2.52) by

$$m(\mu) = \lim_{|x| \to \infty} -\frac{\log G_\mu(0, x)}{|x|}. \tag{2.91}$$

The most convenient way to calculate $m(\mu)$ is by Fourier-transforming $G_\mu(0, x)$. We have

$$\begin{aligned}
\widehat{G}_\mu(p) &= \int dx \, G_\mu(0, x) e^{-ip \cdot x} = \sum_{n=1}^{\infty} \left(\int dx \, e^{-\varphi(|x|) - \mu} e^{-ip \cdot x} \right)^n \\
&= \frac{e^{-\mu} R(p)}{1 - e^{-\mu} R(p)},
\end{aligned} \tag{2.92}$$

where

$$R(p) = \int dx \, e^{-\varphi(|x|) - ip \cdot x}, \tag{2.93}$$

for $p \in \mathbb{R}^d$. The exponential decay rate $m(\mu)$ of $G_\mu(0, x)$ equals the radius of the largest disc centred at 0 in the complex plane, which is contained in the domain of analyticity of $\widehat{G}_\mu$. Assuming that φ increases sufficiently fast at infinity, e.g. if

$$\varphi(x) \geq c \, |x|^\varepsilon$$

for some $c, \varepsilon > 0$, then R is an analytic, even function of $|p|$ in the vicinity of 0 and $R(0) = e^{\mu_0}$. Thus, since

$$1 - e^{-\mu} R(p) = (1 - e^{\mu_0 - \mu}) + \frac{1}{2}\sigma^2 p^2 \, e^{\mu_0 - \mu} + o(p^2) \qquad (2.94)$$

for $|p|$ small, we conclude (in accordance with Eq. (2.87)) that

$$\frac{1}{2}\sigma^2 m^2(\mu) = e^{\mu - \mu_0} - 1 + o(\mu - \mu_0)$$

or

$$m^2(\mu) \sim \mu - \mu_0 \qquad (2.95)$$

for $\mu \to \mu_0$. This should be compared to Eq. (2.50) in the case of hypercubic paths. In particular, it is seen that the critical exponent ν of the mass is $\frac{1}{2}$ in both cases.

We introduce the physical momentum by replacing p by ap. It follows that the first two terms on the right-hand side of Eq. (2.94) are both of order a^2 if we choose $\mu = \mu(a)$ such that

$$\mu - \mu_0 \sim a^2.$$

The continuum limit is obtained as

$$\widehat{G}(p) = \lim_{a \to 0} a^2 \widehat{G}_{\mu(a)}(ap) = \frac{c}{m^2 + p^2}, \qquad (2.96)$$

where $c > 0$ and $m > 0$ are constants. Not surprisingly, we recognize $\widehat{G}(p)$ as the Fourier transform of $G(x, y)$ up to a constant factor.

Analogues of Eqs. (2.42) and (2.90), and more generally Eq. (2.85), will form the starting point of the discussion in subsequent chapters of discrete approximations to measures on spaces of manifolds of dimension 2 or higher. For that purpose it will be important to keep in mind the geometric significance of the variables introduced. In particular, n labels the orbit under the action of $Diff_+[0, 1]$ on metrics defined on $[0, 1]$ with volume n, and X may be thought of as a mapping $X: \{0, \frac{1}{n}, \ldots, \frac{n-1}{n}, 1\} \mapsto \mathbb{R}^d$ specifying a piecewise linear path from x to y parametrized by $[0, 1]$.

Any metric on $[0, 1]$ with volume n specifies in an obvious way a "triangulation", i.e. a subdivision of $[0, 1]$ into intervals of length 1 with respect to the metric. Any two triangulations obtained from metrics in the orbit labelled by n are equivalent, which in this simple case just means that they have an equal number of links (or vertices). The variable n therefore labels equivalence classes of triangulations of $[0, 1]$ and we can view X as a mapping from the set of vertices of such a triangulation into $\mathbb{R}^d$. We shall find it convenient to adopt this point of view for later generalizations.

2.3 Rigid random walks

2.3.1 *Curvature-dependent action*

The requirement of reparametrization invariance led us in the previous section to consider action functionals expressible in terms of the length of paths or approximations to the length. In addition to the metric, or length element, there are other natural geometric quantities associated to one-dimensional manifolds embedded in $\mathbb{R}^d$, for instance the curvature or extrinsic curvature since one-dimensional manifolds have no intrinsic curvature. In differential geometric language the metric and the extrinsic curvature are often referred to as the first and second fundamental form, respectively.

In this section we consider a class of random path models whose action functionals depend on the extrinsic curvature as well as the path length. The introduction of curvature-dependent terms into the action allows us to adjust the coupling constant λ in front of the extrinsic curvature term and obtain in the continuum limit a model where the generic paths are smoother than those described by the Wiener measure. This, however, will require a tuning of λ to infinity. For finite λ we show that the extrinsic curvature term is irrelevant and that the Wiener measure is recovered in the continuum limit. Here we generalize the results of Subsection 2.2.2 to include extrinsic curvature. Similar considerations apply to the lattice models.

We are aiming at generalizing the expression (2.90). For this purpose we first recall that the extrinsic curvature $k(s)$ of a smooth path

$$X : [0,1] \mapsto \mathbb{R}^d$$

is given by

$$k = \frac{[\dot{X}^2 \ddot{X}^2 - (\dot{X} \cdot \ddot{X})^2]^{1/2}}{\dot{X}^2}, \tag{2.97}$$

and transforms according to

$$k(s) \mapsto \dot{\xi}(s) k(\xi(s)), \tag{2.98}$$

under the action of diffeomorphisms $\xi \in \mathit{Diff}_+[0,1]$. It follows that the expression

$$S(X,e) = \int_0^1 ds\, e(s)\, \varphi_0 \left(\frac{|\dot{X}(s)|}{e(s)} \right) + \int_0^1 ds\, e(s)\, \psi_0 \left(\frac{k(s)}{e(s)} \right) \tag{2.99}$$

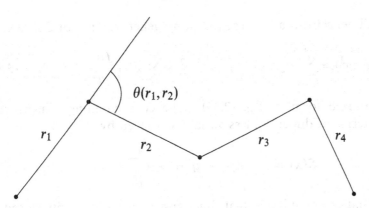

Fig. 2.3. The angle $\theta(r_i, r_{i+1})$ between successive tangent vectors in a piecewise linear path.

represents a natural reparametrization invariant generalization of the action given by Eq. (2.78), where φ_0 and ψ_0 are suitable real functions.

In order to discretize S we proceed as in the previous section and set $e(s) = na^2$, where $n \in \mathbb{N}$ and $a > 0$ is an ultraviolet cutoff, and consider paths $X \in \Omega_{1,n}(x, y)$. From Eq. (2.97) it is evident that the curvature of X is concentrated at the points $s = \frac{1}{n}, \frac{2}{n}, \ldots, \frac{n-1}{n}$. Since extrinsic curvature is naturally related to bending of the path we define the regularized curvature at $\frac{i}{n}$ to be a function of the angle $\theta_i = \theta(r_i, r_{i+1})$ between the ith and the $(i+1)$st step (see Fig. 2.3). The angle $\theta_i \in [0, \pi]$ is determined by

$$\cos \theta(r_i, r_{i+1}) = \frac{r_i \cdot r_{i+1}}{|r_i|\,|r_{i+1}|}, \tag{2.100}$$

where

$$r_i = x_i - x_{i-1}, \tag{2.101}$$

and $x_i = X(\frac{i}{n})$. The discrete approximations to higher-derivative terms in action functionals are never uniquely determined. We can smooth out each corner $X\left(\frac{i}{n}\right)$ of a piecewise linear path by circular arcs tangent to r_i and r_{i+1}. An easy calculation shows that the integral of the curvature along this arc equals θ_i, and the average curvature over the interval $\left[\frac{i-1}{n}, \frac{i}{n}\right]$ is therefore $n\theta_i$. In view of the ambiguity of the discretization of k we choose the general form $f(\theta_i)a^{-2}$, for the average value of ke^{-1} over the interval $\left[\frac{i-1}{n}, \frac{i}{n}\right]$, where we used $e(s) = na^2$ and f is a real valued function on $[0, \pi]$ such that $f(\theta) = \theta + o(\theta)$ for θ small.

We shall henceforth adopt the following generalization of Eq. (2.80):

$$S(\underline{x}, na^2) = \sum_{i=1}^{n} \varphi_0 \left(\frac{|x_i - x_{i-1}|}{a^2} \right) a^2 + \sum_{i=1}^{n-1} \psi_0 \left(\frac{f(\theta_i)}{a^2} \right) a^2 \qquad (2.102)$$

as a discretized form of Eq. (2.99). The corresponding dimensionless action in terms of dimensionless variables is given by

$$S(\underline{x}) = \sum_{i=1}^{n} \varphi(|x_i - x_{i-1}|) + \lambda \sum_{i=1}^{n-1} \psi(\theta_i), \qquad (2.103)$$

where φ and ψ are suitable real functions, and we have introduced the extrinsic curvature coupling λ. It should be noted that Eq. (2.103) inherits the scale invariance of the curvature term in S in the sense that both $k(s)$ and θ_i are invariant under scale transformations $x \mapsto \Omega x$.

2.3.2 The two-point function

The two-point function associated with the action (2.103) is defined in analogy with Eq. (2.90) by

$$\begin{aligned} G_{\mu,\lambda}(x, y) &= \sum_{n=1}^{\infty} \int \prod_{i=1}^{n-1} dx_i \, \exp\left(-S(\underline{x}) - \mu n \right) \qquad (2.104) \\ &= \sum_{n=1}^{\infty} \int \prod_{i=1}^{n-1} dx_i \exp\left(-\sum_{i=1}^{n} \varphi(|x_i - x_{i-1}|) - \lambda \sum_{i=1}^{n-1} \psi(\theta_i) - \mu n \right). \end{aligned}$$

We assume that $\psi(\theta)$ is a continuous non-negative function on $[0, \pi]$, whose only zero is at $\theta = 0$. This property ensures that smooth paths will be favoured as λ grows large and positive. In the following we show that for λ finite the continuum limit of the model defined by S is described by the Wiener measure, i.e. the local rigidity represented by the ψ-term of Eq. (2.103) becomes irrelevant at large distances with respect to the metric e. However, we construct a different continuum limit for $\lambda \to \infty$. This is the main purpose of the rest of this section.

We define the Fourier transformed two-point function by

$$\widehat{G}_{\mu,\lambda}(p) = \int dx \, G_{\mu,\lambda}(0, x) e^{-ip \cdot x}, \qquad (2.105)$$

and the *susceptibility* $\chi(\mu, \lambda)$ is defined in analogy with a spin system as the integrated two-point function, i.e.

$$\chi(\mu, \lambda) = \widehat{G}_{\mu,\lambda}(0) = \int dx \, G_{\mu,\lambda}(0, x). \qquad (2.106)$$

We have

$$
\widehat{G}_{\mu,\lambda}(p) = \sum_{n=1}^{\infty} \int \prod_{i=1}^{n} dx_i \, e^{-\mu n} \tag{2.107}
$$

$$
\times \exp\left(-\sum_{i=1}^{n} \varphi(|x_i - x_{i-1}|) - \lambda \sum_{i=1}^{n-1} \psi(\theta_i) - ip \cdot \sum_{i=1}^{n}(x_i - x_{i-1})\right)
$$

$$
= \sum_{n=1}^{\infty} \int \prod_{i=1}^{n} d\Omega(\hat{r}_i) \, e^{-\mu n} \prod_{i=1}^{n-1}(F(p \cdot \hat{r}_i)K_\lambda(\hat{r}_i, \hat{r}_{i+1}))F(p \cdot \hat{r}_n),
$$

where r_i is defined by Eq. (2.101), $\hat{r} = \frac{r}{|r|}$ for $r \neq 0$, $d\Omega$ is the uniform measure on the sphere S^{d-1} induced by dr and

$$
F(p \cdot \hat{r}) = \int_0^\infty d|r| \, |r|^{d-1} e^{-\varphi(|r|)} e^{-ip \cdot \hat{r}|r|}, \tag{2.108}
$$

$$
K_\lambda(\hat{r}, \hat{r}') = e^{-\lambda\psi(\theta(\hat{r}, \hat{r}'))}. \tag{2.109}
$$

It will be convenient to rewrite Eq. (2.107) in operator form. We denote by F_p the operator on $L^2(S^{d-1}, d\Omega)$ given by multiplication by the function $F(p \cdot \hat{r})$, and by K_λ the operator on $L^2(S^{d-1}, d\Omega)$ whose kernel is $K_\lambda(\hat{r}, \hat{r}')$. Then Eq. (2.107) assumes the form

$$
\widehat{G}_{\mu,\lambda}(p) = \left\langle 1, (1 - e^{-\mu}F_pK_\lambda)^{-1}e^{-\mu}F_p \, 1\right\rangle
$$

$$
= \left\langle 1, e^{-\mu}F_p(1 - e^{-\mu}K_\lambda F_p)^{-1} \, 1\right\rangle, \tag{2.110}
$$

where 1 denotes the constant function 1 on S^{d-1} and $\langle \cdot, \cdot \rangle$ denotes the inner product on $L^2(S^{d-1}, d\Omega)$.

Since $\psi \geq 0$ and $\theta(\hat{r}, \hat{r}')$ is symmetric in $\hat{r}$ and $\hat{r}'$ it follows that K_λ is a self-adjoint Hilbert–Schmidt operator. Since the kernel of K_λ is positive the Perron–Frobenius theorem implies that $\|K_\lambda\|$ is a non-degenerate eigenvalue of K_λ and all the other eigenvalues are numerically strictly smaller than $\|K_\lambda\|$. It follows that

$$
\|K_\lambda\| = \int d\Omega(\hat{r}') \, e^{-\lambda\psi(\theta(\hat{r}, \hat{r}'))},
$$

and the eigenfunctions corresponding to $\|K_\lambda\|$ are the constant functions. In order to prove this suppose that $K_\lambda f = \alpha f$, $\alpha \in \mathbb{R}$ and $f \in L^2(S^{d-1}, d\Omega)$. Then we have

$$
|\alpha| \int d\Omega(\hat{r}) \, |f(\hat{r})| = \int d\Omega(\hat{r}) \, |\int d\widehat{\Omega}(\hat{r}') \, K_\lambda(\hat{r}, \hat{r}')f(\hat{r}')|
$$

$$
\leq \int d\Omega(\hat{r})d\Omega(\hat{r}')|K_\lambda(\hat{r}, \hat{r}')| \, |f(\hat{r}')|
$$

$$
= \int d\Omega(\hat{r}) \, e^{-\lambda\psi(\theta(\hat{r}, \hat{r}'))} \int d\Omega(\hat{r}') \, |f(\hat{r}')|,
$$

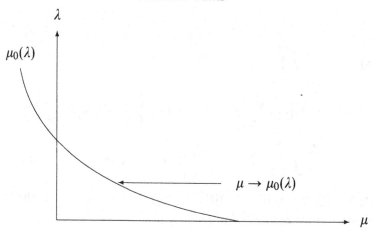

Fig. 2.4. The critical curve (2.114) and the domain of definition of $\widehat{G}_{\mu,\lambda}(p)$ as given by Eq. (2.113). Note that $\mu_0 \to -\infty$ as $\lambda \to \infty$.

i.e.

$$|\alpha| \leq \int d\Omega(\hat{r}) \, e^{-\lambda\psi(\theta(\hat{r},\hat{r}'))},$$

and the claim follows, since $\int d\Omega(\hat{r}) \, e^{-\lambda\psi(\theta(\hat{r},\hat{r}'))}$ is clearly an eigenvalue with the constants as eigenfunctions. The operator F_p for $p = 0$ is a constant given by

$$F_0 = \int_0^\infty d|r| \, |r|^{d-1} e^{-\varphi(|r|)}. \tag{2.111}$$

If ω_d denotes the volume of S^{d-1}, we conclude from Eq. (2.110) that

$$\chi(\mu,\lambda) = \frac{\omega_d \, e^{-\mu} F_0}{1 - e^{-\mu}\|K_\lambda\|F_0} \tag{2.112}$$

for

$$e^\mu > \|K_\lambda\| \, F_0, \tag{2.113}$$

and that $\chi(\mu,\lambda)$ diverges otherwise. One can show that the same is true for $\widehat{G}_{\mu,\lambda}(p)$ and $G_{\mu,\lambda}(x,y)$ for all p and x, y and we take the equation

$$e^{\mu_0} = \|K_\lambda\|F_0 \tag{2.114}$$

as the definition of the critical point $\mu_0 = \mu_0(\lambda)$.

2.3.3 The scaling limits

We next investigate the critical behaviour of the model for fixed finite value of λ. We let $\mu \to \mu_0(\lambda)$ while scaling p as in Eq. (2.96). If φ increases sufficiently fast at infinity (e.g. $\varphi(x) \geq cx^\varepsilon$ for some $\varepsilon > 0$) then F is three times continuously differentiable and by Taylor's theorem we have

$$\frac{F(ap \cdot \hat{r}) - (1 + ic_1 ap \cdot \hat{r})F_0}{a^2} \to -F_0 c_2 (p \cdot \hat{r})^2 \qquad (2.115)$$

uniformly in $\hat{r} \in S^{d-1}$ as $a \to 0$, where

$$c_1 = -iF_0^{-1}F'(0), \qquad c_2 = -\frac{1}{2}F_0^{-1}F''(0)$$

and $c_1, c_2 > 0$ by Eq. (2.108). Regarding both sides of Eq. (2.115) as multiplication operators on $L^2(S^{d-1}, d\Omega)$ this means that we have convergence in norm. The idea is now to expand F_p in Eq. (2.110) in powers of $p \cdot \hat{r}$. Since the operators K_λ and F_p do not commute we need to exercise some care in manipulating the expansion. The basic steps of the calculation are as follows.

We write the operator K_λ in the form

$$K_\lambda = \|K_\lambda\| \bar{K}_\lambda = \|K_\lambda\|(P + R_\lambda), \qquad (2.116)$$

where P is the orthogonal projection onto the constant functions, and $R_\lambda = (1 - P)\bar{K}_\lambda(1 - P)$. Note that the largest eigenvalue of $\bar{K}_\lambda$ is 1 and all the eigenvalues of R_λ are strictly smaller than 1. Since the operator F_{ap} tends to the constant F_0 as $a \to 0$ it follows from Eq. (2.114) that the operator $1 - e^{-\mu}K_\lambda F_{ap}$ approaches

$$1 - e^{\mu_0 - \mu}(P + R_\lambda) = (1 - e^{\mu_0 - \mu})P + (1 - e^{\mu_0 - \mu}R_\lambda)(1 - P).$$

As $\mu \to \mu_0$ the eigenvalue $1 - e^{\mu_0 - \mu}$ of the first term on the right-hand side above tends to zero whereas all eigenvalues of the second term are bounded away from zero. Thus, we anticipate that only the former contributes to the divergence of Eq. (2.110) and that we may replace the operators occurring in Eq. (2.110) by their projections onto the constant functions, i.e. we replace 1 by P, K_λ by $\|K_\lambda\|P$ and F_{ap} by

$$PF_{ap}P = F_0(P - c_2 a^2 P(p \cdot \hat{r})^2 P) + o(a^2),$$

where we have used that $Pp \cdot \hat{r}P = 0$ due to the fact that $\langle 1, p \cdot \hat{r} \rangle = 0$. Accordingly, the operator $1 - e^{-\mu}K_\lambda F_{ap}$ is replaced by

$$(1 - e^{\mu_0 - \mu})P - e^{\mu_0 - \mu}c_2 a^2 P(p \cdot \hat{r})^2 P.$$

We see that the two terms in this expression are both of order a^2 if we choose $\mu = \mu(a)$ such that

$$\mu - \mu_0 = m_1^2 a^2 + o(a^3)$$

for some positive constant m_1. Finally, using the fact that the coordinates $\hat{r}_\mu$, $\mu = 1, \ldots, d$ of $\hat{r}$ fulfil

$$\langle 1, \hat{r}_\mu \hat{r}_\nu \rangle = \delta_{\mu\nu} c, \qquad (2.117)$$

where c is a positive constant, we obtain a finite continuum limit given by

$$\widehat{G}(p) = \lim_{a \to 0} a^2 \widehat{G}_{\mu(a),\lambda}(ap) = \frac{c'}{m^2 + p^2}, \tag{2.118}$$

where $c' > 0$ is a constant and

$$m^2 = \omega_d (c_2 c)^{-1} m_1^2. \tag{2.119}$$

The proof of the convergence in Eq. (2.118) is somewhat technical but straight forward. It involves using the spectral representation of $\bar{K}_\lambda$ to prove the convergence

$$\frac{a^2}{1 - e^{\mu_0 - \mu(a)} \bar{K}_\lambda^{\frac{1}{2}} \frac{F_{ap}}{F_0} \bar{K}_\lambda^{\frac{1}{2}}} \to P \frac{m_1^2}{1 + c_2 m_1^{-2} P(p \cdot \hat{r})^2 P} P. \tag{2.120}$$

We omit the details. Note that the scaling of variables and coupling constants, as well as the scaling limits in Eq. (2.118) and in Eq. (2.96), are identical.

The fact that the continuum limit constructed above is independent of the value of λ leads to the question of whether it is possible to enforce an effective rigidity of continuum paths by tuning λ to infinity as μ approaches $\mu_c(\lambda)$ (see Fig. 2.4). As a preparation for investigating this question we need to analyse the asymptotic behaviour of K_λ as $\lambda \to +\infty$.

First, we observe that

$$K(O\hat{r}_1, O\hat{r}_2) = K(\hat{r}_1, \hat{r}_2)$$

for all $\hat{r}_1, \hat{r}_2 \in S^{d-1}$ and any orthogonal transformation $O : \mathbb{R}^d \mapsto \mathbb{R}^d$. This means that K_λ commutes with the action of the orthogonal group $O(d)$ on $L^2(S^{d-1}, d\Omega)$, which is given by $f \mapsto f \circ O$ for $f \in L^2(S^{d-1}, d\Omega)$ and $O \in O(d)$. It is well known that (see e.g. [231]) this representation of $O(d)$ decomposes into an infinite sequence of finite-dimensional irreducible representations, which are pairwise inequivalent and can be labelled by the eigenvalues $0 = \alpha_0 < \alpha_1 < \alpha_2 < \ldots$ of the Laplace–Beltrami operator L on S^{d-1}, in such a way that the representation space of the representation corresponding to α_ℓ is the corresponding eigenspace of L. For later use we also note (see [231]) that for each $\ell = 0, 1, 2, \ldots$ there exists an eigenfunction φ_ℓ of L with eigenvalue α_ℓ, which depends only on the azimuthal angle $\theta(\hat{r})$ with respect to a fixed north pole $\hat{n}$ on S^{d-1}, i.e. $\cos \theta(\hat{r}) = \hat{n} \cdot \hat{r}$, and

$$\varphi_\ell(0) \neq 0, \qquad \varphi_\ell'(0) = 0. \tag{2.121}$$

Since K_λ commutes with $O(d)$ it acts by scalar multiplication on each eigenspace of L. Let the eigenvalue of $\bar{K}_\lambda$ corresponding to α_ℓ be denoted by $\beta_\ell(\lambda)$. Then

$$\|K_\lambda\|^{-1} \int d\Omega(\hat{r}') \, e^{-\lambda\psi(\theta(\hat{r},\hat{r}'))} \varphi_\ell(\theta(\hat{r}')) = \beta_\ell(\lambda)\varphi_\ell(\theta(\hat{r})),$$

which for $\hat{r} = \hat{n}$ yields

$$\|K_\lambda\|^{-1} \int d\Omega(\hat{r}) \, e^{-\lambda\psi(\theta(\hat{r}))} \varphi_\ell(\theta(\hat{r})) = \beta_\ell(\lambda)\varphi_\ell(0). \qquad (2.122)$$

Using the fact that the action of L on $\varphi_\ell(\theta(\hat{r}))$ is given by

$$(L\varphi_\ell)(\theta) = -\varphi_\ell''(\theta) - (d-2)\cot\theta \, \varphi_\ell'(\theta)$$

it follows that

$$(L\varphi_\ell)(0) = -(d-1)\varphi_\ell''(0).$$

Hence,

$$\alpha_\ell\varphi_\ell(0) = -(d-1)\varphi_\ell''(0). \qquad (2.123)$$

Taylor expanding the function φ_ℓ around 0 and using Eq. (2.123) we find

$$\begin{aligned}
\varphi_\ell(\theta) &= \varphi_\ell(0) + \frac{1}{2}\varphi_\ell''(0)\theta^2 + \theta^3\eta(\theta) \\
&= \left(1 - \frac{\alpha_\ell}{2(d-1)}\theta^2\right)\varphi_\ell(0) + \theta^3\eta(\theta),
\end{aligned}$$

where $|\eta(\theta)|$ is bounded by a constant. Inserting this expansion on the left-hand side of Eq. (2.122), one arrives at the following expression for the ℓth eigenvalue of $\bar{K}_\lambda$:

$$\begin{aligned}
\beta_\ell(\lambda) &= 1 - \frac{\alpha_\ell}{2(d-1)\|K_\lambda\|} \int d\Omega(\hat{r}) \, e^{-\lambda\psi(\theta(\hat{r}))} \theta(\hat{r})^2 \\
&\quad + \frac{1}{\varphi_\ell(0)\|K_\lambda\|} \int d\Omega(\hat{r}) \, e^{-\lambda\psi(\theta(\hat{r}))} \theta(\hat{r})^3 \eta(\theta(\hat{r})). \qquad (2.124)
\end{aligned}$$

The quantity

$$C(\lambda) = \frac{1}{2(d-1)\|K_\lambda\|} \int d\Omega(\hat{r}) \, e^{-\lambda\psi(\theta(\hat{r}))} \theta(\hat{r})^2 \qquad (2.125)$$

tends to zero as $\lambda \to +\infty$ as a consequence of our assumptions on the function ψ, which imply that the measure $\|K_\lambda\|^{-1} e^{-\lambda\psi(\theta(\hat{r}))} d\Omega(\hat{r})$ tends to the Dirac measure at $\hat{n}$ as $\lambda \to +\infty$. For the same reason the last term on the right-hand side of Eq. (2.124) tends to 0 but faster than $C(\lambda)$ as $\lambda \to +\infty$. Eq. (2.124) shows that all the eigenvalues of $\bar{K}_\lambda$ tend to 1 as

$\lambda \to +\infty$ and it can formally be written as an operator equation

$$\bar{K}_\lambda = 1 + C(\lambda) L + o(C(\lambda)). \tag{2.126}$$

Since L is an unbounded operator whereas $\bar{K}_\lambda$ is bounded, it is clear that this equation has to be interpreted properly. Let us use it together with Eq. (2.115) to expand the denominator in Eq. (2.110) for large λ, with p replaced by ap. We now note that since all eigenvalues of $\bar{K}_\lambda$ tend to 1 as $\lambda \to +\infty$ we cannot disregard the first-order term in Eq. (2.115), contrary to what we did when we took the scaling limit for fixed λ. With the choice $\lambda = \lambda(a)$ such that

$$C(\lambda(a)) = c_1 \lambda_R a, \qquad \lambda_R > 0, \tag{2.127}$$

we obtain to first order in a

$$1 - e^{-\mu} \bar{K}_{\lambda(a)} \bar{F}_{ap} = 1 - e^{\mu_0 - \mu} + a c_1 (e^{\mu_0 - \mu} \lambda_R L + i p \cdot \hat{r}).$$

All terms in the above equation are of order a provided we choose $\mu = \mu(a)$ such that

$$\mu(a) - \mu_0(\lambda(a)) = c_1 \mu_R a, \qquad \mu_R > 0.$$

It follows that

$$a \widehat{G}_{\mu(a),\lambda(a)}(ap) \to \langle 1, (\mu_R + \lambda_R L - i p \cdot \hat{r})^{-1} 1 \rangle. \tag{2.128}$$

This is the scaling limit we were looking for.

The convergence in Eq. (2.128) is a consequence of the strong convergence

$$\frac{1}{1 - H^{\frac{1}{2}}_{\mu(a),\lambda(a)}(F_{ap} - 1)H^{\frac{1}{2}}_{\mu(a),\lambda(a)}} \to \frac{1}{1 - (\mu_R + \lambda_R L)^{-\frac{1}{2}} i p \cdot \hat{r}(\mu_R + \lambda_R L)^{-\frac{1}{2}}}, \tag{2.129}$$

where

$$H_{\mu,\lambda} = \bar{K}_\lambda (1 - e^{\mu_0 - \mu} \bar{K}_\lambda)^{-1}. \tag{2.130}$$

We omit the somewhat technical details of its proof and summarize our findings as follows.

Theorem 2.2 *Let $G_{\mu,\lambda}(x, y)$ be the two-point function of the random walk model defined by Eq. (2.104), and let $\widehat{G}_{\mu,\lambda}(p)$ denote its Fourier transform. Then $G_{\mu,\lambda}$ and $\widehat{G}_{\mu,\lambda}$ are finite for $\mu > \mu_0(\lambda)$, where $\mu_0(\lambda)$ is determined by Eq. (2.114).*

(i) For fixed λ, let the coupling constant μ be a function of the cutoff $a > 0$ such that

$$\mu(a) - \mu_0(\lambda) = m_1^2 a^2 + o(a^3). \tag{2.131}$$

Then the continuum limit $\widehat{G}(p)$ *defined by*

$$\widehat{G}(p) = \lim_{a \to 0} a^2 \widehat{G}_{\mu(a), \lambda}(ap) \tag{2.132}$$

exists and equals (up to a multiplicative constant)

$$\widehat{G}(p) = \frac{1}{m^2 + p^2}, \tag{2.133}$$

where $m > 0$ is a constant proportional to m_1.

(ii) *Given $\mu_R, \lambda_R > 0$, there exist functions $\mu(a)$ and $\lambda(a)$ of $a > 0$, where $\mu(a) > \mu_0(\lambda(a))$, $\lambda(a) \to +\infty$ and $\mu(a) \to -\infty$ as $a \to 0$, such that the continuum limit $\widehat{G}$ defined by*

$$\widehat{G}(p) = \lim_{a \to 0} a \widehat{G}_{\mu(a), \lambda(a)}(ap) \tag{2.134}$$

exists and equals (up to a multiplicative constant)

$$\widehat{G}(p) = \left\langle 1, (\mu_R + \lambda_R L - ip \cdot \hat{r})^{-1} 1 \right\rangle, \tag{2.135}$$

where L is the Laplace–Beltrami operator on S^{d-1} and $\langle \, , \, \rangle$ denotes the scalar product in $L^2(S^{d-1}, d\Omega)$.

2.3.4 *The tangent–tangent correlation function*

We have discussed at length the fact that the continuum measure corresponding to the propagator (2.133) is the Wiener measure. Furthermore, for finite λ, no trace of the local stiffness of the discretized paths is left in the scaling limit. If $\lambda \to +\infty$ we observe a different behaviour and it is an interesting unsolved problem to determine the continuum measure corresponding to Eq. (2.135). We conjecture that the support of such a measure consists of smoother paths than those of the Wiener measure. To corroborate this conjecture it is convenient to consider the tangent–tangent correlation function of paths. This function is defined as the average value of the inner product of two normalized tangent vectors $\hat{r}_i$ and $\hat{r}_j$, as a function of the number of steps from i to j. The average is computed in the ensemble of paths of arbitrary length and free endpoint. By translational invariance the average only depends on $|i - j|$, so we choose $i = 1$, $j = n + 1$, and the contributions of the remaining steps in paths cancel in the average. Hence, we can express the average as

$$
\begin{aligned}
\langle \hat{r}_1 \cdot \hat{r}_{n+1} \rangle_\lambda &= \|K_\lambda\|^{-n} \int d\Omega(\hat{r}_1) \ldots d\Omega(\hat{r}_n) \, \hat{r}_1 \cdot \hat{r}_{n+1} \\
&\qquad \times K_\lambda(r_1, r_2) \ldots K_\lambda(\hat{r}_n, \hat{r}_{n+1}) \\
&= \sum_{i=1}^{d} \langle \hat{r}^i, \bar{K}_\lambda^n \hat{r}^i \rangle \tag{2.136}
\end{aligned}
$$

where $\hat{r}^i$ denotes the ith coordinate function of $\hat{r}$. The last expression in Eq. (2.136) can be evaluated by observing that the function $\varphi_v(\hat{r}) = v \cdot \hat{r}$ is an eigenfunction (in fact one corresponding to the eigenvalue $\beta_1(\lambda)$ in Eq. (2.124)) of $\bar{K}_\lambda$ for each $v \in \mathbb{R}^d$. Indeed,

$$\left(\bar{K}_\lambda(\varphi_v)\right)(\hat{r}) = \|K_\lambda\|^{-1} \int d\Omega(\hat{r}')e^{-\lambda\psi(\theta(\hat{r},\hat{r}'))}v \cdot \hat{r}'$$

is a linear function of v, invariant under simultaneous rotations of v and $\hat{r}$, and hence proportional to φ_v. From this we obtain

$$\langle \hat{r}_1 \cdot \hat{r}_{n+1} \rangle_\lambda = \beta_1(\lambda)^n. \tag{2.137}$$

In particular, the tangent–tangent correlation length is given by

$$\xi(\lambda) = -\frac{\log\langle \hat{r}_1 \cdot \hat{r}_{n+1} \rangle_\lambda}{n} = \frac{1}{-\log|\beta_1(\lambda)|}.$$

It diverges if and only if $|\beta_1(\lambda)| \to 1$ which is the case exactly if $\lambda \to +\infty$. If $\lambda(a)$ is chosen as in Eq. (2.127), we obtain the continuum limit T of the tangent–tangent correlation function as

$$T(s) = \lim_{a \to 0} \langle \hat{r}_1 \cdot \hat{r}_{s/a} \rangle_{\lambda(a)} = e^{-c_1\alpha_1\lambda_R s} \tag{2.138}$$

where we have used Eq. (2.124) with $\ell = 1$. To the extent that the tangent–tangent correlation function measures the stiffness of paths we see that λ_R is a measure of the rigidity of continuum paths.

It is important to note that the scaling $n \to na = s$ in Eq. (2.138) of the internal distance on paths is identical to the scaling $x \to xa$ of coordinates in embedding space $\mathbb{R}^d$. This shows that the tangents to paths are correlated not only as functions of intrinsic distance n, but also in embedding space $\mathbb{R}^d$. If we define the correlation function

$$\langle \hat{r}_{x_1} \cdot \hat{r}_x \rangle = \frac{1}{G_{\mu,\lambda}(0,x)} \sum_{n=1}^{\infty} \int \prod_{i=1}^{n-1} dx_i \tag{2.139}$$

$$\times \exp\left(-\sum_{i=1}^{n} \varphi(|x_i - x_{i-1}|) - \lambda\sum_{i=1}^{n-1} \psi(\theta_i) - \mu n\right) \hat{r}_{x_1} \cdot \hat{r}_x,$$

where $\hat{r}_{x_i} = (x_i - x_{i-1})/|x_i - x_{i-1}|$, $x_0 = 0, x_n = x$, then it falls off exponentially with $|x|$ in the scaling limit $a \to 0$. This strongly suggests that generic continuum paths are "close" to being straight. In order to exhibit this feature in a quantitative form we introduce the concept of *Hausdorff dimension* d_H of paths.

Let us first define the expectation value of an observable $\mathcal{O}$ on paths with a free endpoint by

$$\langle \mathcal{O} \rangle = \frac{1}{\chi(\mu, \lambda)} \sum_{n=1}^{\infty} \int \prod_{i=1}^{n} dx_i \, \mathcal{O} \tag{2.140}$$

$$\times \exp\left(-\sum_{i=1}^{n} \varphi(|x_i - x_{i-1}|) - \lambda \sum_{i=1}^{n-1} \psi(\theta_i) - \mu n\right).$$

We define the mean square extent $\langle x^2 \rangle_{\mu, \lambda}$ of paths by Eq. (2.140) with $\mathcal{O} = x_n^2$, and the average internal length of paths, $\langle n \rangle_{\mu, \lambda}$, by Eq. (2.140) with $\mathcal{O} = n$. Note that

$$\langle x^2 \rangle_{\mu, \lambda} = -\chi(\mu, \lambda)^{-1} \nabla^2 \widehat{G}_{\mu, \lambda}(0) \tag{2.141}$$

and

$$\langle n \rangle_{\mu, \lambda} = -\chi(\mu, \lambda)^{-1} \frac{\partial}{\partial \mu} \chi(\mu, \lambda). \tag{2.142}$$

We define the *Hausdorff dimension* by the asymptotic relation

$$\langle n \rangle_{\mu(a), \lambda(a)} \sim \langle x^2 \rangle_{\mu(a), \lambda(a)}^{\frac{1}{2} d_H} \tag{2.143}$$

as $a \to 0$.

For dimensional reasons we have

$$\langle x^2 \rangle_{\mu(a), \lambda(a)} \sim a^{-2} \tag{2.144}$$

as $a \to 0$, in the absence of infrared divergences. The asymptotic behaviour of $\langle n \rangle_{\mu(a), \lambda(a)}$ is most easily obtained from that of the susceptibility $\chi(\mu, \lambda)$ for μ close to $\mu_0(\lambda)$ which to leading order in $\mu - \mu_0(\lambda)$ is given by

$$\chi(\mu, \lambda) \sim (\mu - \mu_0(\lambda))^{-\gamma} \tag{2.145}$$

as $\mu \to \mu_0(\lambda)$, where γ is called the *susceptibility exponent*, which here is assumed to be positive. It follows from Eq. (2.131) and Eq. (2.132) with $p = 0$ that

$$\gamma = 1 \tag{2.146}$$

for all finite values of λ. From Eq. (2.142) we obtain

$$\langle n \rangle_{\mu, \lambda} \sim (\mu - \mu_0(\lambda))^{-1} \tag{2.147}$$

as $\mu \to \mu_0(\lambda)$. In particular,

$$\langle n \rangle_{\mu(a), \lambda(a)} \sim (\mu(a) - \mu_0(\lambda(a)))^{-1} \tag{2.148}$$

as $a \to 0$ provided the asymptotic form (2.147) holds uniformly in a as $a \to 0$, in case (ii) of Theorem 2.2. Combining Eq. (2.148) with Eq. (2.131) and Eq. (2.127), it follows that

$$\langle n \rangle_{\mu(a), \lambda(a)} \sim \begin{cases} a^{-2} & \text{in case (i)} \\ a^{-1} & \text{in case (ii)} \end{cases} .$$

By Eq. (2.143) and Eq. (2.144) this means that

$$d_H = \begin{cases} 2 & \text{in case (i)} \\ 1 & \text{in case (ii)} \end{cases}. \tag{2.149}$$

For later reference it is useful to note that the above discussion of the Hausdorff dimension applies to the more general situation where the Fourier transformed two-point function is given as a scaling limit of the form

$$\widehat{G}(p) = \lim_{a \to 0} a^{2-\eta} \widehat{G}_{\mu(a)}(ap), \tag{2.150}$$

where $\widehat{G}_\mu$ is a regularized two-point function parametrized by the coupling constant μ (and possibly other coupling constants) such that $\mu(a) - \mu_0(a) \to 0$ as $a \to 0$ for some given function μ_0. The exponent η, which is assumed to be ≤ 2, is called the *anomalous dimension* of the two-point function. With the susceptibility defined as above it follows by inserting $p = 0$ that, provided $\widehat{G}(0)$ is finite, we have

$$\chi(\mu(a)) \sim a^{\eta-2} \tag{2.151}$$

as $a \to 0$. By definition, any mass parameter $m(\mu)$ associated with the scaling limit given by Eq. (2.150) behaves as

$$m(\mu(a)) \sim a \tag{2.152}$$

as $a \to 0$. Thus, if the critical exponents γ and v of the susceptibility and the mass, respectively, are defined by

$$\chi(\mu(a)) \sim |\mu(a) - \mu_0(a)|^{-\gamma} \quad \text{and} \quad m(\mu(a)) \sim |\mu(a) - \mu_0(a)|^v \tag{2.153}$$

as $a \to 0$, we obtain Fisher's scaling relation

$$\gamma = v(2 - \eta) \tag{2.154}$$

from Eq. (2.151) and Eq. (2.152). *The exponent v is related to the geometry of paths* by the scaling relation

$$d_H = v^{-1}. \tag{2.155}$$

This relation is established by the following argument. From Eq. (2.152) and Eq. (2.153) we see that $a \sim |\mu(a) - \mu_0(a)|^v$. Therefore, by Eq. (2.144), $\langle x^2 \rangle \sim |\mu(a) - \mu_0(a)|^{-2v}$. It follows from Eq. (2.147) and Eq. (2.143) that $\langle x^2 \rangle \sim |\mu(a) - \mu_0(a)|^{-2/d_H}$, which implies Eq. (2.155). In particular, we see that

$$v = \frac{1}{2} \quad \text{and} \quad \eta = 0 \quad \text{in case (i)},$$

and

$$v = 1 \quad \text{and} \quad \eta = 1 \quad \text{in case (ii)},$$

and $\gamma = 1$ in both cases.

We shall henceforth refer to case (ii) as the rigid random walk and to case (i) as the Wiener random walk. A different example of a random walk model with Hausdorff dimension 1 is the Ornstein–Uhlenbeck process [120]. Although this process is generated by a non-geometric action it can nevertheless be treated by the techniques developed in this section. For details we refer to [30], where a lattice version of the rigid random walk is also discussed. On the lattice one can compute the propagator of the rigid walk explicitly, and in the scaling limit rotational invariance is not recovered, contrary to the lattice Wiener walk which has a Euclidean invariant scaling limit.

As a byproduct of the analysis presented here we mention an application to one-dimensional spin chains. Let us view the directions $\hat{r}_i \in S^{d-1}$ of the steps in a walk as classical spin variables with nearest neighbour interaction $\psi(\theta(\hat{r}_i, \hat{r}_{i+1}))$ and with $\lambda = \frac{1}{kT}$, where T is the temperature. The corresponding transfer matrix is $\bar{K}_\lambda$ and for bounded functions $f_1, \ldots, f_n$ of the spin variables $\hat{r}_1, \ldots, \hat{r}_n$ the expectation value of $f_1(\hat{r}_1) \ldots f_n(\hat{r}_n)$ is given by

$$\langle f_1(\hat{r}_1) \ldots f_n(\hat{r}_n) \rangle_\lambda = \int d\Omega(\hat{r}_1) \ldots d\Omega(\hat{r}_n) f_1(\hat{r}_1) \ldots f_n(\hat{r}_n)$$
$$\times \bar{K}_\lambda(\hat{r}_1, \hat{r}_2) \ldots \bar{K}_\lambda(\hat{r}_{n-1}, \hat{r}_n)$$

which can also be written as

$$\langle f_1(\hat{r}_1) \ldots f_n(\hat{r}_n) \rangle_\lambda = \langle 1, f_1 \bar{K}_\lambda f_2 \ldots \bar{K}_\lambda f_n 1 \rangle, \qquad (2.156)$$

where $f_1, \ldots, f_n$ are treated as multiplication operators on $L^2(S^{d-1}, d\Omega)$. From Eq. (2.126) we obtain, for $t \geq 0$,

$$\bar{K}_{\lambda(n)}^{tn} \varphi \rightarrow e^{-tL} \varphi \qquad (2.157)$$

as $n \rightarrow +\infty$ for all $\varphi \in L^2(S^{d-1}, d\Omega)$, if $\lambda(n)$ is chosen such that

$$nC(\lambda(n)) \rightarrow 1 \quad \text{as} \quad n \rightarrow \infty. \qquad (2.158)$$

Applying Eq. (2.157) to the expectation values (2.156) we find that

$$\lim_{i \to \infty} \langle f_1(\hat{r}_{t_1 n_i}) f_2(\hat{r}_{(t_1 + t_2) n_i}) \ldots f_k(\hat{r}_{(t_1 + \ldots + t_k) n_i}) \rangle_{\lambda(n_i)}$$
$$= \langle 1, f_1 e^{-t_2 L} f_2 \ldots f_{k-1} e^{-t_k L} f_k 1 \rangle \qquad (2.159)$$

for arbitrary rational values of $t_1, t_2, \ldots, t_k \geq 0$ and any increasing sequence $n_1 < n_2 < n_3 < \ldots$, such that $t_1 n_i, \ldots, t_k n_i$ are integers for all i, which is required for the left-hand side of Eq. (2.159) to make sense. Since the probability density associated with diffusion on the sphere S^{d-1} is given by the kernel of e^{-tL}, where t denotes time, we see that Eq. (2.159) expresses the fact that the scaling limit of the spin chain at the critical point at zero temperature equals Brownian motion on S^{d-1}. This result was first proved for the Heisenberg spin chain, i.e. for

$$\psi(\theta) = 1 - \cos\theta,$$

in [235]. The result (2.159) exhibits its universal character.

2.4 Fermionic random walks

In this section we present a random walk representation of the Dirac operator, valid in any dimension. It is a natural generalization of the Wiener random walk and has certain features in common with the rigid random walk, but is quite different since paths are endowed with an internal structure associated to a spin degree of freedom. To each piecewise linear path we assign a matrix valued weight, which is given by the product of the rotation matrices describing the turns of the particle at each vertex of the path.

In order to define the model we start with an irreducible representation γ_μ, $\mu = 1, \ldots, d$, of the d-dimensional Clifford algebra, i.e.

$$\gamma_\mu = \gamma_\mu^* \tag{2.160}$$

and

$$\{\gamma_\mu, \gamma_\nu\} = 2\delta_{\mu\nu}\,1, \tag{2.161}$$

where $\{\cdot\,,\,\cdot\}$ denotes the anticommutator and 1 is the unit matrix. We recall [91] that this representation is unique up to unitary equivalence and its dimension is

$$M = 2^{[\frac{d}{2}]}, \tag{2.162}$$

where $[\frac{d}{2}]$ denotes the integer part of $d/2$. Then the matrices $s_{\mu\nu}$ defined by the commutator

$$s_{\mu\nu} = \frac{i}{4}\,[\gamma_\mu, \gamma_\nu] \tag{2.163}$$

form a representation of the Lie algebra of the d-dimensional rotation group $SO(d)$.

Given two (linearly independent) vectors $r, r' \in \mathbb{R}^d$ we let n be the normalized antisymmetric tensor of rank two associated with the oriented plane in $\mathbb{R}^d$ spanned by r and r'. In coordinates n reads

$$n_{\mu\nu} = \frac{\hat{r}_\mu \hat{r}'_\nu - \hat{r}_\nu \hat{r}'_\mu}{\sin\theta}, \tag{2.164}$$

where $\hat{r} = r/|r|$, $\hat{r}' = r'/|r'|$ and $\theta = \theta(\hat{r}, \hat{r}')$ is the angle between $\hat{r}$ and $\hat{r}'$. In three dimensions n is the (normalized) cross-product of r and r'. The matrix

$$K(\hat{r},\hat{r}') = \exp\left(-\frac{i}{2}\sum_{\mu,\nu} n_{\mu\nu} s_{\mu\nu}\theta(\hat{r},\hat{r}')\right) \qquad (2.165)$$

represents the rotation in the plane spanned by r and r' which rotates $\hat{r}'$ to $\hat{r}$. In particular

$$K(\hat{r},\hat{r}')\gamma \cdot \hat{r}' K(\hat{r},\hat{r}')^{-1} = \gamma \cdot \hat{r}. \qquad (2.166)$$

Of course, $K(\hat{r},\hat{r}')$ is only determined up to a sign by this property. Our choice of sign ensures that $K(\hat{r},\hat{r}')$ is continuous in $\hat{r}$ and $\hat{r}'$ and equals the identity matrix 1 for $\hat{r} = \hat{r}'$. Let

$$\mathscr{H} = L^2(S^{d-1}) \otimes \mathbb{C}^M \qquad (2.167)$$

denote the space of square integrable functions on S^{d-1} with values in $\mathbb{C}^M$. We can regard K as an operator on $\mathscr{H}$ whose kernel is given by Eq. (2.165).

We now define the *fermionic random walk* by replacing the operator K_λ, entering the expression (2.107) for the propagator of the rigid random walk, by the operator K. Thus instead of the smoothing factor associated with extrinsic curvature we attribute a spin rotation matrix to each vertex of a piecewise linear path. *This unitary matrix factor causes cancellations among paths such that, effectively, smooth paths dominate the two-point function and the Hausdorff dimension of paths in this model equals one, as for the rigid random walk. This does not require any fine tuning of coupling constants as is required in the case of rigid paths.*

Before writing down the two-point function let us note that according to the arguments given in the previous section (see Eq. (2.120)) only the constant mode on S^{d-1} propagates in the continuum limit representing the Wiener random walk. In other words, the continuum limit of the operator, whose matrix element defines the two-point function as given by Eq. (2.110), vanishes on the orthogonal complement to 1. This should be contrasted with the rigid random walk for which the corresponding operator, by Eq. (2.129), equals

$$(\mu_R + \lambda_R L - ip \cdot \hat{r})^{-1}. \qquad (2.168)$$

If the Fourier transformed propagator in the rigid case is defined as a function of p as well as of the directions $\hat{r}$ and $\hat{r}'$ of the initial and final steps of the walk, then the continuum limit equals the kernel of the operator (2.168).

In the present case it is crucial to take into account the $\hat{r}$-degree of freedom as it turns out that the propagating modes constitute a finite-dimensional subspace of $\mathscr{H}$ carrying a representation of the Clifford algebra. Fixing the directions $\hat{r}_1 = \hat{r}$ and $\hat{r}_n = \hat{r}'$ of the initial and final steps, the two-point function $G_\mu(\hat{r}, \hat{r}'; x)$ is defined by

$$
\begin{aligned}
G_\mu(\hat{r}, \hat{r}'; x) &= \sum_{n=1}^{\infty} e^{-\mu n} \int \prod_{i=2}^{n-1} dr_i |r_1|^{d-1} d|r_1| |r_n|^{d-1} d|r_n| \\
&\quad \times \prod_{i=1}^{n} e^{-\varphi(|r_i|)} \prod_{i=1}^{n-1} K(\hat{r}_i, \hat{r}_{i+1}) \delta\left(x - \sum_{i=1}^{n} r_i\right), \quad (2.169)
\end{aligned}
$$

where φ is a real function, and the last product is ordered along the path. Due to the matrix character of K the definition (2.169) is not written in terms of an action functional as in the previously considered models. Neither do any fermionic fields appear in Eq. (2.169), so the title of this section might seem a misnomer. However, in addition to the fact that (2.169) converges to the Dirac propagator in the scaling limit, it is possible to argue [328] that the continuum version of Eq. (2.169) can be obtained from a supersymmetric (non-local) action after integrating over the fermionic degrees of freedom.

The Fourier transformed two-point function can be written as

$$
\begin{aligned}
\widehat{G}_\mu(\hat{r}, \hat{r}'; p) &= \int dx\, G_\mu(\hat{r}, \hat{r}'; x)\, e^{-ip \cdot x} \quad (2.170) \\
&= \sum_{n=1}^{\infty} e^{-\mu n} \int \prod_{i=2}^{n-1} d\Omega(\hat{r}_i) \prod_{i=1}^{n} F(p \cdot \hat{r}_i) \prod_{i=1}^{n-1} K(\hat{r}_i, \hat{r}_{i+1}),
\end{aligned}
$$

where $F(p \cdot \hat{r})$ is given by Eq. (2.108). Thus, $\widehat{G}_\mu(\hat{r}, \hat{r}'; p)$ is the kernel of the operator

$$
\widehat{G}_\mu(p) = (1 - e^{-\mu} F_p K)^{-1} e^{-\mu} F_p, \quad (2.171)
$$

where F_p denotes the multiplication operator on $\mathcal{H}$ corresponding to the function $F(p \cdot \hat{r})$.

We now proceed to determine a continuum limit of $\widehat{G}_\mu(p)$ as in the previous section. We note that K is a self-adjoint Hilbert–Schmidt operator on $\mathcal{H}$. Using

$$
\left(\sum_{\mu,\nu} n_{\mu\nu} s_{\mu\nu}\right)^2 = 1
$$

we have

$$
K(\hat{r}, \hat{r}') = \cos \frac{1}{2}\theta(\hat{r}, \hat{r}')\, 1 - \sin \frac{1}{2}\theta(\hat{r}, \hat{r}') \sum_{\mu,\nu} n_{\mu\nu} s_{\mu\nu},
$$

from which it follows that

$$
\int_{S^{d-1}} d\Omega(\hat{r}')\, K(\hat{r}, \hat{r}') = C_d\, 1, \quad (2.172)
$$

where

$$C_d = \int_{S^{d-1}} d\Omega(\hat{r}') \, \cos \frac{1}{2}\theta(\hat{r},\hat{r}'). \tag{2.173}$$

We normalize K by setting

$$\bar{K} = C_d^{-1}K \tag{2.174}$$

and similarly F_p is normalized to $\bar{F}_p = F_0^{-1}F_p$ as in the previous section, where F_0 is given by Eq. (2.111).

It follows from Eq. (2.172) that constant spinors, i.e. constant functions in $\mathscr{H}$, are eigenvectors of $\bar{K}$ corresponding to the eigenvalue 1, and similarly Eq. (2.166) and Eq. (2.172) imply that the columns of the matrix $\hat{r} \cdot \gamma$ are eigenvectors of $\bar{K}$, also with eigenvalue 1. We shall need the following lemma.

Lemma 2.3 *Let V denote the subspace of $\mathscr{H}$ spanned by the constant spinors and the columns of the matrix $\hat{r} \cdot \gamma$. Then V is the eigenspace of $\bar{K}$ corresponding to eigenvalue 1 and all other eigenvalues of $\bar{K}$ are numerically smaller than 1.*

Proof Denoting by Tr the trace on $\mathscr{H}$ and by tr the trace on $\mathbb{C}^M$, the square of the Hilbert–Schmidt norm of the operator $\bar{K}$ is given by

$$\begin{aligned}
\mathrm{Tr}\,\bar{K}^2 &= \int d\Omega(\hat{r})d\Omega(\hat{r}') \, \mathrm{tr}\left[\bar{K}(\hat{r},\hat{r}')\bar{K}(\hat{r}',\hat{r})\right] \\
&= C_d^{-2} \int d\Omega(\hat{r})d\Omega(\hat{r}') \, \mathrm{tr}\left[K(\hat{r},\hat{r}')K(\hat{r}',\hat{r})\right] \\
&= M(C_d^{-1}\omega_d)^2, \tag{2.175}
\end{aligned}$$

where M and C_d are given by Eqs. (2.162) and (2.173) and we have used the fact that

$$K(\hat{r}',\hat{r}) = K(\hat{r},\hat{r}')^{-1}.$$

A simple calculation yields

$$C_d = 2^{d-2}\frac{\Gamma\left(\frac{d-1}{2}\right)\Gamma\left(\frac{d}{2}\right)}{\Gamma\left(d-\frac{1}{2}\right)}\omega_{d-1},$$

which implies

$$\mathrm{Tr}\,\bar{K}^2 = M\left(\frac{\Gamma\left(d-\frac{1}{2}\right)\sqrt{\pi}}{\Gamma\left(\frac{d}{2}\right)^2 2^{d-2}}\right)^2. \tag{2.176}$$

Now let $\lambda_1, \lambda_2, \ldots$ denote the distinct eigenvalues of $\bar{K}$ restricted to $V^\perp$ and let $m_1, m_2, \ldots$ denote their multiplicities. We then have

$$\text{Tr}\,\bar{K}^2 = 2M + \sum_{i=1}^{\infty} m_i \lambda_i^2,$$

which by Eq. (2.176) implies

$$\sum_{i=1}^{\infty} m_i \lambda_i^2 M^{-1} = \left(\frac{\Gamma\left(d - \frac{1}{2}\right)\sqrt{\pi}}{\Gamma\left(\frac{d}{2}\right)^2 2^{d-2}} \right)^2 - 2. \qquad (2.177)$$

Let $\pi : Spin(d) \mapsto SO(d)$ denote the canonical covering homomorphism. The action of $Spin(d)$ on $\mathcal{H}$ is given by

$$(Of)(\hat{r}) = \varphi(O)f(\pi(O)^{-1}\hat{r}),$$

where $O \in Spin(d)$ and φ denotes the representation whose generators are given by Eq. (2.163). The operator K commutes with the action of $Spin(d)$ on $\mathcal{H}$. This fact follows from

$$\varphi(O)^{-1}K(\hat{r}, \hat{r}')\varphi(O) = K(\pi(O)^{-1}\hat{r},\ \pi(O)^{-1}\hat{r}').$$

Decomposing $\mathcal{H}$ into irreducible components under $Spin(d)$ it follows that $\bar{K}$ is a multiple of the identity on each of these, so the multiplicities m_i are sums of integer multiples of dimensions of irreducible representations of $Spin(d)$. The lowest possible dimensions of these are known [91] to be $2^{[\frac{d}{2}]}$ if d is odd, and $2^{\frac{d}{2}-1}$ if d is even. Combining this observation with Eq. (2.177) and recalling that $M = 2^{[\frac{d}{2}]}$ we conclude that it suffices to verify that the right-hand side of Eq. (2.177) is less than $\frac{1}{2}$. This is proved by induction from d to $d+2$ after having checked it explicitly for $d = 2$ and $d = 3$. This concludes the proof of the lemma.

Lemma 2.3 implies, by Eq. (2.171), that $\widehat{G}_\mu(p)$ is well defined for $\mu > \mu_0$, where the critical coupling μ_0 is given by

$$e^{\mu_0} = C_d F_0, \qquad (2.178)$$

in analogy with Eq. (2.114). We can then rewrite Eq. (2.171) in the form

$$\begin{aligned}
\widehat{G}_\mu(p) &= e^{-\mu}\left(1 - e^{\mu_0 - \mu}\bar{F}_p\bar{K}\right)^{-1} F_p & (2.179) \\
&= e^{-\mu}H_\mu^{-\frac{1}{2}}\left(1 - e^{\mu - \mu_0}H_\mu^{-\frac{1}{2}}(\bar{F}_p - 1)\bar{K}H_\mu^{-\frac{1}{2}}\right)^{-1} H_\mu^{-\frac{1}{2}}F_p,
\end{aligned}$$

where we have introduced

$$H_\mu = 1 - e^{\mu_0 - \mu}\bar{K}. \qquad (2.180)$$

Denoting by Q the orthogonal projection onto the subspace V it follows from the lemma that

$$(\mu - \mu_0)^{\frac{1}{2}} H_\mu^{-\frac{1}{2}} \to Q \tag{2.181}$$

in norm as $\mu \to \mu_0$. In analogy with Eq. (2.115) we have

$$\frac{\bar{F}(ap \cdot \hat{r}) - 1}{a} \to i c_1 p \cdot \hat{r} \tag{2.182}$$

uniformly in $\hat{r}$ on S^{d-1} as $a \to 0$. We conclude from Eqs. (2.179), (2.181) and (2.182) that if we choose $\mu(a) = \mu_0 + m_1 a + o(a)$, then

$$a\widehat{G}_{\mu(a)}(ap) \to e^{-\mu_0} Q(m_1 - c_1 Qip \cdot \hat{r}Q)^{-1} Q \cdot F_0 \tag{2.183}$$

as $a \to 0$. In other words, the scaling limit of $\widehat{G}_\mu(p)$ vanishes on the orthogonal complement of V, whereas on V it is given by

$$\widehat{G}(p) = C_d^{-1}(m_1 - c_1 Qip \cdot \hat{r}Q)^{-1}. \tag{2.184}$$

We now show that the operators

$$\gamma_\mu' = dQ\hat{r}_\mu Q, \tag{2.185}$$

form a representation of the d-dimensional Clifford algebra, in fact a direct sum of two irreducible ones. As before, the operator $\hat{r}_\mu$ acts on $\mathscr{H}$ by multiplication by the coordinate function $\hat{r}_\mu$. First, we note that by Eqs. (2.160) and (2.161) the operators

$$P^\pm = \frac{1}{2}(1 \pm \hat{r} \cdot \gamma),$$

are orthogonal projections in $\mathscr{H}$ fulfilling

$$P^+ P^- = 0 \quad \text{and} \quad P^+ + P^- = 1.$$

By Eq. (2.166) these projections commute with K so the eigenspace V is invariant under $P^\pm$ and we have

$$V = V^+ \oplus V^-,$$

where

$$V^\pm = P^\pm V.$$

Letting $e_i^\pm$, $i = 1, \dots, M$, denote the ith column of the matrix-valued function $\frac{1}{2}(1 \pm \hat{r} \cdot \gamma)$, it follows that these vectors span V. It is, furthermore, easily checked that

$$\langle e_i^+, e_j^+ \rangle = \langle e_i^-, e_j^- \rangle = \frac{1}{2}\omega_d \delta_{ij}$$

and

$$\langle e_i^+, e_j^- \rangle = 0$$

for all $i, j = 1, \ldots, M$, where $\langle \cdot, \cdot \rangle$ denotes the standard inner product on $\mathscr{H}$ and ω_d is the volume of S^{d-1}. Thus the vectors $f_i^\pm = (\tfrac{1}{2}\omega_d)^{-1/2} e_i^\pm$, $i = 1, \ldots, M$, form an orthonormal basis for $V^\pm$. The matrix elements of the operators $Q\hat{r}_\mu Q$ with respect to this basis are

$$
\begin{aligned}
\langle f_i^+, \hat{r}_\mu f_j^+ \rangle &= \frac{1}{2\omega_d} \int d\Omega(\hat{r}) \left((1 + \hat{r} \cdot \gamma)\hat{r}_\mu (1 + \hat{r} \cdot \gamma) \right)_{ij} \\
&= \frac{1}{\omega_d} \int d\Omega(\hat{r}) (\hat{r} \cdot \gamma)_{ij} \hat{r}_\mu = \frac{1}{\omega_d} (\gamma_\mu)_{ij} \int d\Omega(\hat{r}) \, \hat{r}_\mu^2 \\
&= \frac{1}{d} (\gamma_\mu)_{ij},
\end{aligned}
$$

where we have used the fact that $d\Omega$ is invariant under $\hat{r} \to -\hat{r}$. Similarly, we obtain

$$
\langle f_i^-, \hat{r}_\mu f_j^- \rangle = -\frac{1}{d} (\gamma_\mu)_{ij}
$$

and

$$
\langle f_i^+, \hat{r}_\mu f_j^- \rangle = 0.
$$

This proves our claim that the operators $dQ\hat{r}_\mu Q$ form a direct sum of two irreducible representations of the Clifford algebra. Moreover, expressing $\hat{G}_\mu(p)$ on V with respect to the basis $f_i^\pm$, $i = 1, \ldots, M$, we have proved the following result.

Theorem 2.4 *Let $\hat{G}_\mu(\hat{r}, \hat{r}'; p)$ be defined by Eq. (2.170) and let $\hat{G}_\mu(p)$ denote the operator on $\mathscr{H} = L_2(S^{d-1}) \otimes \mathbb{C}^M$ whose kernel is $\hat{G}_\mu(\hat{r}, \hat{r}'; p)$. Then $\hat{G}_\mu(p)$ is well defined for $\mu > \mu_0$, where μ_0 is given by Eq. (2.178).*

If the coupling constant μ is chosen as a function of the cutoff $a > 0$ such that

$$
\mu(a) - \mu_0 = m_1 a + o(a) \tag{2.186}
$$

for some constant $m_1 > 0$, then the scaling limit $\hat{G}(p)$ defined by

$$
\hat{G}(p) = \lim_{a \to 0} a \hat{G}_{\mu(a)}(ap) \tag{2.187}
$$

exists in the sense of norm convergence and vanishes on the orthogonal complement of V. On V it is given by

$$
\hat{G}(p) = \begin{pmatrix} \frac{c}{m - ip \cdot \gamma} & 0 \\ 0 & \frac{c}{m + ip \cdot \gamma} \end{pmatrix} \tag{2.188}
$$

with respect to a suitable orthonormal basis for V, where c and m are positive constants.

We recognize $\hat{G}(p)$ as the direct sum of two copies of the d-dimensional Euclidean Dirac propagator. This doubling phenomenon is related to the Nielsen–Ninomiya theorem [311] and cannot be avoided.

We also note that the scaling used in Eq. (2.134) for the rigid random walk is identical to the one used here in Eq. (2.187). In other words, the critical exponents γ and ν defined by Eq. (2.153) are identical and are given by

$$\gamma = 1, \qquad \nu = 1,$$

which by Eq. (2.155) leads to

$$d_H = 1.$$

The susceptibility, as defined in Eq. (2.106), is an operator in the present case. In order to define γ it suffices to take a matrix element of this operator which does not vanish in the continuum limit, or, alternatively, take its Hilbert–Schmidt norm, since all non-vanishing matrix elements scale in the same way.

One can modify the random walk model considered here by including, in addition, the smoothing factor K_λ of the previous section and considering its rigid limit $\lambda \to +\infty$ and $\mu \to -\infty$. The Hausdorff dimension is again 1 and the continuum limit of the two-point function is given by

$$(\mu_R + \lambda_R(J^2 - s^2) - ip \cdot \hat{r})^{-1},$$

where J^2 is the square of the total angular momentum operator and $s^2 = \sum_{\mu,\nu} s_{\mu\nu}^2$ is the square of the spin operator on $\mathcal{H}$. Letting the effective rigidity λ_R tend to $+\infty$, only the zero modes of $J^2 - s^2$ survive. Since these exactly constitute the subspace V introduced in the lemma we recover in this limit the Dirac propagator (2.188). We refer to [32, 247] for details of the argument.

2.5 Branched polymers

We close this chapter by a brief discussion of one more generalization of the Wiener random walk, which turns out to be of importance for the understanding of the critical behaviour of random surface models, as well as higher-dimensional quantum gravity models to be introduced in the subsequent chapters.

2.5.1 Extrinsic properties

Rather than modifying the weights of paths as in the generalizations discussed up to now, the model we have in mind is obtained by modifying the geometry of the objects under consideration by allowing branchings of the paths at their vertices. The objects obtained in this fashion are called *branched polymers* and are defined more precisely as follows.

Let $\mathbb{R}^2$ be the plane with a fixed orientation. By a graph G on $\mathbb{R}^2$ we mean a finite collection $L(G)$ of line segments in $\mathbb{R}^2$, called *links*, such that any two links are either disjoint or share one endpoint. An endpoint of a link in G is called a *vertex* of G and the set of vertices of G is denoted by $V(G)$. The number σ_i of links sharing the vertex $i \in V(G)$ is called the *order* of i, and these σ_i links are ordered cyclically in a positive direction around i. Now let G_1 and G_2 be two graphs and assume that there exists a bijective mapping $\varphi : L(G_1) \mapsto L(G_2)$ such that two links $\ell, \ell' \in L(G_1)$ share a vertex $i \in V(G_1)$ if and only if $\varphi(\ell)$ and $\varphi(\ell')$ share a vertex in G_2. Denoting the vertex in $V(G_2)$ corresponding to $i \in V(G_1)$ by $\varphi'(i)$, it follows that φ' is a well-defined bijective mapping from $V(G_1)$ onto $V(G_2)$ and that the order of i equals the order of $\varphi'(i)$ for each vertex i of G_1. If φ also preserves the cyclic ordering of links at each vertex we identify G_1 and G_2.

Similarly, given two graphs G_1 and G_2 with n (ordered) marked vertices $v_1^1, \ldots, v_n^1$ and $v_1^2, \ldots, v_n^2$, respectively, we identify G_1 and G_2 provided there exists a mapping φ as above such that $\varphi'(v_i^1) = v_i^2$, $i = 1, \ldots, n$.

Under these identifications we define a *planar branched polymer* b with n marked vertices to be a connected *tree graph* with n marked vertices, i.e. any two vertices in b can be connected by a sequence of links in b and no closed loop can be formed by any sequence of (pairwise different) links in b. We shall henceforth denote the set of planar branched polymers with n marked vertices by B_n and we usually drop the adjective "planar" from the nomenclature.

An embedding of a branched polymer b into $\mathbb{R}^d$ is a mapping $\omega : V(b) \mapsto \mathbb{R}^d$. In the special case where b is a path, i.e. all vertices in b except two of them have order 2, such an embedding is nothing but a random walk in the language of previous sections. We shall often refer to an embedding $\omega : V(b) \mapsto \mathbb{R}^d$ as a branched polymer.

We consider action functionals on branched polymers ω as above of the form

$$S(\omega) = \sum_{\langle ij \rangle} \varphi(|x_i - x_j|), \tag{2.189}$$

where φ is a real valued function on $\mathbb{R}$ and $x_i = \omega(i)$ for $i \in V(b)$ and the summation is over all nearest neighbour pairs $\langle ij \rangle$ of vertices in b. Vertices i and j are called neighbours if they are connected by a link in b.

Associated with the new degree of freedom represented by the order of vertices we introduce a weight factor $w_m \geq 0$ for each vertex of order $m = 1, 2, 3, \ldots$. Setting

$$D_b\omega = \prod_{i \notin \{i_1, \ldots, i_n\}} dx_i$$

for $b \in B_n$, where $i_1, \ldots, i_n$ denote the marked vertices in b, we define the n-point function $G_\mu^{(n)}$ for $n \geq 1$ by

$$G_\mu^{(n)}(x_1, \ldots, x_n) = \sum_{b \in B_n} \prod_{i \in V(b)} w_{\sigma_i} \int_{(x_1, \ldots, x_n)} D_b \omega \, e^{-S(\omega)} e^{-\mu |b|}, \qquad (2.190)$$

where $|b|$ denotes the number of links in b and $\int_{(x_1, \ldots, x_n)}$ indicates that ω is constrained such that $\omega(i_\ell) = x_\ell$, $\ell = 1, \ldots, n$. We shall see that there exists a critical coupling μ_0 such that (i) $G_\mu^{(n)}$ is a well-defined and finite function on $(\mathbb{R}^d)^n$ for $\mu > \mu_0$ and all $n = 1, 2, 3, \ldots$, (ii) $G_\mu^{(n)}$ depends analytically on μ for $\mu > \mu_0$ and is singular at μ_0.

Let us start by considering the one-point function $G_\mu^{(1)} \equiv G(\mu)$ which is independent of the fixed point x_1 due to the translational invariance of the action (2.189). Setting

$$\alpha_b = \prod_{i \in V(b)} w_{\sigma_i} \int_{(x_1)} D_b \omega \, e^{-S(\omega)} \qquad (2.191)$$

for $b \in B_1$ and

$$\beta_n = \sum_{b \in B_1, |b| = n} \alpha_b, \qquad (2.192)$$

we can rewrite Eq. (2.190) as

$$G(\mu) = \sum_{n=1}^{\infty} \beta_n e^{-\mu n}. \qquad (2.193)$$

$G(\mu)$ is an analytic function of $e^{-\mu}$ in the domain of convergence of the power series (2.193). We define μ_0 such that $e^{-\mu_0}$ is the radius of convergence of this power series. Since the coefficients β_n are non-negative it follows that $G(\mu)$ is non-analytic at $\mu = \mu_0$.

We choose the function φ in Eq. (2.189) and the weights w_m such that $\mu_0 < \infty$. Obviously, this requires that

$$\int dx \, e^{-\varphi(|x|)} < \infty \qquad (2.194)$$

since for $b \in B_1$ we have

$$\int_{(x_1)} D_b \omega \, e^{-S(\omega)} = \left(\int dx \, e^{-\varphi(|x|)} \right)^{|b|}. \qquad (2.195)$$

Using Eq. (2.195) it follows that we must require the convergence radius r_0 of the power series $\sum_m w_m z^m$ to be non-vanishing. Note that $r_0 = \infty$ in all cases where we only allow branching of finite order, i.e. where $w_m = 0$ for m sufficiently large. On the other hand, the existence of an $r_0 > 0$ together with Eq. (2.194) is sufficient to ensure that $\mu_0 < \infty$. This can be seen as follows. First note that by a simple counting argument the number N_n of branched polymers b in B_1 with $|b| = n$ fulfils

$$N_n \leq 4^n. \tag{2.196}$$

We give an exact evaluation of N_n in Section 4.2. The assumption $r_0 > 0$ implies that

$$|w_m| \leq C^m$$

for some finite positive constant C. Because of the relation

$$\sum_{i \in V(b)} \sigma_i = 2|b|$$

it follows that

$$\prod_{i \in V(b)} |w_{\sigma_i}| \leq C^{2|b|}. \tag{2.197}$$

Using Eqs. (2.195), (2.196) and (2.197) in Eqs. (2.191) and (2.192) we conclude that

$$|\beta_n| \leq \left(4C^2 \int dx \, e^{-\varphi(|x|)} \right)^n$$

and hence,

$$e^{\mu_0} \leq 4C^2 \int dx \, e^{-\varphi(|x|)} < \infty.$$

Before proceeding further we note that since the number of vertices in a branched polymer equals $|b| + 1$, a scaling $w_m \to \lambda w_m$, where $\lambda > 0$, can be compensated for by a redefinition of μ so we may without loss of generality assume that

$$w_1 = 1$$

in the following. Similarly, we can assume that φ is normalized such that

$$\int e^{-\varphi(|x|)} dx = 1. \tag{2.198}$$

In order to analyse more closely the critical behaviour of the n-point function it is useful to introduce the *reduced one-point function*

$$Z(\mu) = \sum_{b \in B_1'} \prod_{i \in V(b)} w_{\sigma_i} \int_{(x_1)} D_b \omega \, e^{-S(\omega)} e^{-\mu|b|}, \tag{2.199}$$

where B_1' denotes the subset of B_1 consisting of branched polymers whose marked vertex has order 1. Clearly, $Z(\mu) < G(\mu)$ and there exists a critical point $\mu_0' \leq \mu_0$ such that $Z(\mu)$ is analytic for $\mu > \mu_0'$ and singular at μ_0'. Below we show that, actually, $\mu_0' = \mu_0$.

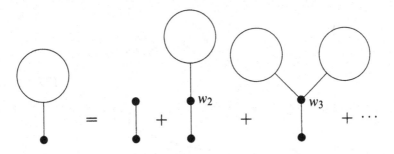

Fig. 2.5.   Graphical illustration of the identity (2.202).

Splitting the sum defining $G(\mu)$ into contributions corresponding to a fixed order of the marked vertex it follows that the mth contribution $G^{(m)}(\mu)$ fulfils

$$\frac{1}{m} w_m Z(\mu)^m \leq G^{(m)}(\mu) \leq w_m Z(\mu)^m$$

and hence

$$\sum_m \frac{1}{m} w_m Z(\mu)^m \leq G(\mu) \leq \sum_m w_m Z(\mu)^m, \qquad (2.200)$$

which in particular shows that $Z(\mu) \leq r_0$ for $\mu > \mu_0$. Moreover, defining the function

$$f(z) = \sum_{m=1}^{\infty} w_m z^{m-1} \qquad (2.201)$$

for $|z| < r_0$ one sees that $Z(\mu)$ satisfies the identity

$$Z(\mu) = e^{-\mu} + e^{-\mu} f(Z(\mu)), \qquad (2.202)$$

which is graphically illustrated in Fig. 2.5. Combining Eq. (2.202) with the second inequality in (2.200) we obtain $G(\mu) \leq Z(\mu)(e^{\mu} Z(\mu) - 1)$, which implies $\mu_0 \leq \mu_0'$ and hence $\mu_0 = \mu_0'$, as claimed.

Using the fact that $Z(\mu) \to 0$ as $\mu \to \infty$ we conclude from Eq. (2.202) that $Z(\mu)$ for $\mu > \mu_0$ is the smallest positive solution z to the equation

$$e^{\mu} = \frac{1 + f(z)}{z} \equiv F(z). \qquad (2.203)$$

There are now two cases to consider:

(i) F is decreasing in $(0, r_0)$,

(ii) F has a minimum in $(0, r_0)$.

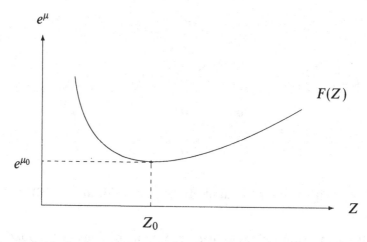

Fig. 2.6. Graphical illustration of the identity (2.203) in the generic case (ii).

In the first case the critical point μ_0 is given by

$$e^{\mu_0} = F(r_0)$$

and $Z(\mu_0) = r_0$. The behaviour of $Z(\mu)$ as $\mu \to \mu_0$ is in this case determined by the behaviour of the function f at the singular point r_0. The second case exhibits a universal character, as we now discuss. This is the case e.g. if $f(z)$ is a polynomial of degree at least 2 and $w_m > 0$ for some $m \geq 3$. In that case $r_0 = \infty$ and $F(z) \to \infty$ as $z \to \infty$. Note that the Wiener random walk is obtained for $w_2 > 0$ and $w_m = 0$ for $m \geq 3$, i.e. it belongs to case (i).

The condition for the existence of a minimum at $Z_0 \in (0, r_0)$ is, by Eq. (2.203), that $F'(Z_0) = 0$. Using that w_m is non-negative one easily verifies that the second derivative of $F(z)$ is positive on $(0, r_0)$. In particular, Z_0 is *unique and the minimum at Z_0 is quadratic* (see Fig. 2.6). We thus conclude that μ_0 is determined by

$$e^{\mu_0} = F(Z_0) \tag{2.204}$$

and

$$Z(\mu_0) - Z(\mu) \sim (\mu - \mu_0)^{\frac{1}{2}} \tag{2.205}$$

as $\mu \to \mu_0$.

Let us next consider the two-point function $G_\mu^{(2)}(x, y)$. In each branched polymer $b \in B_2$ with marked vertices i_1 and i_2 there is a unique shortest path b_0 connecting i_1 and i_2, and we can decompose b uniquely into b_0 and a number of branched polymers belonging to B_1' and rooted in the vertices of b_0 (see Fig. 2.7). Summation over the outgrowths at a given vertex i of b_0 gives rise to a factor $1 + f(Z(\mu))$ if i is an endpoint of b_0 and

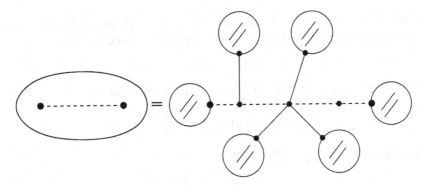

Fig. 2.7. Graphical representation of the two-point function for branched polymers. The dashed line represents the unique shortest path between the two marked vertices. The "blobs" represent the contributions from polymers rooted in vertices of this path.

otherwise a factor $f'(Z(\mu))$. Thus, we can rewrite $G_\mu^{(2)}(x, y)$ in the form

$$G_\mu^{(2)}(x, y) = [1 + f(Z(\mu))]^2 \sum_{n=1}^{\infty} e^{-\mu n} \left(f'(Z(\mu))\right)^{n-1} \qquad (2.206)$$

$$\times \int \prod_{i=1}^{n-1} dx_i \, \exp\left(-\sum_{i=1}^{n} \varphi(|x_i - x_{i-1}|)\right),$$

where $n = |b_0|$.

Comparing this expression with the form of Eq. (2.90), the corresponding two-point function $G_\mu(x, y)$ for the Wiener random walk, we see that

$$G_\mu^{(2)}(x, y) = \frac{(1 + f(Z(\mu)))^2}{f'(Z(\mu))} G_{\bar{\mu}}(x, y), \qquad (2.207)$$

where the "renormalized" coupling $\bar{\mu}$ is given by

$$\bar{\mu} = -\log\left(e^{-\mu} f'(Z(\mu))\right). \qquad (2.208)$$

In analogy with the definition Eq. (2.106) of the susceptibility χ for the ordinary random walks, we define the susceptibility χ_b of the branched polymer model by

$$\chi_b(\mu) = \int dx \, G_\mu^{(2)}(0, x).$$

It follows from Eq. (2.207) that

$$\chi_b(\mu) = \frac{(1 + f(Z(\mu)))^2}{f'(Z(\mu))} \chi(\bar{\mu}). \qquad (2.209)$$

Similarly, if the mass associated with $G_\mu^{(2)}(x,y)$ is denoted by $m_b(\mu)$ and the one corresponding to $G_{\bar\mu}$ is denoted by $m(\bar\mu)$ it follows that

$$m_b(\mu) = m(\bar\mu). \qquad (2.210)$$

Differentiating Eq. (2.203) with respect to μ one finds that

$$e^{-\mu} f'(Z(\mu)) = 1 + \frac{Z(\mu)}{Z'(\mu)}$$

and it follows from Eq. (2.205) that

$$\bar\mu \sim (\mu - \mu_0)^{\frac{1}{2}} \qquad (2.211)$$

as $\mu \to \mu_0$. Thus, in particular, as $\mu \to \mu_0$ we see that $\bar\mu$ tends to the critical coupling of the Wiener random walk model, which is 0 because of the normalization of φ given by Eq. (2.198) (see Eq. (2.82)). Recalling the values 1 and $\frac{1}{2}$ of the critical exponents of the susceptibility and mass, respectively, for the Wiener random walk, we obtain from Eqs. (2.209), (2.210) and (2.211) that

$$\chi_b(\mu) \sim \bar\mu^{-1} \sim (\mu - \mu_0)^{-\frac{1}{2}}$$

and

$$m_b(\mu) \sim \bar\mu^{\frac{1}{2}} \sim (\mu - \mu_0)^{\frac{1}{4}}$$

as $\mu \to \mu_0$. Hence,

$$\gamma = \frac{1}{2} \quad \text{and} \quad v = \frac{1}{4} \qquad (2.212)$$

for the branched polymer models of type (ii).

For the anomalous dimension η (see Eq. (2.150)) we find

$$\eta = 0 \qquad (2.213)$$

as a consequence of Eq. (2.207), which furthermore implies that the scaling limit of the two-point function is identical to that of the Wiener random walk.

The general n-point function can be treated similarly by applying the technique of summing over outgrowths. The result is an expression for $G_\mu^{(n)}(x_1,\ldots,x_n)$ in the form of a sum of terms labelled by connected tree-diagrams in $\mathbb{R}^d$ whose boundary vertices (i.e. vertices of order 1) equal $x_1,\ldots,x_n$. The term corresponding to a given diagram is given by the standard prescription of perturbation theory, with each link being represented by a two-point function $G_{\bar\mu}(x,y)$ with coupling constants associated to vertices given in terms of derivatives of the function f evaluated at $Z(\mu)$. Simple power counting shows that the dominant terms in the scaling limit are those corresponding to diagrams whose internal vertices all have order 3, i.e. they are φ^3-diagrams, and that a finite scaling limit is obtained as

$$G^{(n)}(x_1,\ldots,x_n) = \lim_{a\to 0} a^{(4-d)n+d-6} G_{\mu(a)}^{(n)}(a^{-1}x_1,\ldots,a^{-1}x_n), \qquad (2.214)$$

where $\mu(a)$ is chosen, according to Eq. (2.212), to satisfy

$$\mu(a) - \mu_0 = m_1 a^4 + o(a^4), \qquad (2.215)$$

where $m_1 > 0$ is proportional to the physical mass. It is possible to evaluate the function $G^{(n)}(x_1,\ldots,x_n)$ as a sum over connected φ^3-tree-diagrams in $\mathbb{R}^d$, but now with links represented by standard scalar propagators.

This completes our account of the scaling limits of the n-point functions, the main lesson being that they can all be expressed in a simple way in terms of the ordinary two-point function of the Wiener random walk. Nevertheless, the values of the critical exponents γ and ν are different for branched polymers and Wiener random walks, reflecting the different geometric characteristics of the underlying objects. In particular, it follows from Eqs. (2.212) and (2.155) that the Hausdorff dimension of branched polymers is

$$d_H = 4. \qquad (2.216)$$

Intuitively this is the result one would expect by superposing transversally two Wiener random walks. In explicit terms it is a consequence of the non-analytic behaviour at μ_0 of the effective coupling constant $\bar{\mu}$ obtained from the summation over outgrowths, which can be regarded as a renormalization group action by which the branched polymer model is mapped to the Wiener random walk. In subsequent chapters we shall see several applications of similar renormalization group arguments.

2.5.2 *Intrinsic properties*

An interesting geometric characteristic of the branched polymer model is the *intrinsic Hausdorff dimension* d_h, which measures the average extent of polymers in terms of an intrinsic distance scale (as opposed to the extrinsic one defining d_H). In a given polymer b the *distance* between vertices i_1 and i_2 is defined as the number of links in the unique shortest path in b connecting i_1 and i_2. Let $B_2(n)$ denote the set of branched polymers with two marked vertices i_1 and i_2 separated a distance n. The intrinsic two-point function $G_\mu^I(n)$ is defined as

$$G_\mu^I(n) = \sum_{b \in B_2(n)} \prod_{i \in V(b)} w_{\sigma_i} \int dx_2 \int_{(x_1,x_2)} D_b\omega \, e^{-S(\omega)-\mu|b|}. \qquad (2.217)$$

The intrinsic mass $m_I(\mu)$ is defined as the exponential decay rate of G_μ^I, i.e.

$$m_I(\mu) = -\lim_{n \to \infty} n^{-1} \log G_\mu^I(n). \tag{2.218}$$

It is easy to check that this limit exists. From the representation (2.206) of $G_\mu^{(2)}(x, y)$ it follows that

$$G_\mu^I(n) = \frac{(1 + f(Z(\mu)))^2}{f'(Z(\mu))} (e^{-\mu} f'(Z(\mu)))^n \tag{2.219}$$

so

$$m_I(\mu) = -\log(e^{-\mu} f'(Z(\mu))) = \bar{\mu}. \tag{2.220}$$

Hence, by Eq. (2.211),

$$m_I(\mu) \sim (\mu - \mu_0)^{\frac{1}{2}}$$

as $\mu \to \mu_0$, i.e. the critical exponent ν_I of m_I is

$$\nu_I = \frac{1}{2}. \tag{2.221}$$

In order to construct a continuum limit

$$G^I(r) = \lim_{a \to 0} a^{1 - \eta_I} G_{\mu(a)}^I(a^{-1} r), \tag{2.222}$$

where η_I by definition is the intrinsic anomalous dimension, we have to choose

$$\mu(a) - \mu_0 \sim a^2 \tag{2.223}$$

according to Eq. (2.221). From Eq. (2.219) we see that

$$\eta_I = 1 \tag{2.224}$$

and

$$G^I(r) = C\, e^{-m_I r},$$

where $C > 0$ is a constant, and where m_I is the intrinsic physical mass.

We may now define d_h by

$$\langle |b| \rangle_{\mu(a)} \sim \langle n \rangle_{\mu(a)}^{d_h} \tag{2.225}$$

as $a \to 0$, where

$$\langle |b| \rangle_\mu = -\frac{1}{\chi(\mu)} \frac{d}{d\mu} \chi(\mu) \tag{2.226}$$

is the average value of $|b|$ and

$$\langle n \rangle_\mu = \chi(\mu)^{-1} \sum_{n=1}^{\infty} n\, G_\mu^I(n) \tag{2.227}$$

is the average distance between the two marked vertices in $b \in B_2$. Here we have used the fact that the intrinsic susceptibility

$$\chi_I(\mu) = \sum_{n=1}^{\infty} G_\mu^I(n) \qquad (2.228)$$

equals the ordinary susceptibility $\chi(\mu)$. In order to evaluate d_h we note that

$$\langle |b| \rangle_\mu \sim |\mu - \mu_0|^{-1} \sim a^{-2},$$

by Eq. (2.223), and

$$\langle n \rangle_\mu \sim a^{-1}.$$

It follows that

$$d_h = 2. \qquad (2.229)$$

This is consistent with the scaling relation

$$d_h = \nu_I^{-1},$$

which can be established in the same manner as Eq. (2.155). Likewise, Fisher's scaling relation, given by Eq. (2.154), holds for the intrinsic critical exponents.

In later chapters we shall take up a detailed discussion of intrinsic geometric properties of more interesting and less trivial statistical models, but it will be clear in the study of these models that *branched polymers represent a fractal structure that is generic and can only be avoided by making special choices of coupling constants in the models.*

2.5.3 *Generalizations*

Finally, we mention that the assumption that the vertex-weights w_m are non-negative is not essential for the exact solubility of the models considered in this section. As the reader may verify, the evaluation of $Z(\mu)$, and hence of all the critical exponents, is only based on the behaviour of the function F in Eq. (2.203) at its smallest positive minimum Z_0 inside the interval $(0, r_0)$. It is clear that if the weights w_m are allowed to take negative values they can be tuned such that

$$F'(Z_0) = \ldots = F^{(m-1)}(Z_0) = 0, \quad F^{(m)}(Z_0) \neq 0 \qquad (2.230)$$

for any given $m = 2, 3, 4, \ldots$. One then obtains

$$Z(\mu) - Z_0 \sim (\mu - \mu_0)^{\frac{1}{m}} \qquad (2.231)$$

and consequently

$$\bar{\mu} \sim (\mu - \mu_0)^{\frac{m-1}{m}} \qquad (2.232)$$

as $\mu \to \mu_0$. From this we obtain

$$\gamma = \frac{m-1}{m}, \quad v = \frac{m-1}{2m} \tag{2.233}$$

and, of course,

$$\eta = 0. \tag{2.234}$$

For the "intrinsic" exponents one obtains

$$\gamma_I = \gamma, \quad v_I = \frac{1}{2}v, \quad d_h = \frac{1}{2}d_H, \quad \eta_I = 1. \tag{2.235}$$

In Chapter 4 we study the so-called "multi-critical" random surface models which are analogous to this branched polymer model in that surfaces do not necessarily have positive weight factors.

2.6   Notes

In this chapter we have considered various random walk models with local action functionals. Section 2.1 is meant as an introduction to some basic results in the theory of the Wiener integral without entering into mathematical details. For a thorough treatment of the mathematical results related to path integrals we refer to [85, 348].

The principal reason for the rather detailed discussion of parametrization independent aspects in Section 2.2 is that it is possible in this context to give a precise account of how to obtain reparametrization invariant regularizations. Hopefully, it motivates and illustrates the parallel discussion of higher-dimensional cases in subsequent chapters. In those cases rigorous continuum results are not available and the discretized models are taken as the basic entities, their form being dictated mainly by geometric considerations. The form (2.58) of the scalar particle action first appeared in [99]. Its path integral quantization has been discussed in [148, 328]. The discretized version of this action was first discussed in [23]. The action (2.31) is of standard use in the theory of the lattice random walk. A classic treatise on the lattice random walk is [349].

The theory of random walks with curvature-dependent action has a long history in polymer physics [232, 340]. The discussion of generalized random walks in Section 2.3 is based on [30], see also [28], and was originally motivated by parallel work in the theory of random surfaces described in Chapter 3. Continuum functional integral approaches to rigid walks can be found in [323, 13, 276]. Canonical quantization of rigid random walks has been discussed in [324, 325].

The content of Section 2.4 is essentially taken from [32]. The continuum version of the model considered was first proposed by Polyakov in [328], where a corresponding supersymmetric action was also discussed. We are not aware of any discretization of that action preserving supersymmetry. Further references to path integral methods for spinning paricles are given in [32], see also [244] and references therein.

The treatment of branched polymers in Section 2.5 follows [33] closely and is mainly motivated by their applications in the study of higher-dimensional manifolds in later chapters.

There is a multitude of applications of random walks in quantum physics, which we have not touched upon. We refer to, e.g., [348, 204, 239, 240] for accounts of such applications. Perhaps the most direct link between quantum field theory and random walk is the one given by Symanzik in [355], where a scalar field theory with quartic interaction is represented as a gas of random paths with local repulsive interaction. For a lattice version of this representation see [105]. The fact that this theory is interacting in less than four dimensions but non-interacting in more than four dimensions is related in this picture to the fact that two Wiener paths originating at different points intersect with positive probability in dimension less than four but with zero probability in dimension four or more. Naively, these intersection properties can be understood as a consequence of the value 2 of their Hausdorff dimension, the dimension four being a marginal case. Rigorous proofs can be found in [164, 283]. Rigorous proofs of the triviality results for the scalar field theory are given in [10, 186]. For further applications of random walks in lattice field theory see [176].

Random walk models with non-local interactions such as self-avoiding random walks are naturally of great importance in condensed matter physics. Analytical results are scarce for this case. For a review of both analytical and numerical work see [144]. The existence of a critical point as well as entropy estimates for lattice self-avoiding walks were obtained by Hammersley [218], see also [273]. In view of the intersection properties of the Wiener random walk mentioned above, one would naively expect that the self-avoidance constraint is irrelevant in dimensions higher than four. Partial rigorous results in this direction can be found in [282, 106].

Self-avoiding random walks arise naturally in the $N \to 0$ limit of $O(N)$ vector models. This connection is used in [110, 109] to study the properties of self-avoiding walks.

3

Random surfaces

3.1 Introduction

In the previous chapter we explained how the propagator of a relativistic particle has a natural representation as a sum over random walks. In this chapter we repeat the construction for closed relativistic strings, which constitute a natural generalization of point particles: the zero-dimensional particle is extended to a one-dimensional object, the closed string. The propagator $G(x, y)$ of the particle is obtained by integrating over all paths connecting the points x and y in $\mathbb{R}^d$. Similarly, the propagator $G(\gamma_1, \gamma_2)$ of a string from a loop (i.e. a closed curve) γ_1 to a loop γ_2 in $\mathbb{R}^d$ is obtained by integrating over *all surfaces** embedded in $\mathbb{R}^d$ which have γ_1 and γ_2 as boundary components.

Recall that we discussed two equivalent descriptions of the path integral for the particle. In the first case we integrated over equivalence classes of paths $X : I \mapsto \mathbb{R}^d$ with endpoints x and y. The action of a path was chosen proportional to its length (see Eq. (2.28)). The most obvious generalization to strings is to integrate over equivalence classes (under diffeomorphisms) of surfaces $X : S^1 \times I \mapsto \mathbb{R}^d$ with boundaries γ_1, γ_2 and to choose the action of a surface equal to its *area*, i.e.

$$G(\gamma_1, \gamma_2) = \int \mathscr{D}[X] \, e^{-S_{NG}(X)}, \tag{3.1}$$

where the so called *Nambu–Goto* action is given by

$$S_{NG}(X) = \int dA(X) = \int d^2\xi \sqrt{|\det h|}. \tag{3.2}$$

Here $dA(X)$ denotes the area element of the surface in $\mathbb{R}^d$ and h is the

---

* By *all surfaces* we have in mind surfaces of the simplest topology. In the next section we shall discuss the problem of fluctuating topology.

66

induced metric given by

$$h_{ab} = \frac{\partial X^\mu}{\partial \xi^a} \frac{\partial X^\mu}{\partial \xi^b}, \tag{3.3}$$

where (ξ^1, ξ^2) are local coordinates on the manifold $S^1 \times I$. In the second description of the path integral of the relativistic particle an intrinsic metric was introduced on the one-dimensional manifold I and integration was performed over all equivalence classes of both intrinsic metrics g and paths X (see Eq. (2.29)). The corresponding expression for strings is

$$G(\gamma_1, \gamma_2) = \int \mathscr{D}[X, g] \, e^{-S_P(X,g)}, \tag{3.4}$$

where the reparametrization invariant *Polyakov action* S_P is defined by

$$S_P(X, g) = \frac{1}{2} \int d^2\xi \sqrt{g(\xi)} \, g^{ab}(h_{ab}(X) + \mu g_{ab}), \tag{3.5}$$

and $g(\xi)$ denotes the determinant of the intrinsic metric $g_{ab}(\xi)$. We note that the action S_P can also be viewed as the action of *two-dimensional gravity coupled to d scalar fields*. The classical equations of motion for X which can be derived from $S_P(X, g)$ agree with the ones obtained from $S_{NG}(X)$. In the next chapter we shall return to a discussion of the gravity aspects of Eq. (3.5) and in particular of two-dimensional gravity coupled to certain conformally invariant field theories.

In the previous chapter we have shown by explicit construction that the path integrals Eq. (2.28) and Eq. (2.29), which are the one-dimensional analogues of Eq. (3.1) and Eq. (3.4), yield equivalent descriptions of point particles. In this chapter we address this problem for Eq. (3.1) and Eq. (3.4).

Let us recall briefly the important points in the construction of the path integrals Eq. (2.28) and Eq. (2.29). In both cases the main step was to introduce a reparametrization invariant cutoff a and replacing continuum paths with an ensemble of piecewise linear random paths. In the first case the cutoff a was introduced as the length of the steps of random paths on the hyper-cubic lattice $a\mathbb{Z}^d \subset \mathbb{R}^d$. In the second case the cutoff a^2 was chosen to be a length scale prescribing the subdivision of the one-dimensional manifold I into pieces of length a^2, measured with respect to the intrinsic metric*. In addition to the equivalence of the two corresponding continuum limits we observed important features of universality: the scaling limits of a wide variety of discretized versions of Eq. (2.29) were shown to yield identical continuum limits. In particular, it is worth while mentioning the case where the function φ in Eq. (2.80) is

---

* Recall that a^2 rather than a appears as the intrinsic cutoff, since the Hausdorff dimension of the random walk is two.

chosen to be a δ-function at a. In this case all contributing piecewise linear paths in $\mathbb{R}^d$ have fixed step length a, and it follows that the corresponding regularization of Eq. (2.29) can also be viewed as a regularization of Eq. (2.28).

The definition of the path integral for the string turns out to be considerably more complicated and several new phenomena arise. We are not able to construct the scaling limit of the two-loop function $G(\gamma_1, \gamma_2)$ explicitly, and an analogue of Donsker's theorem is not known in the two-dimensional case. We shall nevertheless argue that scaling limits exist of the regularized versions of both Eq. (3.1) and Eq. (3.4) and we provide some evidence that they should be identical. The relevance of the scaling limits from a physical point of view will then be discussed.

We start by regularizing Eq. (3.4). This requires some discussion of triangulations of surfaces and the framework of *Regge calculus*. A basic idea in the following is to represent equivalence classes of intrinsic metrics on a two-dimensional *orientable* manifold M by triangulations T of M. The model based on this idea will be called a *dynamically triangulated random surface model*, abbreviated to a DTRS-model. We shall establish the basic scaling properties of this model.

Some properties of random surfaces are analysed more easily by using a regularization of Eq. (3.1) on a hyper-cubic lattice. We shall see that the scaling properties of this model, which we call a *lattice random surface model*, abbreviated to an LRS-model, agree with those of the dynamically triangulated random surface model whenever they can be determined in both models.

The motivation for studying the random surface models described above originates mainly from relativistic quantum field theory. At the end of this chapter we discuss more general random surface models. The study of these models is partly motivated by the results obtained for the dynamically triangulated and the lattice random surface model, and partly by the theory of membranes as we know it from solid state physics and biology. In the following chapter we see the close relation of these models with quantum gravity.

3.2 The dynamically triangulated random surface model

The main problem in defining a suitable regularization of the path integral in Eq. (3.4) is to find a workable discretization of the space of equivalence classes of metrics on the two-dimensional orientable parameter manifold M. For the one-dimensional manifolds considered in the last chapter we achieved this goal by the use of *fundamental building blocks* in the form of subintervals of I with fixed reparametrization invariant length given by the cutoff $\varepsilon = a^2$, from which we constructed the piecewise linear discretization of the space $\tilde{\Gamma}(x, y)$. We attempt a similar procedure for the surfaces. Any two-dimensional orientable manifold M can be triangulated. At least intuitively it is clear that given a metric on M we can always find a suitable triangulation, such that the lengths of the sides of the triangles are almost equal, if we are allowed to choose the lengths of the sides sufficiently small, and the connectivity of the triangulations freely. However, our task is not to approximate a particular, given, Riemannian manifold, but to integrate over all metrics on M. We propose to approximate this integration by a summation over all triangulations of M which have the property that each triangle is equilateral with sides of length ε. By this construction, *triangles will serve as the fundamental building blocks of two-dimensional Riemannian manifolds* (M, g). It is natural, and in accordance with the role of ε as a short distance cutoff, to regard the metric in the interior of a triangle as being flat. The volume assigned to each triangle is then $dA_\varepsilon = \sqrt{3}\varepsilon^2/4$, and the total volume of such a *piecewise linear* manifold is $N_t\, dA_\varepsilon$, where N_t denotes the number of triangles in the triangulation. For fixed ε the volume of the manifold can only take certain discrete values. In the one-dimensional case considered in the previous chapter the volume was the *only* reparametrization invariant quantity of the Riemannian manifold I and a discretization of the volume therefore entails a discretization of the moduli space of metrics. For a two-dimensional manifold M the volume is not the only invariant. For example, the scalar curvature is a local invariant. An exhaustive characterization of the space of equivalence classes[*] of metrics may be given in terms of the so-called Teichmüller space of metrics $\hat{g}(t_1, \ldots, t_m)$, where $t_1, \ldots, t_m$ are complex parameters with the property that for any given metric g there are unique values of the parameters and a function ϕ on M such that g is equivalent to

$$e^\phi \hat{g}(t_1, \ldots, t_m). \tag{3.6}$$

For later use we note that the complex dimension m of Teichmüller space for a compact two-dimensional orientable manifold $M_{h,b}$ with h handles and b punctures is

$$m = (3h - 3) + b \tag{3.7}$$

if $3h - 3 + b > 0$ and zero if $3h - 3 + b < 0$. Furthermore, $m = 1$ for $h = 0, b = 3$ or $h = 1, b = 0$.

One can think of the degrees of freedom associated with the Teichmüller parameters $t_1, \ldots, t_m$ as reflecting the freedom in gluing together triangles

---

[*] Two metrics g_1 and g_2 on M are *equivalent* if there is an orientation preserving diffeomorphism $\varphi : M \mapsto M$ such that $\varphi_* g_1 = g_2$.

to construct triangulations, whereas the *conformal factor* e^ϕ reflects the presence of additional invariants such as the curvature and the volume.

In Chapter 4 we show that the number of triangulations of $M_{h,b}$ with A triangles is given, for large A, by

$$N_{h,b}(A) \sim A^{\gamma_h+b-3} e^{\mu_0 A}(1 + O(1/A)) \tag{3.8}$$

where

$$\gamma_h = \frac{5h-1}{2} \tag{3.9}$$

and $\mu_0 > 0$ is a number independent of h and b.

In the next section we discuss in some detail the class of triangulations to be considered and a natural way of defining a *metric and curvature* on piecewise linear manifolds. At the same time there is assigned an equivalence class of metrics to a given triangulation. The integration over equivalence classes of metrics is then replaced by a summation over triangulations of a given topology.

For each triangulation T we define a piecewise linear surface X in $\mathbb{R}^d$ as a map of the vertices i in T to points $X(i) = x_i \in \mathbb{R}^d$. If i, j, k are vertices belonging to a triangle in T we view x_i, x_j, x_k as corners of a triangle in $\mathbb{R}^d$. Similarly, if i, j are vertices belonging to a link in T we view x_i, x_j as endpoints of a link in $\mathbb{R}^d$. We say that *the surface X is based on T*. If T is a triangulation of a manifold with boundary, the boundary vertices and the links connecting them will be mapped into polygonal boundaries of the piecewise linear surface X. Other vertices will be called *interior* vertices and the set of these is denoted by $V_I(T)$. The next step, analogous to Eq. (2.68) in the previous chapter, is to calculate the Polyakov action $S_P(X)$, defined by Eq. (3.5), for the piecewise linear surface X. As a result of the specific metric assignment given to the triangulation T, the free field action in Eq. (3.5) becomes

$$S_T(X) = \frac{1}{2} \sum_{(ij)} (x_i - x_j)^2 + \mu N_t(T), \tag{3.10}$$

where the summation is over all links (i, j) in T. We verify this in detail in the next section. The integration over X in Eq. (3.4) for a given metric is now replaced by the integration $\int \prod_{i \in V_I(T)} dx_i$ over positions of the images $x_i \in \mathbb{R}^d$ of the interior vertices, exactly as in the random walk case.

We can write the regularization of Eq. (3.4) as

$$G_\mu(\gamma_1, \gamma_2) = \sum_{T \in \mathcal{T}(n_1, n_2)} \int \prod_{i \in V_I(T)} dx_i \, e^{-S_T(X)}, \tag{3.11}$$

where γ_1, γ_2 are two polygonal boundary loops in $\mathbb{R}^d$, i.e. closed loops consisting of finite numbers n_1 and n_2 of linear pieces, the vertices of which are the images of the boundary vertices of the triangulations, and where the sum runs over triangulations $\mathcal{T}(n_1, n_2)$ of a surface (world-sheet) with two boundary components. More generally, we define the b-loop functions $G_\mu(\gamma_1, \ldots, \gamma_b)$ by

$$G_\mu(\gamma_1, \ldots, \gamma_b) = \sum_{T \in \mathcal{T}(n_1, \ldots, n_b)} \int \prod_{i \in V_I(T)} dx_i \, e^{-S_T(X)}, \qquad (3.12)$$

where the sum is over a class $\mathcal{T}(n_1, \ldots, n_b)$ of triangulations with b boundary components, the ith component consisting of n_i links, which are mapped onto polygonal loops γ_i, $i = 1, \ldots b$.

We have not yet discussed the summation over manifolds with different topologies. Such a summation is natural in Eq. (3.1) or Eq. (3.4), at least if we consider three- and higher-loop Green functions. If a string can split into two and two strings can join into one, we should also allow for world-sheets where one string first splits into two which subsequently join into one, i.e a world-sheet which has the topology of a torus with two boundary components. Iteration of this process leads to manifolds with an arbitrary genus.

The summation over genus is not included in a natural way in the continuum treatment since the Riemannian metrics are defined on a fixed manifold, and in continuum string theory the "string loop" amplitudes are defined (formally) only in a perturbative sense, i.e. as an expansion in the genus h of the world-sheet of the string:

$$G(\gamma_1, \ldots, \gamma_b) = \sum_{h=0}^{\infty} w(h, b) \int \mathcal{D}[X, g] \, e^{-S_P(X, g)}, \qquad (3.13)$$

where the integration is over equivalence classes of metrics on $M_{h,b}$ and embeddings $X : M_{h,b} \mapsto \mathbb{R}^d$. We shall occasionally refer to $M_{h,b}$ as the *parameter manifold*. Two such manifolds M_{h_1,b_1} and M_{h_2,b_2} are homeomorphic if and only if $h_1 = h_2$ and $b_1 = b_2$. Since b is fixed in Eq. (3.13), the only topological degree of freedom is the genus h. We have introduced a weight factor $w(h, b)$ associated with the summation over h. One can show that the only form of this topological weight that is compatible with the combinatorial unitarity alluded to above is

$$w(h, b) = e^{\chi(h,b)/G}, \qquad (3.14)$$

where $\chi(h, b) = 2 - 2h - b$ is the Euler characteristic of $M_{h,b}$ and G is a constant. In string theory $e^{-1/G}$ is called the string coupling constant for the obvious reason that, by Eq. (3.14), a factor $e^{-1/G}$ can be associated with each string-splitting or -joining. In addition, the term $e^{\chi/G}$ fits nicely into the interpretation of $S_P(X)$ as the action of two-dimensional gravity coupled to d scalar fields $X^\mu(\xi)$ since the Einstein–Hilbert action is a topological invariant in two dimensions as a consequence of the Gauss–Bonnet formula

$$\int_{M_{h,b}} d^2\xi \sqrt{g}\, R = 4\pi \chi(h,b). \tag{3.15}$$

It is then natural to redefine the Polyakov action by

$$S_P(X,g) = \int d^2\xi \sqrt{g(\xi)} \left(\frac{1}{2} g^{ab}(\xi) \frac{\partial X^\mu}{\partial \xi^a} \frac{\partial X^\mu}{\partial \xi^b} + \left[\mu - \frac{1}{4\pi G} R(\xi) \right] \right), \tag{3.16}$$

where the expression in the square bracket is the two-dimensional Einstein–Hilbert action with cosmological constant μ.

The curvature term in Eq. (3.16) is readily included in the discretized expressions since the Euler characteristic can be defined combinatorially for any triangulation T of a manifold $M_{h,b}$ by Euler's formula

$$\chi(T) = N_t(T) - N_l(T) + N_v(T), \tag{3.17}$$

where $N_\alpha(T)$, $\alpha = t, l, v$, denotes the number of triangles, links and vertices, respectively, in the triangulation T. The regularized expressions Eq. (3.11) and Eq. (3.12) seem to open up the possibility of defining the sum over topologies by simply evaluating the sum (3.13) with triangulations of arbitrary topology. This will not work. If we consider all triangulations with a fixed number N_t of triangles, then the expressions given by Eqs. (3.11) and (3.12) are finite and well defined even though no constraint is placed on the topology. However, we are interested in summing over the volume. One can show that the total number of non-isomorphic triangulations* constructed from N_t triangles has a *lower* bound of the form

$$(cN_t)!, \tag{3.18}$$

where c is a positive constant. Any reasonable action functional of the matter fields will result in a free energy F which for fixed volume V fulfils

$$F \le fV, \tag{3.19}$$

for some constant f. This is certainly the case for the regularized version Eq. (3.10) of S_P, as will be shown later. It follows that *the entropy factor (3.18) renders the regularized loop-functions, given by Eqs. (3.11) and (3.12), ill defined*, since we have

$$G_\mu(\gamma_1, \ldots, \gamma_b) \ge \sum_{N_t} (cN_t)!\, C^{N_t}, \tag{3.20}$$

---

* Two triangulations are isomorphic if there is a bijective mapping between them which maps neighbours to neighbours; see next section for a precise definition.

where c and C are positive constants and C might depend on the matter couplings and the cutoff ε.

Once the topology is fixed the situation is different. By Eq. (3.8) the number of triangulations then grows exponentially with N_t. As we show in Section 3.4 this implies that the sums in Eqs. (3.11) and (3.12) are well defined for sufficiently large values of the coupling μ and a critical, smallest coupling μ_0 exists at which a scaling limit can be constructed (under some assumptions), in close analogy with the procedure employed for the relativistic particle in the previous chapter. Since the critical value μ_0 is independent of the topology, see Eq. (3.8), we can write a formal sum over topology. In Chapter 4 we discuss some aspects of this perturbative expansion in topology.

Before turning to a detailed discussion of the scaling properties of the dynamically triangulated random surface model defined by Eqs. (3.10)–(3.12), in the next section we review some useful facts about triangulations of surfaces and the concept of Regge calculus on two-dimensional piecewise linear manifolds.

3.3 Triangulations and Regge calculus

We can view triangulations in three different ways: (i) as *combinatorial triangulations*, i.e. abstract triangulations defined via incidence relations between triangles, (ii) as piecewise linear surfaces constructed by gluing together triangles in $\mathbb{R}^D$ for some $D \geq 2$, and finally (iii) as a dissection of a given surface M into triangles, where a triangle in M is defined as the image of a standard triangle (two-dimensional simplex) in $\mathbb{R}^D$, $D \geq 2$, under a homeomorphism. We will appeal to all three interpretations of triangulations at various occasions.

Let us start by defining combinatorial two-dimensional triangulations. The definition we give here will be somewhat more general than the one we give in Chapter 6 for higher-dimensional manifolds. A combinatorial, oriented, connected two-dimensional triangulation is a collection of oriented, closed triangles, i.e. including their links and vertices, with pairwise identifications of a subset of the links subject to the following constraints:

(i) A link can only be identified with an oppositely oriented link. The identification of the links implies identification of the corresponding endpoints.

(ii) Any two triangles can be connected by a "path" of adjacent triangles, i.e. triangles which have a link in common.

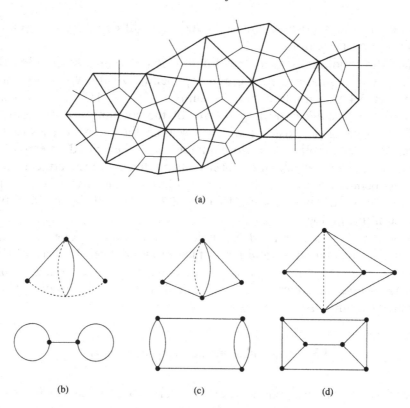

(a)

(b)   (c)   (d)

Fig. 3.1. Various triangulations and their dual ϕ^3 graphs. Fig. 3.1(a) shows the
basic transformation from a triangulation to a ϕ^3 graph. Figs. 3.1(b)–(d) show
the simplest closed type I–type III triangulations and their associated ϕ^3 graphs.

We call this class of triangulations type I triangulations. They can be
rather singular, as indicated in Fig. 3.1. In the following we shall almost
exclusively consider the less singular type II and type III triangulations.
Type II triangulations satisfy the further restriction that single links are
not allowed to form a loop, i.e. the endpoints of a link may not be
identified with each other. One of the simplest triangulations where the
endpoints of a link are identified is shown in Fig. 3.1(b). In type III
triangulations loops, i.e. closed connected paths of links, of length one
and length two, are not allowed.

For a triangulation we define the *boundary* ∂T as the set of links
(including endpoints) which only belong to one triangle. These are the
links that have not been identified with any other link. The number of
connected components of ∂T will usually be denoted by b. The collection
of all links in a triangulation T will be denoted by $L(T)$ and the collection
of all vertices in T will be denoted by $V(T)$. We have already introduced
the notation $N_t(T)$, $N_l(T)$ and $N_v(T)$ for the number of triangles, links

and vertices. They satisfy Euler's relation Eq. (3.17) so the genus h of a triangulation is well-defined. We denote the collection of all combinatorial triangulations with h handles and b boundary components of length $n_1, \ldots, n_b$ by $\mathcal{T}^{(h)}(n_1, \ldots, n_b)$. By the length of a boundary component we mean the number of links in the boundary component. In the rest of the chapter we do not consider the rather singular type I triangulations. Most of our results apply to both type II or type III triangulations so we shall usually not be explicit about the class under consideration. This should not cause any confusion.

Two combinatorial triangulations are by definition *isomorphic* if there is a bijective mapping between triangles, links and vertices which preserves the incidence relations, but not necessarily the orientation.

A combinatorial triangulation T with h handles and b boundary components can (up to isomorphism) be regarded as a graph drawn without link intersections on the manifold $M_{h,b}$ under consideration, with the vertices of the graph being the corners of the triangles and the links being the edges of the triangles. It is often convenient to consider the graph *dual* to a triangulation. It is constructed by placing a vertex in the interior of each triangle and connecting two of these vertices by a link if and only if the corresponding triangles share a link. It follows that the dual graph will be a ϕ^3-graph such that type I–III triangulations correspond to ϕ^3-graphs with tadpoles, without tadpoles and without tadpoles and self-energy diagrams, respectively. We have illustrated this in Fig. 3.1.

Above we have discussed how we wish to use triangulations to construct random surfaces. It is now convenient to take the opposite point of view for a moment and consider piecewise linear surfaces that are embedded without self-intersections in some Euclidean space $\mathbb{R}^D$ of sufficiently high dimension. A piecewise linear embedded two-dimensional manifold is by definition a collection of triangles in $\mathbb{R}^D$ which satisfy the same incidence relations as the combinatorial triangulations of type II or III above. Thus, a piecewise linear manifold defines a triangulation T. For given link lengths the metric structure is well defined on such a piecewise linear manifold if we regard the triangles as flat in the interior, since we can calculate the shortest distance between any two points in the manifold. We next explain how to define the concept of curvature and parallel transport without explicit use of the metric.

Recall that the scalar curvature R of a smooth surface satisfies

$$\frac{1}{2} \int_t dA \, R = \beta_1 + \beta_2 + \beta_3 - \pi \equiv \varepsilon_t, \qquad (3.21)$$

for any *geodesic triangle* t on the surface, i.e. a triangle whose edges are geodesic curves, and $\beta_1, \beta_2, \beta_3$ denote the angles between the curves where they intersect at the corners of the triangle. The quantity ε_t is called

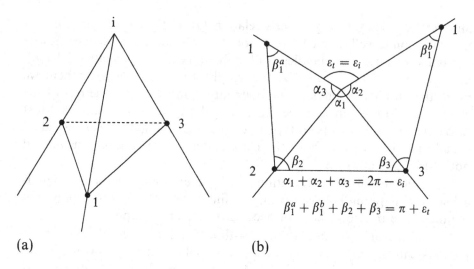

(a)   (b)

Fig. 3.2. A geodesic triangle on a piecewise linear surface. The vertex i is an interior point of the triangle (Fig. 3.2(a)). Fig. 3.2(b) shows the excess angle after the piecewise linear neighbourhood has been cut open along link $(1i)$ and pushed down into the plane.

the *excess angle* of the triangle. This relation has a simple extension to piecewise linear surfaces where it is natural to associate the curvature with the vertices, since the interior of the triangles is flat and curvature is an intrinsic geometric quantity, i.e. bending around an edge in $\mathbb{R}^D$ does not produce curvature. To each vertex i we associate a *deficit angle* ε_i defined by

$$\varepsilon_i = 2\pi - \sum_{t \ni i} \alpha_i(t), \tag{3.22}$$

where the summation is over the i-angles of the triangles to which the vertex i belongs, see Fig. 3.2. The vertex i of the triangulation T is located in the interior of a geodesic triangle 123. From the diagram it follows that $\varepsilon_i = \varepsilon_t$ so we can write

$$\int_t dA\, R = 2\varepsilon_i, \qquad \int dA\, R = \sum_{i \in V_I(T)} 2\varepsilon_i. \tag{3.23}$$

The second equality in Eq. (3.23) generalizes Eq. (3.21) to any surface whose boundary is a piecewise geodesic curve and the sum is over the interior vertices in the triangulation corresponding to the piecewise linear surface. In particular, we shall use Eq. (3.23) for closed surfaces.

On a smooth surface a vector which undergoes parallel transport along a curve enclosing an infinitesimal area dA, will rotate an angle $d\theta = \frac{1}{2} R\, dA$. On the piecewise linear surface parallel transport is trivial if one avoids the vertices. From Fig. 3.2 it can be seen that a vector parallel transported around the vertex i will undergo a rotation through an angle ε_i, since parallel transport between the two points labelled by "1" in Fig. 3.2 will not rotate the vector, but the identification of the two links with vertices 1 and i will effect a rotation through an angle ε_i. This is consistent with the identification in Eq. (3.23), and lends weight to the claim that the deficit angles ε_i assigned to the vertices capture the geometric properties of curvature.

Associating an area

$$A_i = \frac{1}{3} \sum_{t \ni i} A_t \tag{3.24}$$

to the vertex i, i.e. distributing the area of each triangle equally among its three vertices, we are led to define the curvature R_i at the vertex i of a piecewise linear surface by

$$R_i = \frac{2\varepsilon_i}{A_i}. \tag{3.25}$$

With these definitions we have

$$\int dA = \sum_i A_i, \qquad \int dA\, R = \sum_i R_i A_i. \tag{3.26}$$

Regge's original proposal was to use these formulas to construct a sequence of *piecewise linear approximations*, corresponding to triangulations, to a *given smooth manifold M* such that

$$\sum_i A_i f_i \to \int_M dA\, f, \tag{3.27}$$

for any sufficiently smooth function f on the manifold. Here f_i is the value of the function f at the vertex i on the manifold M.

As already mentioned we take a different point of view since our aim is to integrate over all metrics. We fix the length of each link to be ε, and obtain for a given triangulation T a piecewise linear manifold and an associated equivalence class of metrics. For a fixed topology and fixed cutoff ε the combinatorially non-isomorphic triangulations will form a grid of points in the space of equivalence classes of metrics, and the hope is that this grid will become uniformly dense when $\varepsilon \to 0$.

We now describe the approximation to the action $S_P(X, g)$ in the case of piecewise linear manifolds whose links are all of length equal to ε. Let $y_i \in \mathbb{R}^D$ denote the coordinates of vertices i on a piecewise linear manifold M which is a realization of a triangulation T in some space $\mathbb{R}^D$. If the vertices i, j belong to the same link then $|y_i - y_j| = \varepsilon$. If the vertices i, j, k belong to a triangle t we introduce barycentric coordinates (ξ^1, ξ^2) on t

by

$$y = \xi^1 y_i + \xi^2 y_j + (1 - \xi^1 - \xi^2) y_k. \tag{3.28}$$

The metric corresponding to this coordinate system is

$$g_{ab} = \frac{\partial y^\alpha}{\partial \xi^a} \frac{\partial y^\alpha}{\partial \xi^b} = \varepsilon^2 \begin{pmatrix} 1 & \frac{1}{2} \\ \frac{1}{2} & 1 \end{pmatrix}, \tag{3.29}$$

where the index α runs from 1 to D.

Now consider an arbitrary surface $X : i \mapsto x_i$ in $\mathbb{R}^d$ based on the triangulation T. The dimension d is arbitrary and need not be the same as D. The surface X may be regarded as a piecewise linear surface in $\mathbb{R}^d$ obtained by extending the map X to the interior of the triangle (i, j, k) by

$$X(\xi^1, \xi^2) = \xi^1 x_i + \xi^2 x_j + (1 - \xi^1 - \xi^2) x_k. \tag{3.30}$$

Note that this surface has in general a metric structure quite different from that of the piecewise linear surface M even though they correspond to the same triangulation. From Eqs. (3.29) and (3.30) we obtain

$$g^{ab} \frac{\partial X^\mu}{\partial \xi^a} \frac{\partial X^\mu}{\partial \xi^b} \propto \frac{1}{\varepsilon^2} \left[(x_i - x_j)^2 + (x_i - x_k)^2 + (x_j - x_k)^2 \right], \tag{3.31}$$

and since the area of each equilateral triangle $t \in T$ is $\sqrt{3}\,\varepsilon^2/4$ we can write the Polyakov action on the piecewise linear manifold T with the metric assignment given above as

$$\int d^2\xi \sqrt{g}\, g^{ab} \partial_a X \partial_b X = C \sum_{(ij)\in L(T)} (x_i - x_j)^2, \tag{3.32}$$

where C is a constant. We note that

$$\sum_{(ij)\in L(T)} (x_i - x_j)^2 = \sum_{i,j \in V(T)} x_i (-\Delta_T)_{ij} x_j, \tag{3.33}$$

where Δ_T denotes the *combinatorial Laplacian* corresponding to the graph defined by the triangulation T. It is defined as the $V(T) \times V(T)$ matrix whose diagonal entry Δ_{ii} equals minus the *order* of the vertex i, i.e. minus the number of links incident on i, while Δ_{ij}, $i \neq j$, is 1 if i and j belong to the same link, and zero otherwise. This completes the arguments for our claim that Eqs. (3.10)–(3.12) constitute a discretization of the path integral (3.4), with cutoff ε.

3.4   Basic properties of the loop functions

We now turn to the study of the loop functions of the DTRS-model, given by Eq. (3.12). All triangulations are assumed to have the topology of a sphere or a sphere with a number of holes. After proving the existence of a critical point μ_0 we study the scaling behaviour of the loop functions and define the critical indices in analogy with the random walk case studied in Chapter 2. Some properties of the loop functions are more easily proved by considering the regularization of Eq. (3.1) on the hyper-cubic lattice $\mathbb{Z}^d$, the LRS-model, which we study in the next section. For this reason some of the basic properties of loop functions will only be derived for the LRS-model.

3.4.1 Convergence of the loop functions

Our first goal is to prove that the loop functions

$$G_\mu(\gamma_1, \ldots, \gamma_b) = \sum_{T \in \mathscr{T}(n_1, \ldots, n_b)} \rho(T) e^{-\mu |T|} G^{(T)}(\gamma_1, \ldots, \gamma_b), \qquad (3.34)$$

where

$$G^{(T)}(\gamma_1, \ldots, \gamma_b) = \int \prod_{i \in V_I(T)} dx_i \, e^{-S_T(X)}, \qquad (3.35)$$

are convergent for μ larger than a critical value of the coupling μ_0 and divergent for $\mu < \mu_0$, where μ_0 is independent of the number of boundary components as well as their length and shape. Moreover, the loop functions are continuous functions of the coordinates of the boundary loops. In Eq. (3.34) we have changed notation and written

$$S_T(X) \equiv \frac{1}{2} \sum_{(ij) \in T} (x_i - x_j)^2 \qquad (3.36)$$

for the "Gaussian" part of Eq. (3.10), as well as

$$|T| \equiv N_t(T).$$

Further, we have generalized the definition of the discretized loop function, given by Eq. (3.12), slightly by including a weight factor $\rho(T)$ for each triangulation. The only weight factors we shall consider are of the form

$$\rho(T) = \prod_{i \in V(T)} \sigma_i^\alpha, \qquad (3.37)$$

where $\alpha \in \mathbb{R}$. In the following we usually choose $\rho(T) = 1$, unless explicitly stated otherwise, but most results are valid for weights of the form given by Eq. (3.37). The loop function (3.34) can be viewed as the partition function for "discretized" membranes in $\mathbb{R}^d$, which are attached to the boundary loops γ_i. We usually restrict our attention to surfaces with planar (spherical) topology, but many of the proofs presented in the following are valid for surfaces of arbitrary, but fixed, topology. Finally, it is sometimes convenient to consider the "loop function" with no loops at all, which we call the *partition function*. It is defined by

$$Z(\mu) = \sum_{T \in \mathscr{T}} \frac{\rho(T)}{C_T} e^{-\mu |T|} \int \prod_{i \in V(T)} dx_i \, e^{-S_T(X)} \delta \left(\sum_{i \in V(T)} x_i \right), \qquad (3.38)$$

where $\mathscr{T}$ is the collection of all triangulations of S^2 and C_T is a symmetry factor equal to the order of the automorphism group[*] of the graph corresponding to T. The δ-function fixes the centre of mass of the surface X in $\mathbb{R}^d$. Fixing a vertex x_i rather than the centre of mass yields the same partition function $Z(\mu)$.

Let $T \in \mathscr{T}(n_1, \ldots, n_b)$. Regarding T as an abstract graph, a *spanning tree* in T is another graph U with $V(U) = V(T)$ and $L(U) \subset L(T)$ such that the graph U is a connected tree, i.e. it is connected and contains no closed loop. It is easy to see that any connected graph has a spanning tree and it is not unique unless the original graph is a tree, in which case the spanning tree is identical to the original graph. Let us choose a spanning tree U for T. It follows that

$$
\begin{aligned}
G^{(T)}(\gamma_1, \ldots, \gamma_b) &\leq \int \prod_{i \in V_I(T)} dx_i \, \exp \left(-\frac{1}{2} \sum_{(i,j) \in L(U)} (x_i - x_j)^2 \right) \\
&\leq (2\pi)^{V_I(T)d/2}.
\end{aligned}
\qquad (3.39)
$$

Hence, if $\rho(T) = 1$ the loop function $G_\mu(\gamma_1, \ldots, \gamma_b)$ is finite for $\mu > \frac{d}{2} \log 2\pi$ and there is a number $\mu(\gamma_1, \ldots, \gamma_b)$ such that $G_\mu(\gamma_1, \ldots, \gamma_b)$ is finite and analytic in μ for $\mu > \mu(\gamma_1, \ldots, \gamma_b)$ and infinite for $\mu < \mu(\gamma_1, \ldots, \gamma_b)$. By the same argument there exists $\mu_0 > 0$ such that the partition function $Z(\mu)$ is finite and analytic in μ for $\mu > \mu_0$ and divergent for $\mu < \mu_0$.

We now proceed to show that the numbers $\mu(\gamma_1, \ldots, \gamma_b)$ are in fact all equal to μ_0. The argument we give is tailored to the Gaussian action S_T defined by Eq. (3.36), but the result is valid for a large class of local action functionals, in particular the action

$$\sum_{(i,j)} |x_i - x_j|^p,$$

with $p > 0$, and a large class of weight functions $\rho(T)$.

We define the distance between two polygonal loops γ and γ' with vertices $x_1, \ldots, x_n$ and $x_1', \ldots, x_n'$, respectively, listed in cyclic order, by

$$d(\gamma, \gamma') = \min_\pi \max_{1 \leq i \leq n} |x_i - x_{\pi(i)}'|.$$

---

[*] The automorphism group of T is the group of isomorphisms mapping T to T, as defined in Section 3.3.

where the minimum is taken over all cyclic permutations π of $1,\ldots,n$. We begin by establishing a technical lemma. The proof introduces ideas that will be useful in the sequel.

Lemma 3.1 *Let $\mu > \mu(\gamma_1,\ldots,\gamma_b)$ and $\varepsilon > 0$. Then there exists $\delta > 0$ which only depends on ε and $\gamma_1,\ldots,\gamma_b$, such that*

$$G_{\mu+\varepsilon}(\gamma'_1,\ldots,\gamma'_b) \le G_\mu(\gamma_1,\ldots,\gamma_b) \tag{3.40}$$

for any loops γ'_i, which satisfy $d(\gamma'_i,\gamma_i) \le \delta$, $i = 1,\ldots,b$.

Proof For simplicity we write the proof for a surface with one boundary component γ of length n. The same argument holds for surfaces with an arbitrary number of boundary components.

For $T \in \mathcal{T}(n)$ let $X(\gamma,T)$ minimize the Gaussian action functional $S_T(X)$ among all mappings

$$X : T \mapsto \mathbb{R}^d, \quad X(\partial T) = \gamma,$$

where ∂T denotes the boundary of T. If X is an arbitrary surface contributing to $G^{(T)}(\gamma)$, we can write

$$X = X(\gamma,T) + X', \tag{3.41}$$

where $X'(\partial T) = 0$ and

$$S_T(X) = S_T(X(\gamma,T)) + S_T(X'). \tag{3.42}$$

Let O_n be the degenerate loop all of whose n vertices are at 0. Then

$$G^{(T)}(\gamma) = e^{-S_T(X(\gamma,T))} G^{(T)}(O_n) \tag{3.43}$$

and it follows that

$$G_\mu(\gamma) = \sum_{T \in \mathcal{T}(n)} e^{-\mu|T|} e^{-S_T(X(\gamma,T))} G^{(T)}(O_n). \tag{3.44}$$

Now suppose $d(\gamma',\gamma) < \delta$ and let X' be the surface based on T which maps ∂T to γ but is identical to $X(\gamma',T)$ on the interior vertices of T. Then

$$S_T(X(\gamma,T)) \le S_T(X') \le S_T(X(\gamma',T)) + C_\gamma|T|\delta \tag{3.45}$$

for a suitable constant $C_\gamma > 0$. Combining Eq. (3.44) and Eq. (3.45) we obtain

$$G_{\mu+\varepsilon}(\gamma') \le G_\mu(\gamma), \tag{3.46}$$

provided $\delta \le C_\gamma^{-1}\varepsilon$. This completes the proof of the lemma.

We next show that

$$\mu(\gamma) = \mu(\gamma') \tag{3.47}$$

for any pair of loops γ, γ'. By repeated use of the same argument it follows that $\mu(\gamma_1, \ldots, \gamma_b)$ is independent of the loops γ_i for fixed b.

Let n and n' be the number of corners in γ and γ', respectively. Choose a fixed triangulation $T_0 \in \mathcal{T}(n, n')$. For each $T \in \mathcal{T}(n)$ we let $T' \in \mathcal{T}(n')$ be the triangulation obtained by gluing T and T_0 along the boundary component of length n. This can be done in n different ways: we pick one of them. Similarly, we can glue any surface X based on T with boundary γ to any surface X_0 based on T_0 with boundary $\gamma \cup \gamma'$. In this way we obtain a surface X' based on T' with boundary γ' by putting $X'(i) = X(i)$ for $i \in T$ and $X'(i) = X_0(i)$ for $i \in T_0$. Since

$$S_{T'}(X') = S_T(X) + S_{T_0}(X_0) - S(\gamma),$$

where $S(\gamma)$ is the contribution to $S_T(X)$ (and to $S_{T_0}(X_0)$) from the boundary links, it follows that

$$G^{(T')}(\gamma') = \int \prod_{i \in \partial T} dx_i \, G^{(T)}(\gamma) G^{(T_0)}(\gamma, \gamma') e^{S(\gamma)}.$$

The triangulation T' determines T uniquely if T_0 is fixed and $|T'| = |T| + |T_0|$ so

$$G_\mu(\gamma') \geq \int \prod_i dx_i \, G_\mu(\gamma) G^{(T_0)}(\gamma, \gamma') e^{S(\gamma)} e^{-\mu|T_0|},$$

where the integration is over the vertices of γ. If $\mu > \mu(\gamma')$ we conclude that $G_\mu(\gamma) < \infty$ for all values of the coordinates of γ except possibly for a set of measure 0 in $\mathbb{R}^{dn}$. Now Lemma 3.1 implies that for any $\varepsilon > 0$, $G_{\mu+\varepsilon}(\gamma) < \infty$ and hence $\mu(\gamma) \leq \mu(\gamma')$ for all polygonal loops γ. Reversing the roles of γ and γ' in the above argument completes the proof of Eq. (3.47). Clearly, the argument can be extended to a proof that $\mu(\gamma_1, \ldots, \gamma_b)$ is independent of the loops $\gamma_1, \ldots, \gamma_b$ for each fixed $b \geq 1$.

We next show that the value of $\mu(\gamma_1, \ldots, \gamma_b)$ is independent of the number of boundary components b. Let $\mu > \mu(\gamma_1, \ldots, \gamma_b)$ and choose $T_0 \in \mathcal{T}(n_1)$, $|\gamma_1| = n_1$. We may assume that $n_1 = 3$ and that T_0 is a triangle. Applying the same argument as in the proof of Eq. (3.47) we find that

$$-\frac{d}{d\mu} G_\mu(\gamma_2, \ldots, \gamma_b) \geq \int \prod_{i \in \partial T_0} dx_i \, G_\mu(\gamma_1, \ldots, \gamma_b), \qquad (3.48)$$

where the derivative with respect to μ has been inserted to produce a factor $|T'|$ for each triangulation T' which bounds the number of ways a given $T' \in \mathcal{T}(n_2, \ldots, n_b)$ can be obtained by gluing T_0 onto a triangulation $T \in \mathcal{T}(n_1, \ldots, n_b)$ along the boundary component of length n_1. Since $G_\mu(\gamma_2, \ldots, \gamma_b)$ is analytic in μ for $\mu > \mu(\gamma_2, \ldots, \gamma_b)$ we obtain as above that $\mu(\gamma_2, \ldots, \gamma_b) \geq \mu(\gamma_1, \ldots, \gamma_b)$.

Similarly, picking a triangulation $T_0 \in \mathcal{T}(n_1, n_2, n_2)$ and gluing it to a triangulation $T \in \mathcal{T}(n_2, n_3, \ldots, n_b)$ along the boundary component of length n_2 we obtain

$$G_\mu(\gamma_1, \ldots, \gamma_b) \geq \int \prod_i dx_i \, G_\mu(\gamma, \gamma_3, \ldots, \gamma_b) G^{(T_0)}(\gamma_1, \gamma_2, \gamma) e^{-\mu|T_0|} e^{S(\gamma)}, \quad (3.49)$$

where the integration above is over the n_2 vertices of the loop γ. It follows that $\mu(\gamma_1, \ldots, \gamma_b) \geq \mu(\gamma_2, \ldots, \gamma_b)$.

The fact that $\mu_0 = \mu(\gamma_1, \ldots, \gamma_b)$ follows easily from (3.44) and the bound

$$\frac{1}{6} \int dx_1 dx_2 \, G_\mu(\gamma) \leq -\frac{dZ}{d\mu} \leq \int dx_1 dx_2 \, G_\mu(\gamma),$$

where γ is the triangular loop with vertices $0, x_1, x_2$. The factor $1/6$ enters because the symmetry factor $C_T \leq 6$ for $T \in \mathcal{T}(3)$.

We have now completed the proof of the fact that the loop functions have exactly the same interval of convergence as the partition function. The analyticity of $G_\mu(\gamma_1, \ldots, \gamma_b)$ in μ for fixed $\gamma_1, \ldots \gamma_b$ implies, together with Lemma (3.1) that the loop functions are continuous functions of the coordinates of the boundary loops for all $\mu > \mu_0$.

We summarize the results as follows.

Theorem 3.2 *There is a positive number μ_0 such that the loop functions are well defined and continuous in the coordinates of the boundary loops and real analytic in μ for $\mu > \mu_0$. The sums defining the loop functions are all divergent for $\mu < \mu_0$. The partition function is finite and analytic for $\mu > \mu_0$ and divergent for $\mu < \mu_0$.*

It is frequently convenient to consider surfaces with degenerate boundary components consisting of single vertices. For this purpose we define $\mathcal{T}_b$ to be the set of all closed triangulations with b different marked and ordered vertices. The degenerate b-loop function, or the b-point function, is defined as

$$G_\mu(x_1, \ldots, x_b) = \sum_{T \in \mathcal{T}_b} e^{-\mu|T|} G^{(T)}(x_1, \ldots, x_b) \quad (3.50)$$

where

$$G^{(T)}(x_1, \ldots, x_b) = \int \prod_{i \in V(T) \setminus \{i_1, \ldots, i_b\}} dx_i \, e^{-S_T(X)} \quad (3.51)$$

and x_k is the image of i_k under X. It follows from the discussion in the previous section that the degenerate loop functions defined by Eq. (3.50) are convergent for the same values of the coupling constant as the ordinary loop functions. The degenerate loops $x_1, \ldots, x_b$, as well as the marked vertices, are called *punctures*. We draw the reader's attention to the difference between the degenerate loop functions and the loop functions where the vertices of each boundary loop sit at the same point in embedding space. While the piecewise linear embedded surfaces look identical the contributing triangulations are not the same.

The singular behaviour of $G_\mu(\gamma_1,\dots,\gamma_b)$ as $\mu \to \mu_0$ is determined by the large $|T|$ tail of Eq. (3.34). This is of course the reason that the continuum limit, which we define later, is associated with the limit $\mu \to \mu_0$. If the "internal length" of the links in the triangulations, i.e. the cutoff, is ε, we expect, in analogy with the random walk models, that there exists a scaling of the link lengths $\varepsilon(\mu) \to 0$ for $\mu \to \mu_0$ such that *continuum loop functions* can be defined. Consider as an example the two-point function $G_\mu(x_1, x_2)$, which only depends on $x = x_1 - x_2$ by translational invariance. The continuum two-point function should only depend on a "physical" scaled distance $x_{ph} = a(\mu)x$, such that x_{ph} is fixed while $a(\mu) \to 0$ and $|x| \to \infty$ in the continuum limit. We further expect that $a(\mu)$, which defines the distance scale in $\mathbb{R}^d$, will depend on ε. In the random walk case we found $a^{d_H} = \varepsilon$. In order to choose the correct scaling for the random surfaces we have to study the behaviour of the loop functions as $\mu \to \mu_0$. In the limit where $\mu \to \mu_0$ and the "distances" $d_{i,j}$ between γ_i and γ_j go to infinity, we expect the behaviour of the loop functions $G_\mu(\gamma_1,\dots,\gamma_b)$ to be independent of the number of corners n_i of the loops γ_i and of the detailed shape of γ_i as long as the numbers n_i stay finite and the sizes of all γ_i remain bounded. In the following we assume this to be the case.

We introduce the *susceptibility* (sometimes called the *string susceptibility*), analogous to the susceptibility for the random walk, as the integrated two-loop function. More precisely, let γ_0 be a fixed polygonal loop in $\mathbb{R}^d$ and let γ' be a polygonal loop with a fixed number n of corners $x_1,\dots,x_n$. Then the susceptibility is defined by

$$\chi(\mu) = \int dx_1 \cdots dx_n \, G_\mu(\gamma_0, \gamma'). \qquad (3.52)$$

We expect that the singular behaviour of χ at μ_0 is independent of γ_0 and n and this can be proved in special cases. We furthermore assume the existence of a *susceptibility exponent* $\gamma \in \mathbb{R}$, independent of γ_0 and n, such that

$$\chi(\mu) \sim (\mu - \mu_0)^{-\gamma} \qquad (3.53)$$

as $\mu \to \mu_0$.

We shall frequently use the fact that the susceptibility may alternatively be defined as minus the derivative of the one-loop function since

$$\chi(\mu) \sim -\frac{d}{d\mu} G_\mu(\gamma_0) \qquad (3.54)$$

as $\mu \to \mu_0$. This is most easily seen by replacing the boundary loop γ'

in Eq. (3.52) by a puncture and assuming $|\gamma_0| = n_0 \geq 2$. Differentiating $G_\mu(\gamma_0)$ with respect to μ brings down a factor $|T|$ in the formula Eq. (3.34) for $G_\mu(\gamma_0)$. On the other hand, we have $|T| = 2N_v(T) + n_0 - 2$ for $T \in \mathcal{T}(n_0)$, i.e. the number of ways $N_v(T)$ of introducing a puncture on T is of the order of $|T|$. Since the order of the automorphism group of T is $\leq 2n_0$, the number of different punctured triangulations obtained from T is also of the order of $|T|$ and the claim (3.54) follows by comparing the defining formula (3.34) for $G_\mu(\gamma_0)$ with that for $G_\mu(\gamma_0, \gamma')$ integrated over the puncture γ'.

3.4.2 The susceptibility exponent γ

It will be useful for later purposes to note that the Gaussian integration in Eq. (3.35) can be done explicitly by first separating out the minimal action as was done in the proof of Lemma 3.1. One then obtains

$$G^{(T)}(\gamma_1, \ldots, \gamma_b) = e^{-S_T(X(\gamma_1, \ldots, \gamma_b, T))} \left(\frac{(2\pi)^{N_v(T \backslash \partial T)}}{\det C_T^0} \right)^{d/2}, \qquad (3.55)$$

where $X(\gamma_1, \ldots, \gamma_b, T)$ is a surface based on T with boundary $\gamma_1 \cup \cdots \cup \gamma_b$ for which S_T is minimal, $N_v(T \backslash \partial T)$ is the number of interior vertices in T and C_T^0 is the so-called *modified adjacency matrix* of T. In order to define the matrix C_T^0 let C_T be the *adjacency matrix* of the graph T' obtained from T by deleting all boundary links and identifying the boundary vertices with a single vertex, which we denote by 0, i.e. C_T is a $N_v(T') \times N_v(T')$ matrix where the entries are labelled by pairs (i, j) of vertices in $V(T')$ and defined by

$$(C_T)_{ij} = \begin{cases} -\#\{\text{links connecting i and j}\} & \text{if } i \neq j \\ \sigma_i & \text{if } i = j. \end{cases} \qquad (3.56)$$

Then C_T^0 is the matrix obtained from C_T by deleting the row and column indexed by 0.

Note that the expression (3.55) allows us to make an analytic continuation of the loop functions to any real value of d. The study of the theory in unphysical dimensions does shed some light on its analytic structure, as we shall see in Chapter 4, and may even be given physical interpretation.

A standard result from graph theory [219] states that

$$D(T') \equiv \det C_T^0 = \#\{\text{spanning trees in } T'\}. \qquad (3.57)$$

For $T \in \mathcal{T}_1$ we note that $C_T^0 = -\Delta_T'$, where Δ_T' is the matrix representing the Laplacian defined by Eq. (3.32) with the zero mode removed by fixing

$x_0 = 0$. In this case Eq. (3.55) is equivalent to

$$G^{(T)}(x_0) = \left(\frac{(2\pi)^{N_v(T')-1}}{\det(-\Delta_T')} \right)^{d/2}. \tag{3.58}$$

This result also follows immediately from definition (3.50).

We next consider the application of the above results to the two-point function $G_\mu(x_0, x_1)$. Given $T \in \mathcal{T}_2$ we denote the marked points in T mapped to x_0 and x_1 by 0 and 1. It is clear that a spanning tree in T' (which in this case is identical to T with 0 and 1 identified) may be identified with a pair (t_0, t_1) of trees in T which together span T such that 0 belongs to t_0 and 1 to t_1. If $\Gamma_T^{(2)}(0, 1)$ denotes the number of such pairs of trees in T we have

$$G^{(T)}(x_0, x_1) = e^{-S_T(X(x_0,x_1,T))} \left(\frac{(2\pi)^{N_v(T)-2}}{\Gamma_T^{(2)}(0, 1)} \right)^{d/2}. \tag{3.59}$$

Note that this expression shows that $G_\mu(x_0, x_1)$ is a decreasing function of $|x_0 - x_1|$.

The Fourier transform of the two-point function

$$\widehat{G}_\mu(p) = \int dx \, e^{-ipx} \, G_\mu(x_0, x), \tag{3.60}$$

can also be expressed in terms of $\Gamma_T^{(2)}(0, 1)$. We have

$$\widehat{G}_\mu(p) = \sum_{T \in \mathcal{T}_2} e^{-\mu|T|} \, \widehat{G}^{(T)}(p), \tag{3.61}$$

where

$$\widehat{G}^{(T)}(p) = e^{-\frac{1}{2}p^2(-\Delta_T'^{-1})_{11}} \left(\frac{(2\pi)^{N_v(T)-1}}{\det(-\Delta_T')} \right)^{d/2}, \tag{3.62}$$

and $(\Delta_T'^{-1})_{11}$ denotes the matrix element of the inverse matrix $\Delta_T'^{-1}$ labelled by the vertex pair $(1, 1)$ in the notation above. By Cramer's formula we have

$$(\Delta_T'^{-1})_{11} = \frac{\det \Delta_T''}{\det \Delta_T'},$$

where Δ_T'' is the matrix obtained from Δ_T' by deleting the row and column indexed by the vertex 1. Note that here we cannot replace $-\Delta_T'$ by C_T^0 since the latter is labelled by deleting the row and column labelled 0 from the adjacency matrix of the graph T' obtained from T by identifying the vertices 0 and 1. In fact, we have $-\Delta_T'' = C_T^0$, i.e. $\det(-\Delta_T'') = \Gamma_T^{(2)}(0, 1)$

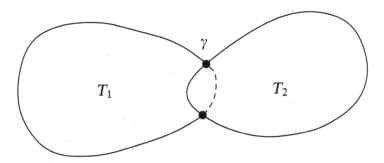

Fig. 3.3. The construction of $T_1 \oplus_\gamma T_2$.

and Eq. (3.62) can be written as

$$\widehat{G}^{(T)}(p) = e^{-\frac{1}{2}p^2\Gamma_T^{(2)}(0,1)/D(T)} \left(\frac{(2\pi)^{N_v(T)-1}}{D(T')} \right)^{d/2}. \qquad (3.63)$$

We now establish some properties of $D(T)$. Let $T_1 \in \mathscr{T}(m)$ and $T_2 \in \mathscr{T}(n)$ and choose two links $\ell_1 \in L(T_1 \setminus \partial T_1)$ and $\ell_2 \in L(T_2 \setminus \partial T_2)$. Cut the two triangulations open along these links so we obtain two new triangulations $T_1' \in \mathscr{T}(m,2)$ and $T_2' \in \mathscr{T}(n,2)$. Now glue the two triangulations together across the boundaries of length 2 and thereby obtain a new triangulation $T \in \mathscr{T}(m,n)$ (see Fig. 3.3 in the case where $m = 0$ and $n = 0$). Let γ be the loop in T where T_1 and T_2 are glued together. We note that T_1 and T_2 can be reconstructed from T if we know γ. Let us denote the triangulation T by $T_1 \oplus_\gamma T_2$. Given a triangulation T and a link $\ell \in L(T)$ we denote by $D_\ell(T)$ the number of trees that span T and include the link ℓ. With the above definitions we have the following result.

Lemma 3.3

$$D(T_1 \oplus_\gamma T_2) = D(T_1)D_{\ell_2}(T_2) + D_{\ell_1}(T_1)D(T_2). \qquad (3.64)$$

Proof Let $D_\ell'(T)$ denote the number of spanning trees of T that do not contain the link ℓ so that

$$D(T) = D_\ell(T) + D_\ell'(T). \qquad (3.65)$$

If t is a tree spanning $T_1 \oplus_\gamma T_2$ containing one of the two links in γ then we can associate to t two trees t_1 and t_2 spanning T_1 and T_2 and containing the links ℓ_1 and ℓ_2, respectively. Since the loop γ consists of two links there is a two to one correspondence between the collection of all trees t as above and pairs of trees (t_1, t_2) spanning T_1 and T_2 and containing ℓ_1 and ℓ_2, respectively. We conclude that the number of trees that span

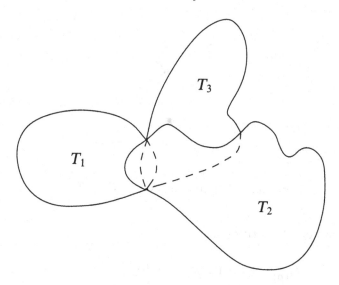

Fig. 3.4.   The construction of $T_1 \oplus_{\gamma_1} T_2 \oplus_{\gamma_2} T_3$.

$T_1 \oplus_\gamma T_2$ and contain one of the links in γ is equal to

$$2D_{\ell_1}(T_1)D_{\ell_2}(T_2). \tag{3.66}$$

If, on the other hand, t is a spanning tree in $T_1 \oplus_\gamma T_2$ containing neither of the two links in γ, then either the part of t lying in T_1 is a spanning tree in T_1 and the part of t lying in T_2 consists of two trees which span T_2, or vice versa. In the former case we may join the two trees in T_2 into a spanning tree of T_2 by adding the link ℓ_2. Similarly, in the latter case we may connect the two trees in T_1 by the link ℓ_1 and obtain a spanning tree of T_1. We conclude that the number of trees spanning $T_1 \oplus_\gamma T_2$ which contain neither of the links in γ is given by

$$D_{\ell_1}(T_1)D'_{\ell_2}(T_2) + D'_{\ell_1}(T_1)D_{\ell_2}(T_2). \tag{3.67}$$

Adding Eq. (3.66) and Eq. (3.67) and using Eq. (3.65) proves the desired result.

Let now $T_1, \ldots, T_n$ be triangulations, $T_i \in \mathcal{T}(n_i)$, and denote by

$$T_1 \oplus_{\gamma_1} T_2 \oplus_{\gamma_2} \ldots \oplus_{\gamma_{k-1}} T_k \tag{3.68}$$

the triangulation in $\mathcal{T}(n_1, \ldots, n_k)$ obtained by cutting the triangulations T_i open along along links ℓ_i, $i = 1, \ldots, n$, and gluing them together successively along pairs of links in the new boundary loops of length two which we denote by γ_i. The loops $\gamma_1, \ldots, \gamma_{k-1}$ have the same vertices and γ_i and γ_{i+1} have one common link; see Fig. 3.4 in the case of $k = 3$. By repeated use of the above lemma we obtain

$$D(T_1 \oplus_{\gamma_1} T_2 \oplus_{\gamma_2} \ldots \oplus_{\gamma_{k-1}} T_k) = \sum_{i=1}^{k} D(T_i) \prod_{j \neq i} D_{\ell_j}(T_j). \qquad (3.69)$$

One of the uses we shall make of Eq. (3.64) is to prove the following theorem.

Theorem 3.4 *The critical exponent of the susceptibility satisfies the inequality $\gamma \leq \frac{1}{2}$.*

Proof According to Eq. (3.52) or Eq. (3.54) we have

$$\chi(\mu) \sim \sum_{T \in \mathcal{T}(2,2)} e^{-\mu|T|} D(T)^{-d/2}.$$

Furthermore, by an argument similar to the one used to relate Eq. (3.52) and Eq. (3.54) one obtains

$$-\frac{d\chi}{d\mu} \sim \sum_{T \in \mathcal{T}(2,2,2)} e^{-\mu|T|} D(T)^{-d/2}. \qquad (3.70)$$

The theorem follows from a lower bound on the right-hand side of Eq. (3.70) which we now establish.

Let T_0 denote a fixed triangulation with a boundary consisting of three loops γ_1, γ_2 and γ_3 of length two. For the sake of concreteness let us choose T_0 as the triangulation defined by a cube whose sides have been divided into two triangles and then cut open along three links which do not share any endpoints. Next, take three triangulations $T_1, T_2, T_3 \in \mathcal{T}(2,2)$ and glue them to T_0 at its three boundary loops which results in a new triangulation T in $\mathcal{T}(2,2,2)$. It is obvious that not all triangulations in $\mathcal{T}(2,2,2)$ can be obtained by this construction, but the "gluing piece" T_0 makes T_1, T_2 and T_3 uniquely determined by T. If T can be obtained in this fashion then

$$D(T) \leq 8D(T_0)D(T_1)D(T_2)D(T_3)$$

by Lemma 3.1. It follows (since we assume $d \geq 0$) that

$$D(T)^{-d/2} \geq 8^{-d/2} D(T_0)^{-d/2} D(T_1)^{-d/2} D(T_2)^{-d/2} D(T_3)^{-d/2}.$$

Summing over the T_i, $i = 1, 2, 3$, we obtain

$$-\frac{d\chi}{d\mu} \geq C\chi^3(\mu) \qquad (3.71)$$

where C is a positive constant. From the assumed singular behaviour of $\chi(\mu)$, described by Eq. (3.53), it follows that $\gamma + 1 \geq 3\gamma$, i.e. $\gamma \leq 1/2$. This completes the proof of Theorem 3.4.

The bound on γ proved above is one of the most general results in the theory of random surfaces and is analogous to mean field bounds on critical exponents in ordinary statistical mechanics. It is straightforward to generalize the proof to the analytically continued theory at $d \leq 0$. Broadly speaking, the bound on γ is expected to hold in any random surface model with local interactions. In the next section we shall give a proof of this inequality for lattice surfaces. For surfaces with non-local interaction, e.g. self-avoiding surfaces, the bound presumably does not hold in general.

3.4.3   Branched polymer surfaces

Recall from Section 2.5 that $\gamma = 1/2$ is the generic value of γ for branched polymers. Let us now consider in some detail how we can saturate the bound in Theorem 3.4 and exhibit the relation of this bound to mean field theory and branched polymers.

Let T be any triangulation in $\mathcal{T}_1$. Choose one link emanating from each vertex in T. Let U be the subgraph of T consisting of these links. The graph U clearly spans T but in general it is not a tree. However, any tree spanning T can be obtained by a construction of this type. We therefore obtain the *upper bound*

$$D(T) \;\leq\; \prod_{j \in T} \sigma_j \leq \left(\frac{\sum_{j \in T} \sigma_j}{N_v(T)} \right)^{N_v(T)}$$

$$= \left(\frac{3|T|}{\frac{1}{2}|T| + \chi} \right)^{\frac{1}{2}|T| + \chi} \leq C_\chi (\sqrt{6})^{|T|}, \qquad (3.72)$$

where we have used Euler's relation $N_v(T) - N_l(T) + |T| = \chi$, as well as $\sum_j \sigma_j = 2N_l(T)$ and $3|T| = 2N_l(T)$. Eq. (3.72) shows that $D(T)$ is exponentially bounded in $|T|$ for a fixed topology when $|T| \to \infty$.

The above bound is surprisingly close to being optimal. Intuitively, one would expect the most regular "two-dimensional" triangulations to have the largest number of spanning trees, i.e. the largest determinants. Let

$$\lambda_{max} = \liminf_{|T| \to \infty} \frac{\log D(T)}{|T|}.$$

There are good arguments (but no proof) that $\lambda_{max} \approx \sqrt{5.03}$, where this particular value originates from the numerical evaluation of determinants $D(T)$ of particular regular triangulations which are of type III [93, 24].

There is an exponential *lower bound* on $D(T)$ [27, 24], but it depends on the class of triangulations used. For triangulations with spherical topology we have

$$D(T) \geq C\,|T|\,\lambda_{min}^{|T|}, \qquad (3.73)$$

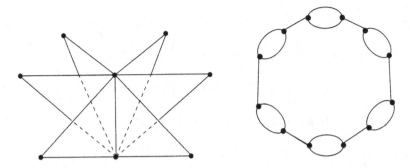

Fig. 3.5. The left-hand diagram shows the type of triangulation which saturates the lower bound for $D(T)$ for type II triangulations. The triangles shown should be viewed as double triangles, glued together along two links, while the central link should be viewed as different links connecting the two vertices of high order. The dual graph is shown on the diagram to the right.

where C is a positive constant and

$$\lambda_{min} = \begin{cases} \sqrt{2} & \text{for type II triangulations} \\ \sqrt{2\sqrt{2}} & \text{for type IIA triangulations} \\ \sqrt{\sqrt{3}+2} & \text{for type III triangulations.} \end{cases} \qquad (3.74)$$

Here we have introduced a new class, type IIA, of triangulations, identical to the type II triangulations, except that we do not allow vertices of order two. It is introduced for calculational convenience, as will become clear below. Rather than proving the bounds we shall exhibit the triangulations which saturate the bound for type II triangulations.

Let T_1 be a triangulation consisting of two triangles glued together along their three boundary links. Take n copies of T_1, cut each of them open along one of the boundary links and glue them together as in Eq. (3.68) to obtain a new triangulation (see Fig. 3.5)

$$T_n = T_1 \oplus_{\gamma_1} T_1 \oplus_{\gamma_2} \ldots \oplus_{\gamma_{n-1}} T_1.$$

Clearly, $D(T_1) = 3$ and $D_\ell(T_1) = 2$ if ℓ is a link in T_1. It follows from Eq. (3.69) that

$$\begin{aligned} D(T_n) &= 3n2^{n-1} \\ &= \frac{3}{4}|T_n|(\sqrt{2})^{|T_n|} \end{aligned} \qquad (3.75)$$

since $|T_n| = 2n$. A similar construction works for type IIA and type III triangulations, although the construction for type III triangulations is slightly more complicated since loops of length two are not allowed.

In statistical mechanics the limit $d \to +\infty$ may often be interpreted as a mean field limit, the reason being that any local interaction will be

Random surfaces

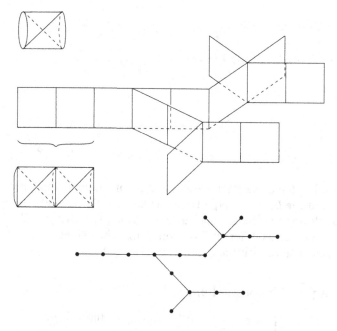

Fig. 3.6.   The construction of branched polymer like triangulations.

dominated by the large number of neighbours present for $d \to \infty$. In the DTRS-model there exists no large d expansion of the loop functions, but it is still possible to obtain a fairly reliable idea of the behaviour of the model in this limit. The results to be derived in the following by use of a certain approximation to the full DTRS-model agree with numerical simulations of the complete model, as well as with strong coupling expansions.

It follows from Eq. (3.58) that $G_\mu(x_0)$ is given by the sum over all triangulations $T \in \mathscr{T}_1$ with weights depending on the triangulations T. The d-dependence of the weight, represented by the the factor $D(T)^{-d/2}$, implies that as $d \to +\infty$ the triangulations with the smallest determinants will be increasingly dominant. For finite d not only the value of $D(T)$ is important, but also the number of triangulations with given values $|T|$ and $D(T)$. This *entropy factor* decreases in importance as d grows. Let us illustrate this, as well as certain aspects of universality, by considering an approximation to the complete model. In order to describe the approximation let T_t be the tetrahedron. We cut T_t open along two links that do not meet and denote the resulting triangulation by T_t'. We shall refer to the boundary links in T_t' as *vertical links*, see Fig. 3.6. We now define the subclass $\mathscr{B}\mathscr{T}_1 \subset \mathscr{T}_1$ to consist of all planar triangulations that can be constructed by gluing together copies of T_t' along the vertical links and closing the triangulation by identifying the unglued boundary links in each T_t'. To each triangulation $B \in \mathscr{B}\mathscr{T}_1$ there is associated a unique rooted *planar branched polymer b* as defined in Section 2.5. The links in b correspond to the tetrahedral building blocks in B; see Fig. 3.6. Vertices of any order can occur in b and the root of b is determined by the marked vertex in B.

The triangulations constructed in this way are type IIA triangulations, but we would like to stress that there is nothing fundamental about the choice of class of triangulations. They are chosen because they do saturate the lower bound on the determinant in the class of type IIA triangulations and their determinant can be calculated rather easily, as can be seen below.

Let T_t' be one of the building blocks in B and i, j its endpoints in the corresponding branched polymer b. Let $x_i, y_i \in \mathbb{R}^d$ be the coordinates of the endpoints of the links corresponding to i and similarly for j. Introducing the variables

$$u_i = \frac{1}{2}(x_i + y_i), \quad v_i = x_i - y_i,$$

it is straightforward to check that the contribution from the links in the building block with vertices x_i, y_i, x_j, y_j to the Gaussian action is given by

$$2(u_i - u_j)^2 + (v_i^2 + v_j^2). \tag{3.76}$$

It follows that the total action of a triangulation $B \in \mathscr{BT}_1$ is

$$S = 2 \sum_{(i,j) \in L(b)} (u_i - u_j)^2 + \sum_{i \in V(b)} \sigma_i v_i^2,$$

where σ_i is the order of the vertex i in b. It is easy to integrate over the variables u_i and v_i. Each v_i integration gives a factor $\pi^{d/2}\sigma_i^{-d/2}$ and integrating over the u_is inward from the vertices of order 1, leaving out the root, yields a factor $(\frac{1}{2}\pi)^{d/2}$ for each integration. It follows that the one-point function $G_\mu(x_0)$ in this reduced DTRS-model is identical to the (reduced) one-point function (see Eq. (2.199)) in the corresponding branched polymer model:

$$G_\mu(x_0) = \sum_{b \in \mathscr{B}_1'} e^{-\mu|b|} \prod_{i \in V(b)} \sigma_i^{-d/2}, \tag{3.77}$$

where we have absorbed the factors of 2 and π into a redefinition of μ. The branching weight factors w_n are given by $w_n = n^{-d/2}$ so the model is of type (ii) in the notation of Section 2.5 if the function

$$F(z) = \frac{1}{z} + \sum_{n=2}^{\infty} \frac{z^{n-2}}{n^{d/2}} \tag{3.78}$$

has a minimum in the interval $0 < z < 1$, since $z = 1$ is the radius of convergence for the sum in Eq. (3.78). This happens when $d \leq 4$. For $d \geq 5$, and in particular for $d \to \infty$, the branched polymer model defined by Eq. (3.77) is a type (i) model. The fact that the reduced model becomes a type (i) branched polymer model could very well be an artifact of the approximation.

Let us mention two examples of modifications of the original random surface model which lead to different "reduced" models where we end up with branched polymers of type (ii) for all d. In both models we change slightly the weights $\rho(T)$ given to the triangulations in the sum (3.34). In the first case we put $\rho(T) = 0$ if the triangulation T contains a vertex i with order $\sigma_i \geq K$, where K is an arbitrarily large, but fixed, number, and $\rho(T) = 1$ otherwise. In this case the sum in Eq. (3.78) is over a finite range of n and $F(z)$ has a quadratic minimum. In the second case set

$$\rho(T) = \prod_{i \in V(T)} \sigma_i^{d/2}. \tag{3.79}$$

In terms of the original tentative correspondence between the integration over equivalence classes of metrics in the path integral and the summation over triangulations this weight corresponds to the inclusion of a so-called *conformal factor*, which can be written formally as

$$\prod_{\xi \in M} g(\xi)^{\frac{d}{4}} \tag{3.80}$$

in the continuum path integral, since the volume density $\sqrt{g(\xi)}$ associated with the vertex i is σ_i and the natural discretized analogue of the formal expression (3.80) is given by Eq. (3.79). The motivation for including this factor is that, formally, $\sqrt{g(\xi)}\, dx(\xi)$ is invariant under the conformal transformation

$$x \mapsto \lambda^{-1}x, \qquad \sqrt{g} \mapsto \lambda^2 \sqrt{g}. \tag{3.81}$$

This, however, is not a strong motivation here since the regularization does not respect the conformal invariance given by Eq. (3.81). If we nevertheless include the conformal factor we see that the factor $\rho(T)$ cancels $\prod_i \sigma_i^{-d/2}$ in Eq. (3.77) and Eq. (3.78) is changed to

$$F(z) = \frac{1}{z} + \sum_{n>1} z^{n-2}, \tag{3.82}$$

which has a minimum in the interval $0 < z < 1$ for all d.

In the cases when the reduced DTRS-model corresponds to branched polymers of type (ii) the susceptibility exponent is

$$\gamma = \frac{1}{2}. \tag{3.83}$$

In the case where Eq. (3.77) holds, one finds for $d \geq 5$ that

$$\gamma(d) = 2 - \frac{d}{2}. \tag{3.84}$$

The above considerations are the first indication that branched polymer-like configurations play an important role in random surface theories. The study of the string tension below, as well as other results presented in the next section, will provide further evidence of this.

3.4.4 Mass and string tension

In the following we introduce two physical parameters which play a fundamental role for the construction of possible scaling limits. The first one is the mass gap $m(\mu)$ which is defined in the same way as for the random walks as the exponential decay rate of the two-point function:

$$m(\mu) = - \lim_{|x| \to \infty} \frac{\log G_\mu(0, x)}{|x|}. \tag{3.85}$$

The existence of the limit (3.85) follows from a subadditive property of $-\log G_\mu(0, x)$ as a function of x. It is technically complicated to prove this property in full detail for the DTRS-model, but somewhat easier to establish in the LRS-model, and we confine ourselves to presenting a proof for the latter model. However, it is not difficult to establish the following weaker result.

Theorem 3.5 *There is a number*

$$m(\mu) \geq c\sqrt{\mu - \mu_0} \tag{3.86}$$

such that

$$G_\mu(0, x) \leq Ce^{-m(\mu)|x|}, \tag{3.87}$$

where c and C are positive constants.

Proof The inequality (3.87) is equivalent to the statement that the Fourier transform of the two-point function $\widehat{G}_\mu(p)$ has an analytic extension in p to the domain

$$\Sigma_{m(\mu)} = \{p \in \mathbb{C}^d : |\mathrm{Im}\, p| < m(\mu)\} \subset \mathbb{C}^d,$$

where $\mathrm{Im}\, p = (\mathrm{Im}\, p^1, \dots, \mathrm{Im}\, p^d)$.

We use the representation (3.63) for $\widehat{G}_\mu(p)$ and note the easily verifiable fact that

$$\Gamma_T^{(2)}(0, 1) \leq \frac{3}{2}|T|D(T)$$

to conclude by Eq. (3.61) that $\widehat{G}_\mu(p)$ is analytic for $p \in \Sigma_{m(\mu)}$ provided

$$\frac{3}{2}m(\mu)^2 \leq (\mu - \mu_0).$$

This completes the proof.

It is not difficult to see that the mass is an increasing concave function of μ, provided it exists. As in the case of the susceptibility it is clear that there is room for some arbitrariness in the definition of the mass. Above, we used the two-point function. We could just as well have used a two-loop function. Let γ be any polygonal loop and let $\gamma(x)$ be the translate of γ by $x \in \mathbb{R}^d$. Then we expect that

$$G_\mu(\gamma, \gamma(x)) \sim e^{-m(\mu)|x|}, \tag{3.88}$$

where $m(\mu)$ is the exponential decay rate of the two-point function defined above.

It is almost obvious that $\chi(\mu)$ cannot diverge as $\mu \to \mu_0$ unless $m(\mu) \to 0$ as $\mu \to \mu_0$. This means that $\gamma > 0$ implies $m(\mu) \to 0$ for $\mu \to \mu_0$. *The vanishing of the mass $m(\mu)$ at $\mu = \mu_0$ is necessary for the scaling limit of the two-point function to exist* for the same reasons as we discussed in the previous chapter for the random walk. We *assume* that the mass scales to zero for $\mu \to \mu_0$ with an exponent ν defined by

$$m(\mu) \sim (\mu - \mu_0)^\nu. \tag{3.89}$$

We now turn to the study of the *string tension*, which is defined as follows. Let $\gamma(L_1, L_2; n_1, n_2)$ be a rectangular loop in $\mathbb{R}^d$ with sides of lengths L_1 and L_2 and $2(n_1 + n_2)$ vertices evenly distributed along the sides of the loop with a vertex at each corner of the loop. We study the behaviour of the loop function $G_\mu(\gamma(L_1, L_2; n_1, n_2))$ as L_1, L_2, n_1 and n_2 tend to infinity with all the ratios L_1/L_2, n_1/L_1 and n_2/L_2, fixed. Hence, we will in the following suppress the arguments n_1 and n_2 in the rectangular boundary loop $\gamma(L_1, L_2; n_1, n_2)$. The string tension $\tau(\mu)$ is defined as

$$\tau(\mu) = \lim_{L_1, L_2 \to \infty} \frac{-\log G_\mu(\gamma(L_1, L_2))}{L_1 L_2}. \tag{3.90}$$

From a continuum, statistical mechanics point of view we have a fixed *frame* in $\mathbb{R}^d$, defined by the loop $\gamma(L_1, L_2)$, and an ensemble of *membranes* X whose boundaries are attached to the frame. The one-loop function $G(\gamma(L_1, L_2))$ then plays the role of a *partition function* for membranes attached to the frame. Let $\mathscr{G}$ denote the Gibbs free energy of the system, i.e.

$$G_\mu(\gamma) = e^{-\mathscr{G}(\gamma)}. \tag{3.91}$$

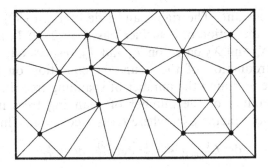

Fig. 3.7. An example of a triangulation fulfilling the boundary condition in the proof of the existence of the string tension.

Then the surface tension is defined as the Gibbs free energy per unit frame area, and we recognize the definition (3.90) of the string tension as being analogous to the definition of the surface tension for a membrane.

It is of some interest to give general arguments which prove the *existence of* $\tau(\mu)$. We first remark that if we do not let n_i increase with L_i ($i = 1, 2$) as L_i is taken to infinity, then $\tau(\mu)$ will obtain a boundary contribution which is independent of μ. If, however, L_i and n_i are taken to be proportional, then $\tau(\mu)$ is independent of the constant of proportionality. In the context of Polyakov's string theory this boundary condition amounts to requiring that the dynamical metric g is proportional, at the boundary, to the metric induced by the embedding in Euclidean space. Let us, further, for purely technical reasons, restrict the boundary of the triangulations so that the boundary vertices have order 4 along the sides and 3 on the corners, see Fig. 3.7. Such a restriction does not influence existence or value of the limit in Eq. (3.90). We denote this class of triangulations by $\mathcal{T}'(L_1, L_2)$. If we glue two triangulations $T_a \in \mathcal{T}'(L_1^a, L_2)$ and $T_b \in \mathcal{T}'(L_1^b, L_2)$ along two vertical sides, we obtain a triangulation $T_a \oplus T_b \in \mathcal{T}'(L_1^a + L_1^b, L_2)$. The technical assumption about the boundary allows us to identify the individual components T_a and T_b unambiguously, given a triangulation $T_a \oplus T_b$. Let the minimal actions associated with T_a, T_b and $T_a \oplus T_b$ be $S_{T_a}^{min}$, $S_{T_b}^{min}$ and $S_{T_a \oplus T_b}^{min}$. We have

$$G_\mu(\gamma(L_1^a + L_1^b, L_2)) \geq \sum_{T_a, T_b} G_\mu^{T_a \oplus T_b}(\gamma(L_1^a + L_1^b, L_2)) \qquad (3.92)$$

$$= \sum_{T_a, T_b} e^{-S_{T_a \oplus T_b}^{min}} \left(\frac{\pi^{N_v(T_a \oplus T_b \setminus \partial(T_a \oplus T_b))}}{\det C_{T_a \oplus T_b}^0} \right)^{d/2} .$$

Next we note that

$$S_{T_a \oplus T_b}^{min} \leq S_{T_a}^{min} + S_{T_b}^{min} + C_1 L_2, \qquad (3.93)$$

where C_1 is a constant. The right-hand side of Eq. (3.93) represents an upper bound on the action of a surface based on $T_a \oplus T_b$ and composed of two surfaces X_a and X_b based on T_a and T_b, respectively, with minimal action. The introduction of the boundary condition on triangulations makes it possible to bound the number of spanning trees in $T_a \oplus T_b$ from above by the product of the number of spanning trees in T_a and T_b, respectively, up to a factor $C_2^{L_2}$, where C_2 is a constant. This gives

$$\det C^0_{T_a \oplus T_b} \leq C_2^{L_2} \det C^0_{T_a} \, \det C^0_{T_b}. \tag{3.94}$$

Using Eq. (3.93) and Eq. (3.94) in Eq. (3.92) we obtain

$$G_\mu(\gamma(L_1^a + L_1^b, L_2)) \geq C_3^{L_2} G_\mu(\gamma(L_1^a, L_2)) \, G_\mu(\gamma(L_1^b, L_2)), \tag{3.95}$$

where C_3 is yet another constant which depends on C_1 and C_2. This shows that the function

$$f(L_1, L_2) = -\log\left(C_3^{L_1+L_2} G_\mu(\gamma(L_1, L_2))\right) \tag{3.96}$$

in *subadditive* in L_1 and by symmetry also in L_2. By a standard result about subadditive functions (see e.g. [338]) it follows that

$$\lim_{L_1,L_2 \to \infty} \frac{f(L_1, L_2)}{L_1 L_2} = \inf_{L_1, L_2} \frac{f(L_1, L_2)}{L_1 L_2}. \tag{3.97}$$

Since $G_\mu(\gamma(L_1, L_2))$ can easily be bounded *from above* by a constant it follows that the right-hand side of Eq. (3.97), and hence the string tension given by Eq. (3.90), exists and is finite.

Note that the finiteness of a string tension implies that the one-loop function cannot decay *faster* than an exponential function of the area enclosed by the boundary loop. A similar statement is known in relativistic field theory for the so-called Wilson loop and implies that the static potential between charged particles cannot rise faster than a linear function. The analogous statement for the relativistic string is that the potential energy of a string can at most grow linearly with the length L of the string.

The final result we wish to establish in this section is the inequality

$$G_\mu(\gamma(L_1, L_2)) \leq C_{L_1, L_2} e^{-L_1 L_2}, \tag{3.98}$$

for all $\mu \geq \mu_0$, where C_{L_1, L_2} grows at most exponentially with $L_1 + L_2$. This inequality implies the following theorem.

Theorem 3.6 *The dimensionless string tension fulfils*

$$\tau(\mu) \geq 1 \tag{3.99}$$

and therefore the string tension cannot vanish as $\mu \downarrow \mu_0$.

Proof Let us for notational simplicity consider square loops, i.e. $L_1 = L_2 = L$ and $n_1 = n_2 = n$ in the notation above. Let $T \in \mathcal{T}(4n)$ and

$$X : T \mapsto \mathbb{R}^d, \quad X(\partial T) = \gamma(L, n). \tag{3.100}$$

With the notation as in the proof of Lemma 3.1 we can split X,

$$X = X(\gamma, T) + X',$$

where $X(\gamma, T)$ minimizes the action S_T among all mappings of the type (3.100) and we obtain

$$G_\mu(\gamma(L, n)) = \sum_{T \in \mathcal{T}(4n)} e^{-\mu|T|} e^{-S_T(X(\gamma,T))} G^{(T)}(O_n).$$

The sum of the squares of the lengths of any two sides of a triangle is larger than, or equal to, four times its area. Hence,

$$S_T(X(\gamma, T)) \geq L^2,$$

from which we deduce that

$$G_\mu(\gamma(L, n)) \leq e^{-L^2} G_\mu(O_n).$$

In order to complete the proof of Eq. (3.98) it suffices to show that

$$G_\mu(O_n) \leq e^{c(\mu)n}. \tag{3.101}$$

Note that if we have a surface X based on a triangulation $T \in \mathcal{T}(n)$ with a polygonal boundary $\gamma(x_1, \ldots, x_n)$ with vertices at $x_1, \ldots, x_n$, then we can construct from X another surface X_0 with a degenerate boundary consisting of a single point, by adding one vertex to T and connecting the new vertex to the boundary vertices of T. It follows that

$$G_\mu(x_0) \geq \sum_{n=2}^{\infty} \sum_{T \in \mathcal{T}(n)} n^{-1} e^{-\mu|T|} \int dx_1 \ldots dx_n \tag{3.102}$$

$$\times G^{(T)}(\gamma(x_1, \ldots, x_n)) \exp\left(-\frac{1}{2}\sum_{i=1}^{n} x_i^2 - \frac{1}{2}\sum_{i-j=1}^{n}(x_i - x_j)^2\right).$$

Now take $\varepsilon > 0$. Then, by Lemma 3.1,

$$G_\mu(\gamma(x_1, \ldots, x_n)) \geq G_{\mu+\varepsilon}(O_n), \tag{3.103}$$

if $|x_i| < \delta$, $i = 1, \ldots, n$, and δ is sufficiently small. It follows, by restricting the range of the x_i-integrations in Eq. (3.103), that

$$G_\mu(x_0) \geq C \sum_{n=2}^{\infty} a^n G_{\mu+\varepsilon}(O_n) \tag{3.104}$$

for positive constants a and C which only depend on ε and μ. Since $G_\mu(x_0) < \infty$ for $\mu > \mu_0$ and $\varepsilon > 0$ is arbitrarily small, the inequality (3.101) follows for any $\mu > \mu_0$.

3.4.5 The Hausdorff dimension

The notion of Hausdorff dimension can be generalized from random walks to random surfaces. The *mean square extent*, also called the *radius of gyration* of a generic surface, is most conveniently defined in analogy with Eq. (2.141) by

$$\langle x^2 \rangle_\mu = \frac{1}{\chi(\mu)} \int dx \, x^2 \, G_\mu(0, x). \tag{3.105}$$

Disregarding irrelevant symmetry factors of the triangulations we have

$$\langle x^2 \rangle_\mu \sim \frac{1}{\chi(\mu)} \sum_{T \in \mathscr{T}_1} e^{-\mu|T|} \int \prod_{i \in V(T) \setminus \{0\}} dx_i \sum_{i \in V(T)} x_i^2 \, e^{-S_T(X)}, \tag{3.106}$$

where 0 is the marked vertex in T and where we have chosen the image $x_0 \in \mathbb{R}^d$ of 0 to be $x_0 = 0$. Performing the integration in Eq. (3.106) yields

$$\langle x^2 \rangle_\mu \sim \frac{1}{\chi(\mu)} \sum_{T \in \mathscr{T}_1} e^{-\mu|T|} \operatorname{Tr}(-\Delta_T') \left(\frac{(2\pi)^{N_v(T)-1}}{D(T)} \right)^{d/2}. \tag{3.107}$$

Similarly, the average number of triangles in a triangulation is defined in analogy with Eq. (2.142) by

$$\langle |T| \rangle_\mu = -\frac{1}{\chi(\mu)} \frac{d}{d\mu} \chi(\mu). \tag{3.108}$$

If there is a number $d_H > 0$ so that, asymptotically,

$$\langle |T| \rangle_\mu \sim \langle x^2 \rangle_\mu^{d_H/2} \tag{3.109}$$

as $\mu \to \mu_0$, then d_H is called the *Hausdorff dimension* or the *extrinsic Hausdorff dimension*. In case $\langle |T| \rangle_\mu$ increases faster than any power of $\langle x^2 \rangle_\mu$ we set $d_H = \infty$.

If we consider a regular $N \times N$ two-dimensional lattice which is triangulated by dividing each square into two triangles, and, say, with periodic boundary conditions on X, the corresponding Hausdorff dimension, as defined by Eq. (3.109), is infinite. This follows since the trace of the inverse Laplacian, with the zero mode removed, diverges logarithmically as one removes the infrared cutoff N. In Section 3.7 we give an alternative, more geometric, derivation of this result.

Assume that the mass $m(\mu)$ scales to zero as $m(\mu) \sim (\mu - \mu_0)^\nu$. Then

$$d_H = \nu^{-1}. \tag{3.110}$$

The proof of this relation is almost identical to the proof of Eq. (2.155) in Chapter 2.

In order to define the *intrinsic Hausdorff dimension* of surfaces, denoted by d_h, we let, for $T \in \mathscr{T}_2$, the *geodesic distance* between the two marked vertices 0 and 1 be denoted by r_T, i.e. r_T is the smallest number of links in a path in T connecting 0 and 1. It should be stressed that there is nothing fundamental about the use of the shortest link path as a measure of the geodesic distance. Another important notion of geodesic distance is the number of triangles in the shortest path of triangles separating the points in question; see Section 4.7 for a further discussion of this concept. The average geodesic distance squared between 0 and 1 is then given by

$$\langle r_T^2 \rangle_\mu = \frac{1}{\chi(\mu)} \sum_{T \in \mathscr{T}_2} r_T^2 \, e^{-\mu|T|} \int dx \, G^T(0, x). \tag{3.111}$$

We define d_h by the asymptotic relation

$$\langle |T| \rangle_\mu \sim \langle r_T^2 \rangle_\mu^{d_h/2}, \tag{3.112}$$

as $\mu \to \mu_0$. It is clear that $d_h \geq 1$ if it exists.

The intrinsic and extrinsic Hausdorff dimensions for ordinary and multicritical branched polymers were calculated in Section 2.5. For the DTRS-model there is not yet any analytic calculation of the Hausdorff dimensions, but in the next section we show under certain assumptions that $d_H = 4$ for the LRS-model. In the next chapter we shall argue that $d_h = 4$ in the case $d = 0$, which corresponds to pure two-dimensional quantum gravity.

There is a general relation between the intrinsic and extrinsic Hausdorff dimensions:

$$d_h \leq d_H, \tag{3.113}$$

i.e. seen from inside the surface the dimension is smaller than in embedding space. Intuitively it is clear that an inequality of this type should hold because the distance between points on the surface is smaller in embedding space than the distance measured along the surface.

In order to prove the inequality (3.113) let $n = r_T$ for a given $T \in \mathscr{T}_2$. As before we let 0 and 1 be the marked vertices in $\mathscr{T}_2$ and choose a geodesic path $j(\alpha)$ from 0 to 1 with $j(0) = 0$, $j(n) = 1$. Then

$$x_1 = \sum_{\alpha=1}^{n} (x_{j(\alpha)} - x_{j(\alpha-1)}),$$

so that

$$x_1^2 \leq n \sum_{\alpha=1}^{n} (x_{j(\alpha)} - x_{j(\alpha-1)})^2.$$

Hence,

$$\langle x_1^2 \rangle_\mu \;\leq\; \chi^{-1}(\mu) \sum_{T \in \mathscr{T}_2} r_T e^{-\mu |T|} \int \prod_{i \in V(T)\backslash\{0\}} dx_i$$

$$\times \sum_{\alpha=1}^{r_T} (x_{j(\alpha)} - x_{j(\alpha-1)})^2 \, e^{-S_T(X)}$$

We now use the fact that

$$(x_{j(\alpha)} - x_{j(\alpha-1)})^2 \, e^{-S_T(X)} \leq C \, e^{-S_{T'}(X)},$$

where C is a constant and the modified action $S_{T'}(X)$ is defined by leaving out the term $(x_{j(\alpha)} - x_{j(\alpha-1)})^2$ in the action, i.e.

$$S_{T'}(X) = \frac{1}{2} \sum_{(k,l) \in L(T')} (x_k - x_l)^2,$$

where T' is the graph obtained by removing the link $(j(\alpha), j(\alpha-1))$ from T. Note that T' is not a triangulation, but the formula (3.57) is still valid for T'. Using this, one can show [246] that $\frac{1}{3} \det C_T \leq \det C_{T'} \leq \det C_T$. It follows that

$$\langle x^2 \rangle_\mu \leq C' \langle r_T^2 \rangle_\mu,$$

where $C' > 0$ is a constant and the inequality (3.113) is thereby proved.

3.4.6 *Scaling and the continuum limit in the DTRS-model*

As already mentioned we *assume* that the mass $m(\mu)$ scales to zero as

$$m(\mu) \sim (\mu - \mu_0)^\nu \tag{3.114}$$

for some $\nu > 0$ as $\mu \to \mu_0$. We expect the following behaviour for the two-point function:

$$G_\mu(0, x) \sim \begin{cases} x^{-\alpha} e^{-m(\mu)x} & \text{for} \quad x \gg m(\mu)^{-1} \\ x^{-d+2-\eta} & \text{for} \quad 0 \ll x \ll m(\mu)^{-1}. \end{cases} \tag{3.115}$$

The exponent η is called the *anomalous scaling dimension*. Since the susceptibility exponent γ is determined from

$$\chi(\mu) = \int dx \, G_\mu(0, x) \sim \frac{1}{(\mu - \mu_0)^\gamma},$$

Fisher's scaling relation

$$\gamma = \nu(2 - \eta) \tag{3.116}$$

follows from Eq. (3.115) in the same way as in the case of random paths.

In order to construct a continuum limit of the two-point function that falls off exponentially at large (physical) distances we apply the same procedure as for the random walk and define the continuum two-point function as

$$G(0, x) = \lim_{a \to 0} a^{-(d-2+\eta)} G_{\mu(a)}(0, a^{-1}x), \tag{3.117}$$

where a is the cutoff. Here x is measured in physical units and the (scaling) function $\mu(a)$ is determined by

$$m(\mu(a)) = ma, \tag{3.118}$$

where m is the physical mass which, together with x, is held fixed in the limit (3.117). It follows from Eq. (3.115) that $G(0, x)$ falls off as $\exp(-m|x|)$ at large distances.

In Eq. (3.117) we have fixed the scaling of the two-point function in terms of the scaling parameter a in target space $\mathbb{R}^d$. The relation between a and lattice spacing ε in the dynamical triangulation can now be determined from, e.g., Eq. (3.109). If we introduce the link length ε the left-hand side of Eq. (3.109) is multiplied by a factor ε^2, which in order to match the factor a^{d_H} present on the right-hand side, if x is now the physical length scale, has to satisfy

$$\varepsilon^2 \sim a^{d_H}. \tag{3.119}$$

Having fixed the scaling of μ as $a \to 0$, we may consider continuum limits of other loop functions. In particular, it is of interest to consider the one-loop function $G_\mu(\gamma(L_1, L_2))$ and the string tension, since in a genuine string theory the string tension is a physical observable with the dimension of mass^2. This means that the scaling

$$\tau(\mu(a)) = \tau a^2(\mu) \tag{3.120}$$

is necessary for the continuum limit of $\tau(\mu)$ to make sense. Here we denote the continuum string tension by τ, which is fixed in the limit $a \to 0$. Clearly, it is impossible for both Eq. (3.120) and $\tau(\mu) \geq 1$ to hold unless the physical string tension $\tau = \infty$. Recall that $G_\mu(\gamma_1, \dots \gamma_b)$ can be interpreted as the partition function for a membrane stretched out between the boundaries $\gamma_1, \dots, \gamma_b$. It follows that we can view the surfaces contributing to the loop functions in the scaling limit as surfaces of minimal area with the prescribed boundary since the infinite physical string tension suppresses all area fluctuations. This does not imply that the surfaces are completely rigid, since at any point on the surfaces there can be thin outgrowths of zero area. In particular, the dominant contributions to the two-point function $G(x, x')$ come from thin tubes of (almost) no area connecting x and x'. Anywhere along the tube there can be outgrowths which themselves are thin tubes, i.e. we are led to the picture of *branched polymer dominance*. These heuristic arguments lend support to the conjecture that the scaling limit of the DTRS-model for $d \geq 2$ is that of branched polymers. For large d we were led to the branched polymer picture by independent mean-field considerations. The missing link in the arguments for smaller d is that we have no proof that the mass $m(\mu)$ tends to zero at μ_0. If that is not the case it is not possible to define a scaling limit in any natural way at all. If the scaling limit is that of branched polymers, we have already shown in Chapter 2 that $v = 1/4$, $\eta = 0$ and $d_H = 4$ while $d_h = 2$.

Two questions arise: is the pathology of branched polymer dominance which we have described above an artifact of the particular model we have considered above or is it a generic feature of relativistic string theories? If it is a generic feature how can it be avoided? In the next section we give a partial answer to the first question while the last two sections of this chapter deal with the second question.

3.5 Random surfaces on a lattice

In this section we discuss the theory of random surfaces embedded in the hyper-cubic lattice $\mathbb{Z}^d$, which we called the LRS-model. These surfaces have, roughly speaking, the same relation to the randomly triangulated models discussed above as the lattice paths have to piecewise linear paths, and they should be viewed as different discretizations of the same continuum theory, i.e. they are expected to belong to the same universality class. As mentioned in the first section of the present chapter the main difference from a continuum point of view is that the action of the lattice theory is proportional to the area, such that it may be regarded as a discretization of the Nambu–Goto action (3.2), while the dynamically triangulated random surface model is a discretization of the Polyakov action (3.5). Appealing to universality it is possible to imagine a family of discretized theories with identical scaling behaviour which interpolate between the DTRS-model and the LRS-models. In Chapter 2 we have seen explicitly how this could be done in the random walk case. For surfaces we do not have any strong tool analogous to the central limit theorem at our disposal and there is no constructive proof that an interpolation is possible. However, one would like to prove that more general actions than the Gaussian lead to the same critical behaviour in the framework of DTRS-models. In particular, one could imagine a limiting case with an action that constrains the triangles in the surfaces in $\mathbb{R}^d$ to be equilateral with fixed edge length a, such that the cosmological term $\mu|T|$ in the Polyakov action is converted into the Nambu–Goto action. As an explicit example of such an interpolation we can write down the following partition function:

$$Z(\mu, \beta, a) \;=\; \sum_{T \in \mathscr{T}} e^{-\mu |T|} \, \beta^{N_l(T)d/2} \int \prod_{i \in V(T)} dx_i$$

$$\times \exp\left(-\frac{\beta}{2} \sum_{(ij) \in L(T)} (|x_i - x_j| - a)^2\right) \delta\left(\sum_{i \in V(T)} x_i\right).$$

For $\beta = 1$ and $a = 0$ this is the partition function of the DTRS-model, while in the limit $\beta \to \infty$ it is the partition function of a *dynamically triangulated Nambu–Goto model*. The latter model defined by summing over all surfaces which can be constructed in $\mathbb{R}^d$ from equilateral triangles with a fixed link length a is potentially quite interesting, and in $d = 2$ it is in fact a true lattice model since the triangles are in this case embedded in a regular triangular lattice with link length a. In higher dimensions the structure is richer, but by additional constraints on the action one can force the piecewise linear surface to live on a hyper-cubic lattice $a\mathbb{Z}^d$. In contrast to the random walk case we cannot control the singular actions needed to carry out this interpolation rigorously, but we consider it likely that such an interpolation exists. Here we shall have to jump directly from the DTRS-model of the previous section to the LRS-model defined below, but the results we derive for the LRS-model seem to corroborate the expectation of universality. Moreover, as mentioned previously, some questions are more easily addressed in the lattice model than in the triangulated model so the two discretizations supplement each other in a useful way.

3.5.1 Definition of the lattice surface model

We start by defining the lattice surfaces. We shall refer to the elements of $\mathbb{Z}^d$ as vertices and denote by $\{e_1, \ldots, e_d\}$ the standard basis for $\mathbb{Z}^d$. An ordered pair $(x(x \pm e_i))$, where $x \in \mathbb{Z}^d$, is called an *oriented link* in $\mathbb{Z}^d$. An unoriented link is the corresponding unordered pair. An *elementary plaquette* in $\mathbb{Z}^d$ is a collection of four vertices $\{x, x + e_i, x + e_j, x + e_i + e_j\}$ where $i \neq j$. An orientation of the plaquette is a choice of one of the two cyclic orderings $x(x + e_i)(x + e_i + e_j)(x + e_j)$ or $x(x + e_j)(x + e_i + e_j)(x + e_i)$ (see Fig. 3.8).

In our study of triangulated surfaces we distinguished between the abstract triangulation T and its embedding into $\mathbb{R}^d$ which we denoted by X. In dealing with lattice surfaces it is often more convenient to focus directly on the image of the surface and regard it as a collection of elementary plaquettes from $\mathbb{Z}^d$ that are glued together along their boundary links in such a way that the resulting *complex* is orientable. Since we shall be considering self-overlapping surfaces it is useful to

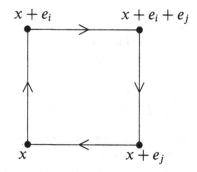

Fig. 3.8. An oriented plaquette.

consider a lattice surface as an embedding of an abstract complex in order to make the concepts clear. For the lattice random walk we did not find this necessary.

An oriented *lattice surface* S is given by the following:

(i) a finite set $V(S)$ called the *vertex set*;

(ii) a family $P(S)$ of cyclically ordered subsets of $V(S)$, called plaquettes, each containing four elements, such that

$$\bigcup_{p \in P(S)} p = V(S);$$

(iii) a mapping

$$S : V(S) \mapsto \mathbf{Z}^d,$$

so that any element of $P(S)$ is mapped onto the corners of an elementary oriented plaquette in $\mathbf{Z}^d$, respecting the cyclic ordering of vertices.

We use the notation $p = (ijkl) \in P(S)$ to indicate the cyclic ordering of the corners i, j, k, l which make up the plaquette p, so $(ijkl) = (jkli)$, etc. We say that (ij) is an oriented link in S if there is a plaquette in S of the form $(ijkl)$. In this case the images of i and j under S form an oriented link in $\mathbf{Z}^d$. Note that (jk), (kl) and (li) are also oriented links in S. Let $L_o(S)$ be the collection of all oriented links in S. We require that any element of $L_o(S)$ is contained in exactly one plaquette. If (ij) is an oriented link then we say that (ji) is the same link with reversed orientation. It follows that if $p_1 = (ijkl)$ and $p_2 = (jinm)$ are two plaquettes in $P(S)$, then both (ij) and (ji) are in $L_o(S)$ and in this case we say that the two plaquettes p_1 and p_2 are glued together along the unoriented link $\{i, j\}$. If $(ij) \in L_o(S)$ but $(ji) \notin L_o(S)$, then we say that (ij) is a boundary link of S. The collection of all unoriented links in S will be denoted $L(S)$. We denote the boundary of S by ∂S and let $|\partial S|$ denote the number of links in ∂S.

Two surfaces S and S' are regarded as being identical if there exists a bijective map $\phi : V(S) \mapsto V(S')$ which induces a bijective map from $P(S)$ onto $P(S')$ such that $S = S' \circ \phi$. In particular, we consider orientable, but not oriented, surfaces, unless otherwise stated explicitly. Note that there is no restriction on the number of distinct elements in $V(S)$ that can be mapped onto a single vertex in $\mathbb{Z}^d$.

A lattice surface is *connected* if, for any two plaquettes $p, p' \in P(S)$, there is a sequence of plaquettes $p_1, p_2, \ldots, p_n$ such that $p_1 = p$, $p_n = p'$ and p_i and p_{i+1} share a link. All lattice surfaces that we consider are assumed to be connected unless we state the opposite explicitly.

We shall use the notation

$$|S| = \#P(S), \quad N_v(S) = \#V(S), \quad N_l(S) = \#L(S).$$

The quantity $|S|$ will be called the area of S. The Euler characteristic of a surface S is defined as in the triangulated model in terms of the number of plaquettes, links and vertices. An orientation of a surface with boundary clearly induces an orientation on the boundary loops. The boundary of a surface is always a union of lattice loops, i.e. of closed lattice paths, which may share some vertices. We leave it to the reader to check that for any loop there is a surface S with $\partial S = \gamma$.

Considering all lattice surfaces without any constraint does not lead to a well defined random surface theory with any natural action functionals because the number of surfaces with a given area A grows faster than any exponential function of A as $A \to \infty$. The reason for this is essentially the same as the reason for the superexponential growth with n of the number of different triangulations made up of n triangles (see Eq. (3.18)).

We shall therefore concentrate on the study of surfaces of genus 0 even though many of our results can be generalized to surfaces with a fixed arbitrary genus. We shall refer to these surfaces as *planar lattice random surfaces*. If $\gamma_1, \ldots, \gamma_b$ are lattice loops we let $\mathscr{S}(\gamma_1, \ldots, \gamma_b)$ denote the collection of all planar lattice surfaces with boundary $\partial S = \gamma_1 \cup \ldots \cup \gamma_b$. We let $\mathscr{S}$ denote the collection of all closed planar surfaces, one of whose vertices is at the origin in $\mathbb{Z}^d$. The model defined below in terms of these surfaces will be called the *planar lattice random surface model* or the PLRS-model.

By analogy with the random walk case and as a regularization of the Nambu–Goto action we choose the action of a surface to be proportional to its area and define the loop functions of the PLRS-model by

$$G_\beta(\gamma_1, \ldots, \gamma_b) = \sum_{S \in \mathscr{S}(\gamma_1, \ldots, \gamma_b)} e^{-\beta|S|}. \tag{3.121}$$

Our first goal is to prove the following basic result analogous to Theorem 3.2.

Theorem 3.7 *There is a positive number β_0 such that the loop functions of the PLRS-model are well defined for $\beta > \beta_0$ and divergent for $\beta < \beta_0$.*

Proof Let us define

$$N_{\gamma_1,\dots,\gamma_b}(A) = \#\{S \in \mathscr{S}(\gamma_1,\dots,\gamma_b) : |S| = A\}.$$

The theorem follows from the existence of the limit

$$\lim_{A\to\infty} \frac{\log N_{\gamma_1,\dots,\gamma_b}(A)}{A} = \beta_0, \tag{3.122}$$

where β_0 is a positive number which is independent of the boundary loops $\gamma_1,\dots,\gamma_b$. The existence of the limit will be established by a subadditivity argument.

Let us denote by $N(A)$ the number of closed lattice surfaces of area A, one of whose vertices is at the origin. We begin by establishing the existence of a positive number β_0 such that

$$\lim_{A\to\infty} \frac{\log N(A)}{A} = \beta_0. \tag{3.123}$$

First we show that there is a constant C_1 such that

$$N(A) \le C_1^A. \tag{3.124}$$

To any lattice surface S we can assign a unique dual ϕ^4-graph. If the surface is planar the corresponding graph is planar, i.e. it can be drawn on the surface of a sphere. The number of planar ϕ^4-graphs with A vertices is bounded by c^A, where c is a constant [279, 82]. This is a consequence of a more general result on the combinatorics of graphs that we derive in the next chapter. Now let $S \in \mathscr{S}$. The surface S induces a labelling of the links of the corresponding dual graph, with each link in the dual graph connecting two dual vertices being labelled by the relative position of the two plaquettes it joins and with each vertex being labelled by the coordinate plane parallel to the corresponding plaquette, see Fig. 3.9. Given this labelling and given which vertex of S sits at the origin, the surface S is uniquely determined once a plaquette containing 0 has been fixed. Since the number of distinct labellings of a given dual graph is bounded by c'^A, where c' is a constant, the claimed bound (3.124) follows.

Let us note in passing that it is simple to show [153] that there exists a constant $C_2 > 1$ such that

$$N(A) \ge C_2^A. \tag{3.125}$$

Now take two surfaces $S_1, S_2 \in \mathscr{S}$ of area A_1 and A_2, respectively. Let ℓ_1 be a link in S_1, both of whose endpoints have a maximal x_1-coordinate.

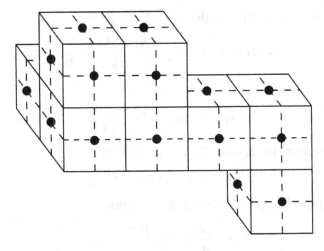

Fig. 3.9. The diagram illustrates a lattice surface and the corresponding ϕ^4-graph constructed by placing dual vertices at the centre of each plaquette and joining them by dual links (the dashed lines) which cross the links of the original surface.

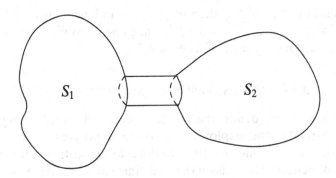

Fig. 3.10. Gluing of S_1 and S_2 by a cylindrical double plaquette surface.

Let ℓ_2 be a link in S_2, both of whose endpoints have a minimal x_1-coordinate. Now rotate S_2 in the lattice $\mathbb{Z}^d$ so that the link ℓ_2 becomes parallel to ℓ_1 and then translate S_2 so that ℓ_2 is moved to $\ell_1 + e_1$. Let us denote the rotated and translated surface by S'_2. Now cut the surfaces S_1 and S'_2 open along ℓ_1 and $\ell_1 + e_1$, respectively, and connect them by gluing them to a cylindrical double plaquette surface whose unoriented boundary consists of two copies of ℓ_1 and two copies of $\ell_1 + e_1$; see Fig. 3.10. In this way we have constructed a new surface $S' \in \mathscr{S}$ of area $A_1 + A_2 + 2$. The surfaces S_1 and S'_2 are uniquely determined by S'. There are at most $2(d-1)(A_2 + 2)$ different surfaces S_2 that can translate and rotate by this construction to the same S_2'. It follows that

$$N(A_1 + A_2 + 2) \geq \frac{N(A_1)N(A_2)}{2(d-1)(A_2+2)}. \tag{3.126}$$

If we define for $A \geq 2$

$$f(A) = -\log\left(\frac{N(A-2)}{4(d-1)A}\right),$$

then, by symmetry in A_1 and A_2, Eq. (3.126) implies

$$f(A_1 + A_2) \leq f(A_1) + f(A_2), \tag{3.127}$$

i.e. the function f is subadditive. It follows that

$$\lim_{A \to \infty} \frac{f(A)}{A} = \inf_A \frac{f(A)}{A} \equiv \beta_0$$

and

$$\log C_2 \leq \beta_0 \leq \log C_1.$$

This proves our first claim Eq. (3.123).

The remaining part of the theorem, i.e. the independence of β_0 on the boundary loops $\gamma_1, \ldots, \gamma_b$, is proved by arguments similar to, but easier than, the corresponding part of Theorem 3.2.

3.5.2 *Mass, susceptibility and string tension*

In this subsection we define the mass, susceptibility and string tension in the PLRS-model and explore some of their properties. Let γ be an arbitrary loop in $\mathbb{Z}^d$ and γ_L its translate by L lattice units in some coordinate direction. The mass $m(\beta)$ is defined in a similar way as in the dynamically triangulated case:

$$m(\beta) = -\lim_{L \to \infty} \frac{\log G_\beta(\gamma, \gamma_L)}{L}. \tag{3.128}$$

If the mass tends to zero as $\beta \downarrow \beta_0$, its critical exponent ν is defined as before by

$$m(\beta) \sim (\beta - \beta_0)^\nu.$$

By the following theorem we have $0 \leq \nu \leq 1$.

Theorem 3.8 *The limit (3.128) defining the mass exists for any γ and $\beta > \beta_0$ and its value is independent of the loop γ. The mass is a positive, concave, increasing function of β fulfilling*

$$2(\beta - \beta_0) \leq m(\beta) \leq 2\beta$$

and

$$G_\beta(\gamma, \gamma_L) \leq C(\beta)e^{-m(\beta)L}.$$

for a suitable constant $C(\beta)$.

Proof We begin by introducing some technical tools that allow us to use subadditivity arguments to prove the existence of the mass. Let us consider surfaces in $\mathscr{S}(\gamma_1, \gamma_2)$ with $\gamma_1 \neq \gamma_2$ and imagine that we have coloured the γ_1 boundary blue, say. We define a restricted surface ensemble $\mathscr{S}'(\gamma_1, \gamma_2)$ consisting of all surfaces $S \in \mathscr{S}(\gamma_1, \gamma_2)$ which do not have any proper subsurface belonging to $\mathscr{S}(\gamma_1, \gamma_2)$, which contains all the blue boundary links. Define the restricted two-loop function by

$$G'_\beta(\gamma_1, \gamma_2) = \sum_{S \in \mathscr{S}'(\gamma_1, \gamma_2)} e^{-\beta|S|}.$$

Now let γ be a loop consisting of two links and denote the two-loop functions $G_\beta(\gamma, \gamma_L)$ and $G'_\beta(\gamma, \gamma_L)$ by $G_\beta(L)$ and $G'_\beta(L)$, respectively. Then there is a function $H(\beta)$ such that

$$G_\beta(L) = H(\beta)G'_\beta(L). \tag{3.129}$$

and

$$G'_\beta(L + K) \geq G'_\beta(L)G'_\beta(K) \tag{3.130}$$

for any $L, K \in \mathbb{Z}^+$. The function $H(\beta)$ is given by the sum with weight $e^{-\beta|S|}$ over all surfaces that can be glued to surfaces in $\mathscr{S}'(\gamma, \gamma_L)$ along one or both links in γ_L in order to obtain surfaces in $\mathscr{S}(\gamma, \gamma_L)$. The inequality (3.130) follows from the fact that if we have a surface $S \in \mathscr{S}'(\gamma, \gamma_{L+K})$ which is constructed by gluing together two surfaces S_1 and S_2 in $\mathscr{S}'(\gamma, \gamma_L)$ and $\mathscr{S}'(\gamma_L, \gamma_{L+K})$, respectively, then S_1 and S_2 are uniquely determined by S.

The smallest surface contributing to $G'_\beta(L)$ has area $2L$ so

$$G'_\beta(L) \geq e^{-2\beta L}. \tag{3.131}$$

By Eq. (3.130) it follows that the function $-\log G'_\beta(L)$ is subadditive so with the aid of Eq. (3.131) we conclude that there is a non-negative number $m(\beta) \leq 2\beta$ such that

$$G'_\beta(L) = e^{-m(\beta)L + o(L)}. \tag{3.132}$$

In fact,

$$m(\beta) = -\inf_L \frac{\log G'_\beta(L)}{L}.$$

By Eq. (3.129), Eq. (3.132) also holds for $G_\beta(L)$. The independence of the mass of the boundary loops follows from the inequalities

$$C_1 G_\beta(\gamma', \gamma'_L) \le G_\beta(\gamma, \gamma_L) \le C_2 G_\beta(\gamma', \gamma'_L),$$

which are valid for an arbitrary loop γ' with constants which only depend on γ and γ'.

In order to establish the lower bound on the mass we note that any surface in $\mathscr{S}(\gamma, \gamma_L)$ can be closed up along γ_L by gluing the two links together so that we obtain a surface in $\mathscr{S}(\gamma)$. A given surface S in $\mathscr{S}(\gamma)$ can arise from, at most, $2|S|$ different surfaces in $\mathscr{S}(\gamma, \gamma_L)$ by this construction. Hence, we obtain the inequality

$$G_\beta(L) \le 2 \sum_{A=2L}^{\infty} A\, N_\gamma(A) e^{-\beta A},$$

which, by Eq. (3.122), implies the lower bound.

If we define

$$m_L(\beta) = -\frac{1}{L} \log G_\beta(L),$$

then

$$\frac{d}{d\beta} m_L(\beta) \ge 2 \quad \text{and} \quad \frac{d^2}{d\beta^2} m_L(\beta) < 0.$$

Since $\lim_{L\to\infty} m_L(\beta) = m(\beta)$ it follows that $m(\beta)$ is an increasing, concave function. This completes the proof.

As in the case of dynamically triangulated surfaces there is no unique way of defining the susceptibility so we choose a definition that is technically convenient to work with. Let γ_0 be the loop which consists of two copies of a link ℓ in $\mathbf{Z}^d$. We shall denote this loop by $\gamma(\ell)$ when we wish to emphasize that the loop consists of two copies of a specific link ℓ. The susceptibility is defined as

$$\chi(\beta) = \sum_{x \in \mathbf{Z}^d} G_\beta(\gamma_0, \gamma_x), \tag{3.133}$$

where γ_x denotes the translate of γ_0 by the vector $x \in \mathbf{Z}^d$. Other possible definitions of the susceptibility will have the same singularity as $\chi(\beta)$ as $\beta \to \beta_0$. The critical exponent of the susceptibility will be denoted by γ as previously.

It is true here as for the DTRS-models that *if the susceptibility diverges as $\beta \to \beta_0$ then the mass vanishes at β_0*. In order to see this in the present case suppose that $m(\beta) \ge M > 0$ for all $\beta > \beta_0$. Then it can be shown [154] that there are positive constants C and c such that

$$G_\beta(\gamma_0, \gamma_x) \le C e^{-cM|x|}.$$

It follows that $\chi(\beta)$ is bounded from above by a constant independent of β.

Closing up the loop γ_x (by identifying the two links in γ_x) in a surface S contributing to $G_\beta(\gamma_0, \gamma_x)$ for some x, we obtain a surface $S' \in \mathscr{S}(\gamma)$. A given surface S' can be obtained in this way from at least $\frac{1}{2}|S'|$ and at most $2|S'|$ different surfaces S. It follows that

$$-\frac{1}{2}\frac{d}{d\beta}G_\beta(\gamma_0) \le \chi(\beta) \le -2\frac{d}{d\beta}G_\beta(\gamma_0); \qquad (3.134)$$

see Eq. (3.54). The asymptotic growth of the number of surfaces with one boundary loop is therefore given by

$$N_{\gamma_0}(A) \sim A^{\gamma-2}e^{\beta_0 A},$$

for any loop γ_0. An argument analogous to the one used to establish Eq. (3.134) shows that

$$N_{\gamma_1,...,\gamma_b}(A) \sim A^{\gamma-3+b}e^{\beta_0 A}.$$

We next establish the analogue of Theorem 3.4 for the PLRS-model. This model was actually the first in which the inequality

$$\gamma \le \frac{1}{2} \qquad (3.135)$$

was obtained [152]. Let S_c be a lattice surface which is an ordinary three-dimensional cube consisting of six plaquettes with one edge parallel to the link ℓ defining γ_0. Therefore, S_c has four edges parallel to ℓ and we cut the cube open along three of these edges so it becomes a surface with a boundary consisting of three non-intersecting loops $\gamma_{x_1}, \gamma_{x_2}$ and γ_{x_3}. Let x_1', x_2', x_3' be three arbitrary points in $\mathbf{Z}^d$ and let $S_i \in \mathscr{S}(\gamma_{x_i}, \gamma_{x_i'})$, $i = 1, 2, 3$. For each i we glue the surface S_i to S_c along the loop γ_{x_i}. In this way we obtain a surface $S \in \mathscr{S}(\gamma_{x_1'}, \gamma_{x_2'}, \gamma_{x_3'})$. Moreover, the subsurfaces S_i are uniquely determined by S. The purpose of the gluing cube S_c is to ensure this uniqueness. It follows, by summing over all translates of S_c, that

$$G_\beta(\gamma_{x_1'}, \gamma_{x_2'}, \gamma_{x_3'}) \ge C \sum_{y \in \mathbf{Z}^d} \prod_{i=1}^{3} G_\beta(\gamma_{x_i+y}, \gamma_{x_i'}). \qquad (3.136)$$

By an argument similar to the one used to establish Eq. (3.134) we find that

$$-\frac{d}{d\beta}\chi(\beta) \sim \sum_{x_2', x_3' \in \mathbf{Z}^d} G_\beta(\gamma_{x_1'}, \gamma_{x_2'}, \gamma_{x_3'}).$$

It follows by summing over x_2' and x_3' in Eq. (3.136) that

$$-\frac{d}{d\beta}\chi(\beta) \ge C\,\chi(\beta)^3, \qquad (3.137)$$

where C is a positive constant, and the bound (3.135) follows. For a discussion of the bound in more general lattice models, see [155], where a family of bounds on loop functions of the type illustrated by Eq. (3.136) are also derived.

Note that the bound we have derived on the critical exponent of the susceptibility, Eq. (3.135), implies that all one-loop functions are finite at the critical point and bounded by a constant which only depends on the boundary loop.

Next, let us consider the string tension in the PLRS-model. Let $\gamma_{R,L}$ be a rectangular lattice loop lying in a coordinate plane with edges of length R and L. The string tension is defined as

$$\tau(\beta) = -\lim_{R\to\infty} \frac{\log G_\beta(\gamma_{R,R})}{R^2}. \tag{3.138}$$

Below we show that this limit exists and establish its basic properties. It can easily be seen that $\tau(\beta) \leq \beta$ by considering the contribution of the minimal surface to $G_\beta(\gamma_{R,R})$. The critical exponent μ of the string tension is defined by

$$\tau(\beta) - \tau(\beta_0) \sim (\beta - \beta_0)^\mu,$$

where we anticipate that $\tau(\beta_0) \neq 0$. From dimensional arguments it follows that the critical exponents of the mass and the string tension are related by

$$\mu = 2\nu. \tag{3.139}$$

The following theorem implies that $0 \leq \mu \leq 1$.

Theorem 3.9 *The limit (3.138) defining the string tension exists for any $\beta > \beta_0$. The string tension is a positive, concave, increasing function of β bounded from below by $\beta - \beta_0$.*

Proof The argument for the existence of the limit (3.138) is a slight variation of the one used for the DTRS-model in the previous section so we shall be brief. For any R and T we assume that the loop $\gamma_{R,T}$ lies in the positive quarter of the RT-plane with one vertex at 0. Given two surfaces S_1 and S_2 with boundaries γ_{R,T_1} and γ_{R,T_2}, respectively, we can construct a surface $S \in \mathscr{S}(\gamma_{R,T_1+T_2+1})$ by first translating S_2 by $T_1 + 1$ units in the positive T-direction and then connecting S_1 to this translated surface by gluing both of them to an intermediate strip along the R-axis of width 1, see Fig. 3.11. The presence of the strip guarantees that S_1 and S_2 are uniquely determined by S. It follows that

$$G_\beta(\gamma_{R,T_1+T_2+1}) \geq e^{-\beta R} G_\beta(\gamma_{R,T_1}) G_\beta(\gamma_{R,T_2}), \tag{3.140}$$

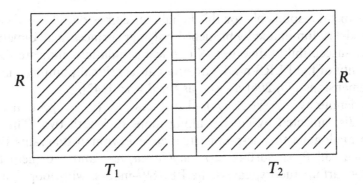

Fig. 3.11. The gluing of two surfaces with frames $R \times T_1$ and $R \times T_2$ along a strip of plaquettes.

which implies that the function

$$f_\beta(R, T) = -\log\left(G_\beta(\gamma_{R-1,T-1})e^{-\beta(R+T-1)}\right)$$

is a subadditive function of T for fixed R, and also, by symmetry, a subadditive function of R with fixed T. We conclude that the limit (3.138) exists and that

$$\tau(\beta) = \inf_R \frac{f_\beta(R, R)}{R^2}. \tag{3.141}$$

In order to prove that $\tau(\beta) \geq (\beta - \beta_0)$, take two surfaces $S_1, S_2 \in \mathcal{S}(\gamma_{R,T})$, translate one lattice spacing in a direction perpendicular to the RT-plane (in two dimensions one argues slightly differently) and glue the resulting surfaces together by adding a ribbon of area $2(R + T)$ to connect them, but leave one link ℓ unglued. If we denote the new surface in $\mathcal{S}(\gamma(\ell))$ that arises by this construction S, then S_1 and S_2 are uniquely determined by S and

$$e^{-2\beta(R+T)}(G_\beta(\gamma_{R,T}))^2 \leq \sum_{A=2RT}^{\infty} N_{\gamma(\ell)}(A)e^{-\beta A}.$$

The desired lower bound follows.

The facts that τ is increasing and concave are proved by mimicking the corresponding argument for the mass. This completes the proof.

3.5.3 *Critical behaviour and continuum limit*

We saw above that if the susceptibility of the PLRS-model diverges, then the mass vanishes at the critical point. We will now show that up to a certain universality assumption the divergence of the susceptibility implies that all the critical exponents of the PLRS-model are calculable.

The mechanism responsible for this behaviour comes out in the wash: the PLRS-model is dominated by branched polymer surfaces in all dimensions, where the susceptibility diverges. Numerical simulations indicate that the susceptibility of the PLRS-model diverges at its critical point in a number of low dimensions [79, 266] as well as for $d = \infty$ [147].

Let us denote by $\mathscr{S}^{(2)}(\gamma_1, \ldots, \gamma_b)$ all surfaces in $\mathscr{S}(\gamma_1, \ldots, \gamma_b)$ that cannot be made disconnected by cutting along loops of length two. These surfaces are analogous to one-point irreducible Feynman diagrams in field theory. We call these surfaces *irreducible surfaces* and use them to define a new surface theory, called the PLRS$^{(2)}$-model, with loop functions defined by restricting the summation to irreducible surfaces, denoted by $G^{(2)}(\gamma_1, \ldots, \gamma_b)$, and other quantities are denoted as before with a superscript (2). All the results we have proved above for the PLRS-model carry over to the PLRS$^{(2)}$-model. In particular there is a critical value $\beta_0^{(2)}$ for the coupling and $\tau^{(2)}(\beta) > 0$ if $\beta > \beta_0^{(2)}$. It is clear that $\beta_0^{(2)} \leq \beta_0$. Let us denote by $\mathscr{S}_2(\gamma_1, \ldots, \gamma_b)$ the set of surfaces in $\mathscr{S}(\gamma_1, \ldots, \gamma_b)$ which can be obtained by scaling surfaces in $\mathscr{S}(\gamma_1/2, \ldots, \gamma_b/2)$ by a factor 2, where we assume that the loops $\gamma_1/2, \ldots, \gamma_b/2$ obtained by scaling $\gamma_1, \ldots, \gamma_b$ by a factor $1/2$ are in the lattice $\mathbf{Z}^d$. It is then clear that

$$\mathscr{S}_2(\gamma_1, \ldots, \gamma_b) \subseteq \mathscr{S}^{(2)}(\gamma_1, \ldots, \gamma_b) \subseteq \mathscr{S}(\gamma_1, \ldots, \gamma_b),$$

which suggests that the critical behaviours of the PLRS- and the PLRS$^{(2)}$-models are identical, since this is true for the model built on $\mathscr{S}_2$ and the PLRS-model. Taking this universality for granted we only have to assume that the susceptibility of the PLRS-model diverges in the theorem given below. Note also that the argument given above implies that $\frac{1}{4}\beta_0 \leq \beta_0^{(2)}$. The main result of this section is the following.

Theorem 3.10 *If the susceptibility of the PLRS-model and the susceptibility of the PLRS$^{(2)}$-model both diverge at their respective critical points, then the mass m of the PLRS-model tends to zero at β_0. All the critical exponents of the PLRS-model have their mean field theory values and the string tension does not vanish at β_0. Furthermore, the scaling limit of the two-loop function of the PLRS-model exists and equals that of a free scalar field theory.*

Proof We let γ denote a loop of length two consisting of two copies of a link ℓ. It will be convenient to denote the one-loop function $G_\beta(\gamma)$ by $G(\beta)$ and $G_\beta^{(2)}(\gamma)$ by $G_2(\beta)$. Of course, the values of these functions are independent of the link ℓ.

Let us consider a surface $S \in \mathscr{S}^{(2)}(\gamma_{R,R})$, $R \geq 1$ and let ℓ be an interior link in S. If we cut the surface S open along ℓ and glue to it a surface S_ℓ from $\mathscr{S}(\gamma(\ell))$, we obtain a surface in $\mathscr{S}(\gamma_{R,R})$. Similarly, if ℓ is a boundary link in S, then we can glue a surface from $\mathscr{S}(\gamma(\ell))$ to S along one copy of ℓ and obtain a surface in $\mathcal{S}(\gamma_{R,R})$. For any subset of links $\{\ell_1, \ldots, \ell_n\}$ in S we can perform a gluing of the kind described above and obtain a surface $S' \in \mathcal{S}(\gamma_{R,R})$. It is clear that the *outgrowths* S_ℓ and the underlying irreducible surface S are uniquely determined by S'. Conversely, given any surface $S' \in \mathcal{S}(\gamma_{R,R})$, it can be decomposed into maximal outgrowths and an irreducible surface S as above. We shall refer to the minimal loops of length two, where the outgrowths meet an irreducible surface, as *bottlenecks*. We can assume that there is an outgrowth emanating from every link in S by allowing empty outgrowths. It follows that

$$
\begin{aligned}
G_\beta(\gamma_{R,R}) &= \sum_{S \in \mathcal{S}^{(2)}(\gamma_{R,R})} e^{-\beta|S|}(1 + G(\beta))^{N_l(S)} \\
&= (1 + G(\beta))^{2R} \sum_{S \in \mathcal{S}^{(2)}(\gamma_{R,R})} e^{-\beta|S|}(1 + G(\beta))^{2|S|} \\
&= (1 + G(\beta))^{2R} G_{\bar{\beta}}^{(2)}(\gamma_{R,R}),
\end{aligned} \tag{3.142}
$$

where the *renormalized coupling* $\bar{\beta}$ is given by

$$
\bar{\beta} = \beta - 2\log(1 + G(\beta)) \tag{3.143}
$$

and we have used the fact that $N_l(S) = 2|S| + 2R$ for $S \in \mathcal{S}^{(2)}(\gamma_{R,R})$. Note that the rewriting Eq. of (3.142) is analogous to Eq. (2.207) for branched polymers.

By taking R to infinity in Eq. (3.142) it follows that

$$
\tau(\beta) = \tau^{(2)}(\bar{\beta}). \tag{3.144}
$$

Let $\mathcal{S}'(\gamma)$ consist of those surfaces in $\mathcal{S}(\gamma)$ that cannot be cut along a single link so that one obtains two surfaces in $\mathcal{S}(\gamma)$. Such surfaces are said to have *no pockets at* γ. Define

$$
K(\beta) = \sum_{S \in \mathcal{S}'(\gamma)} e^{-\beta|S|}.
$$

Then

$$
G(\beta) = \frac{K(\beta)}{1 - K(\beta)}
$$

and

$$
K(\beta) = \frac{G^{(2)}(\bar{\beta})}{1 + G(\beta)},
$$

since no outgrowths sit on the boundary links of the surfaces contributing to K. It follows that

$$
G(\beta) = G^{(2)}(\bar{\beta}). \tag{3.145}
$$

Taking Eq. (3.134) into account we work in the remainder of this proof with susceptibilities χ and $\chi^{(2)}$ given by

$$\chi(\beta) = -\frac{d}{d\beta}G(\beta)$$

and

$$\chi^{(2)}(\bar{\beta}) = -\frac{d}{d\bar{\beta}}G^{(2)}(\bar{\beta}).$$

Differentiating Eq. (3.145) and using

$$\frac{d\bar{\beta}}{d\beta} = 1 + \frac{2\chi(\beta)}{1 + G(\beta)},$$

we find that

$$\chi(\beta) = \frac{(1 + G(\beta))\chi^{(2)}(\bar{\beta})}{1 + G(\beta) - 2\chi^{(2)}(\bar{\beta})}. \tag{3.146}$$

The susceptibilities are positive, $G(\beta_0) < \infty$ (by Eq. (3.143)) and $\chi^{(2)}(\bar{\beta})$ is assumed to diverge as $\bar{\beta} \downarrow \beta_0^{(2)}$, so we conclude from Eq. (3.146) that

$$\bar{\beta}(\beta_0) > \beta_0^{(2)}. \tag{3.147}$$

From Eq. (3.144) and the lower bound $\bar{\beta} - \beta_0^{(2)}$ on the string tension $\tau^{(2)}(\bar{\beta})$ it follows that $\tau(\beta_0) > 0$.

The susceptibility of the PLRS-model diverges when the denominator on the right-hand side of Eq. (3.146) vanishes, and therefore

$$1 + G(\beta_0) = 2\chi^{(2)}(\bar{\beta}(\beta_0)).$$

In order to determine the critical exponent γ we differentiate Eq. (3.146) with respect to β. Noting that the the derivative of $\chi^{(2)}(\bar{\beta})$ with respect to $\bar{\beta}$ is finite at $\bar{\beta}(\beta_0)$ we find after a short calculation that

$$\chi'(\beta) \sim \chi(\beta)^3.$$

Hence, $\gamma = \frac{1}{2}$.

In order to prove that $\nu = \frac{1}{4}$ we need to decompose the two-point function of the PLRS-model and write it as a sum of convolutions of the two-point function of the PLRS$^{(2)}$-model, which we next proceed to explain.

If γ is a loop in $\mathbf{Z}^d$ and x a vector in $\mathbf{Z}^d$, we let γ_x denote the translate of γ by x, as before, and define the Fourier transform of the two-loop function $G_\beta(\gamma, \gamma_x)$ as

$$\hat{G}_\beta(\gamma; p) = \sum_{x \in \mathbf{Z}^d} e^{-ip \cdot x} G_\beta(\gamma, \gamma_x). \tag{3.148}$$

For $\beta > \beta_0$ the sum is absolutely convergent for any $p \in \mathbb{R}^d$ since the mass is positive. In fact, the Fourier transform extends to an analytic function on a small neighbourhood of the real subspace of $\mathbb{C}^d$ for any $\beta > \beta_0$.

Let γ be a loop of length two with endpoints x and $x + e_j$ where $x \in \mathbb{Z}^d$ and e_j, $j \in \{1, \ldots, d\}$, is a vector in an orthonormal basis for $\mathbb{Z}^d$. Note that there is a one to one correspondence between loops of length two and pairs (x, j) as above. We denote this correspondence by $\gamma \leftrightarrow (x, j)$. If γ and γ' are loops of length two we can express the corresponding two-loop function as

$$G_\beta(\gamma, \gamma') = G_\beta(x - x'; j, j') \qquad (3.149)$$

where γ and γ' correspond to (x, j) and (x', j'), respectively, and we have indicated in our notation that $G_\beta(\gamma, \gamma')$ only depends on the difference $x - x'$. In this way one can view $G_\beta(\gamma, \gamma')$ as a $d \times d$-matrix valued function which we denote by $\mathbb{G}(x - x')$. Its Fourier transform,

$$\widehat{\mathbb{G}}_\beta(p) = \sum_{x \in \mathbb{Z}^d} \mathbb{G}(x) e^{-ip \cdot x},$$

is then again a matrix valued function with matrix elements

$$\widehat{G}_\beta(p; i, j) = \sum_{x \in \mathbb{Z}^d} G_\beta(x; i, j) e^{-ip \cdot x}. \qquad (3.150)$$

Let γ_1 and γ_2 be loops of length two. Consider a surface $S \in \mathscr{S}(\gamma_1, \gamma_2)$. Either we can decompose S into two surfaces $S_1 \in \mathscr{S}(\gamma_1, \gamma)$ and $S_2 \in \mathscr{S}(\gamma, \gamma_2)$ by cutting along a bottleneck in S that sits at some loop γ of length two or this is not possible for any such γ. In the first case we see that any path that lies in the surface S and connects the two boundary components crosses γ. Bottlenecks that separate S into two cylindrical surfaces in this way will be called *intermediate bottlenecks*. Let $\mathscr{S}'(\gamma_1, \gamma_2) \subset \mathscr{S}(\gamma_1, \gamma_2)$ be the collection of all surfaces with no intermediate bottlenecks and no pockets at γ_2, where pockets are defined as above. We can decompose any surface with boundary $\gamma_1 \cup \gamma_2$ into a surface in $\mathscr{S}'(\gamma_1, \gamma)$ and a surface in $\mathscr{S}(\gamma, \gamma_2)$ by cutting along the first intermediate bottleneck which one encounters in travelling from γ_1 to γ_2 along the surface. The second surface is empty in case there is no intermediate bottleneck. It follows, by an argument similar to the one that lead to Eq. (3.142), that we have the decomposition

$$G_\beta(\gamma_1, \gamma_2) = G_{\bar{\beta}}^{(2)}(\gamma_1, \gamma_2)(1 + G(\beta))^2 + \sum_\gamma G_{\bar{\beta}}^{(2)}(\gamma_1, \gamma) G_\beta(\gamma, \gamma_2), \qquad (3.151)$$

where the sum runs over all loops of length two and the factor $(1 + G(\beta))^2$

on the right hand side comes from the sum over all pockets at γ_2. Hence,

$$G'_\beta(\gamma_1, \gamma_2) = G^{(2)}_{\bar{\beta}}(\gamma_1, \gamma_2) + \sum_\gamma G^{(2)}_{\bar{\beta}}(\gamma_1, \gamma) G'_\beta(\gamma, \gamma_2), \qquad (3.152)$$

where

$$G'_\beta(\gamma_1, \gamma_2) = \frac{G_\beta(\gamma_1, \gamma_2)}{(1 + G(\beta))^2}$$

is a two-loop function, defined by restricting the sum over surfaces to those that have no pockets at γ_2.

Assuming that $\gamma_1 \leftrightarrow (0, i)$ and $\gamma_2 \leftrightarrow (x, j)$ and Fourier transforming both sides of Eq. (3.152) in the variable x we obtain

$$\widehat{\mathbb{G}}'_\beta(p) = \widehat{\mathbb{G}}^{(2)}_{\bar{\beta}}(p) + \widehat{\mathbb{G}}^{(2)}_{\bar{\beta}}(p)\widehat{\mathbb{G}}'_\beta(p), \qquad (3.153)$$

where $\widehat{\mathbb{G}}^{(2)}_{\bar{\beta}}$ and $\widehat{\mathbb{G}}'_\beta$ are the Fourier transforms of the matrices corresponding to the two-loop functions $G^{(2)}_{\bar{\beta}}$ and G'_β defined analogously to $\mathbb{G}_\beta$. Eq. (3.153) can be rewritten

$$\widehat{\mathbb{G}}'_\beta(p) = \frac{\widehat{\mathbb{G}}^{(2)}_{\bar{\beta}}(p)}{1 - \widehat{\mathbb{G}}^{(2)}_{\bar{\beta}}(p)} \qquad (3.154)$$

where the 1 in the denominator denotes the unit $d \times d$-matrix.

Note that all the matrix elements of $\widehat{\mathbb{G}}'_\beta(0)$ diverge like $\chi(\beta)$ as $\beta \to \beta_0$. On the other hand, it follows from Eq. (3.147) and the positivity of the mass in the PLRS$^{(2)}$-model that all the matrix elements of $\widehat{\mathbb{G}}^{(2)}_{\bar{\beta}}(p)$ are real analytic in the components of p in a neighbourhood of $p = 0$ for $\bar{\beta} \geq \bar{\beta}(\beta_0)$. For simplicity let us take $p = (p_1, 0, \dots, 0)$. Then $\widehat{\mathbb{G}}'_\beta(p)$ is analytic in p_1 in a disc of radius $m(\beta)$ around 0 in the complex p_1-plane and the divergence of $\widehat{\mathbb{G}}'_\beta(0)$ at $\beta = \beta_0$ is due to the non-invertibility of the matrix $1 - \widehat{\mathbb{G}}^{(2)}_{\bar{\beta}(\beta_0)}(0)$. We now analyse this non-invertibility more closely.

We expand the matrix valued function $\widehat{\mathbb{G}}^{(2)}_{\bar{\beta}}(p)$ in Taylor series around $p = 0$. Since $G^{(2)}_{\bar{\beta}}(x; i, j)$ is even in each component of x we obtain

$$\widehat{\mathbb{G}}^{(2)}_{\bar{\beta}}(p) = \widehat{\mathbb{G}}^{(2)}_{\bar{\beta}}(0) + p_1^2 \, \mathbb{K} + O(p_1^4), \qquad (3.155)$$

where, due to the symmetry of the hypercubic lattice, the matrices $\widehat{\mathbb{G}}^{(2)}_{\bar{\beta}}(0)$ and $\mathbb{K}$ have the structure

$$\widehat{G}^{(2)}_{\bar{\beta}}(0; i, j) = \begin{cases} g_1 & \text{if } i = j \\ g_2 & \text{if } i \neq j \end{cases} \qquad (3.156)$$

and

$$\mathbb{K} = \begin{pmatrix} k_1 & k_2 & k_2 & \cdots & \cdots & k_2 \\ k_2 & k_3 & k_4 & k_4 & \cdots & k_4 \\ k_2 & k_4 & k_3 & k_4 & \cdots & k_4 \\ \vdots & \vdots & & \ddots & & \vdots \\ \vdots & \vdots & & & \ddots & k_4 \\ k_2 & k_4 & \cdots & \cdots & k_4 & k_3 \end{pmatrix}. \tag{3.157}$$

The matrix elements are real analytic in $\bar\beta$ in a neighbourhood of $\bar\beta(\beta_0)$. Let us now sum over the second index of the matrices on both sides of Eq. (3.154) and put $p = 0$. Then we obtain

$$\psi_\beta = \frac{\psi_{\bar\beta}^{(2)}}{1 - \psi_{\bar\beta}^{(2)}} \tag{3.158}$$

where

$$\psi_\beta \equiv \sum_{j=1}^d \widehat{G}'_\beta(0; i, j) \tag{3.159}$$

is independent of i. Since ψ_β diverges as $\beta \to \beta_0$ we conclude by Eq. (3.156) that

$$g_1 + (d - 1)g_2 = 1 \tag{3.160}$$

at $\bar\beta = \bar\beta(\beta_0)$. It follows that the vector, all of whose entries are equal, is annihilated by $1 - \widehat{\mathbb{G}}_{\bar\beta(\beta_0)}^{(2)}(0)$. It is easy to check that all the other eigenvalues are non-zero since $g_1, g_2 > 0$. For $\beta > \beta_0$ the matrix $1 - \widehat{\mathbb{G}}_{\bar\beta(\beta)}^{(2)}(0)$ is symmetric and invertible, so we conclude that its inverse can be expressed as

$$\left(1 - \widehat{\mathbb{G}}_{\bar\beta(\beta)}^{(2)}(0)\right)^{-1} = \frac{c}{\sqrt{\beta_0 - \beta}}\,\mathbb{I} + \mathbb{A}, \tag{3.161}$$

since $\gamma = \frac{1}{2}$. Here $\mathbb{I}$ is the matrix with all entries equal to 1, c is a positive constant and $\mathbb{A}$ is a matrix depending on β which has a finite well-defined limit as $\beta \to \beta_0$. Using now the expansion (3.155) in Eq. (3.154) we obtain

$$\widehat{\mathbb{G}}'_\beta(p) = \widehat{\mathbb{G}}_{\bar\beta}^{(2)}(p)\left(1 - \widehat{\mathbb{G}}_{\bar\beta}^{(2)}(0)\right)^{-1}\left(1 + p_1^2\,\mathbb{K}\left(1 - \widehat{\mathbb{G}}_{\bar\beta}^{(2)}(0)\right)^{-1} + O(p_1^4)\right)^{-1}. \tag{3.162}$$

Furthermore,

$$\mathbb{K}\left(1 - \widehat{\mathbb{G}}_{\bar\beta}^{(2)}(0)\right)^{-1} = \frac{1}{\sqrt{\beta_0 - \beta}}\,\mathbb{M} + \mathbb{B}, \tag{3.163}$$

where $\mathbb{M}$ is a constant matrix of the form

$$\mathbb{M} = \begin{pmatrix} a & a & \cdots & a & a \\ b & b & \cdots & b & b \\ \vdots & \vdots & & \vdots & \vdots \\ b & b & \cdots & b & b \end{pmatrix} \qquad (3.164)$$

and $\mathbb{B}$ is a matrix which depends on β but has a finite limit as $\beta \to \beta_0$. It follows that $\widehat{\mathbb{G}}'_\beta(p)$ is analytic in p_1 in a disc of radius $r \sim (\beta_0 - \beta)^{\frac{1}{4}}$ around 0. This means that

$$m(\beta) \sim (\beta_0 - \beta)^{\frac{1}{4}} \qquad (3.165)$$

and we have established that $\nu = \frac{1}{4}$.

We finally turn to the construction of the scaling limit and proceed in a similar way as before. Let a be a lattice cutoff and choose the coupling $\beta(a)$ to be a decreasing function of a such that

$$\lim_{a \to 0} a^{-1} m(\beta(a)) = m > 0. \qquad (3.166)$$

This is of course always possible if the mass vanishes at the critical point. To be more precise, we choose a sequence $\{a_n\}$ of lattice cutoffs tending to zero through rational numbers as $n \to \infty$ and let x_n be a sequence of lattice points tending to infinity in such a way that

$$\lim_{n \to \infty} a_n x_n = X \in \mathbb{R}^d.$$

The scaling limit $G(X)$ of the two-loop function is then given by

$$\lim_{n \to \infty} a_n^{-(d-2+\eta)} G_{\beta_n}(\gamma, \gamma'_{x_n}) = G(X), \qquad (3.167)$$

where $\beta_n = \beta(a_n)$, provided the limit exists for some η and is nonzero. Here the anomalous scaling dimension η is defined as in Eq. (3.115) with $G_\mu(0, x)$ replaced by $G_\beta(\gamma, \gamma'_x)$ and the mass of the DTRS-model $m(\mu)$ replaced by the mass of the lattice theory $m(\beta)$. If the limit exists for some loops γ and γ' then we expect it to exist for any loop and have the same value. Note that the only adjustable parameter in the limit is the continuum mass m and an overall constant factor.

In order to construct the scaling limit it is convenient to work with the Fourier transform of the two-point function. Note that

$$\lim_{n \to \infty} a_n^{2-\eta} \widehat{G}_{\beta_n}(P a_n; i, j) = \widehat{G}(P), \qquad (3.168)$$

where $\widehat{G}(P)$ is the Fourier transform of $G(X)$, is equivalent to Eq. (3.167) for $\gamma \leftrightarrow (0, i)$, $\gamma' \leftrightarrow (0, j)$. With $a_n = (\beta_0 - \beta_n)^{\frac{1}{4}}$, as $\beta_n \to \beta_0$, we can take the scaling limit of Eq. (3.162). With $p_1 = P_1 a_n$ sufficiently small the

Neumann series for

$$\left(1 + p_1^2\, \mathbb{K}\left(1 - \widehat{\mathbb{G}}_{\hat{\beta}}^{(2)}(0)\right)^{-1} + O(p_1^4)\right)^{-1}$$

converges. Using Eq. (3.163) and noting that

$$\mathbb{M}^{l+1} = (a + (d-1)b)^l\, \mathbb{M} \tag{3.169}$$

for any $l \geq 0$, we find that as $n \to \infty$

$$\left(1 + p_1^2\, \mathbb{K}\left(1 - \widehat{\mathbb{G}}_{\hat{\beta}}^{(2)}(0)\right)^{-1} + O(p_1^4)\right)^{-1} \to \sum_{l=0}^{\infty} (-P_1^2)^l\, \mathbb{M}^l$$

$$= 1 - \frac{P_1^2\, \mathbb{M}}{1 + P_1^2(a + (d-1)b)}.$$

Similarly, by Eq. (3.161),

$$a_n^2 \left(1 - \widehat{\mathbb{G}}_{\hat{\beta}(\beta_n)}^{(2)}(0)\right)^{-1} \to c\mathbb{I} \tag{3.170}$$

as $n \to \infty$. It follows that $\eta = 0$ and the scaling limit of the matrix valued two-point function is, up to an overall constant factor, equal to

$$\mathbb{I}\left(1 - \frac{P_1^2}{1 + P_1^2(a + (d-1)b)}\mathbb{M}\right) = \frac{m^2}{P_1^2 + m^2}\mathbb{I} \tag{3.171}$$

where the continuum mass m is given by

$$m = \frac{1}{\sqrt{a + (d-1)b}}. \tag{3.172}$$

A slight generalization of the above calculations shows that we obtain exactly the same scaling limit even though p is not assumed to have the simple form $p = (p_1, 0 \ldots, 0)$. We see therefore that the scaling limit of the two-loop function has full Euclidean invariance and is independent of the relative orientation of the two boundary loops, as expected. We have thus arrived at the conclusion that the continuum limit of the two-point function is the propagator of a free scalar particle and this completes the proof of the theorem.

3.6 Rigid surfaces

3.6.1 Motivation

In this section we study random surface models with an action functional which depends on the extrinsic curvature of the surfaces. Recall that the extrinsic curvature for a piecewise linear random walk was expressed as a function of the angle between successive steps; see Eq. (2.103). For piecewise linear surfaces one should think of the extrinsic curvature in terms of the angle between neighbouring triangles or plaquettes.

Complete definitions will be given in the following subsections. The results in the two preceding sections show that the string tension does not scale to zero at the critical coupling, neither in the dynamically triangulated case nor in the lattice model, presumably due to the dominance of branched polymer-like configurations. One way to suppress spiky outgrowths of surfaces is to add an extrinsic curvature term to the action. Adding such a term has the effect of making the surfaces smoother at short distances, but it is a non-trivial question whether this effect persists at longer distances and in a scaling limit where short distance details should be unimportant. In this section we summarize what is presently known about models of random surfaces with an extrinsic curvature term in the action.

Let us first provide some further motivation for studying such models. In Nature we find a number of physical systems which may approximately be described as fluctuating membranes with both surface tension and bending rigidity. In some of these models the surface tension can be quite small and, correspondingly, the fluctuations in area large. The random surface models we will consider could possibly be regarded as toy models of such membranes as far as their universal properties are concerned. This requires the existence of critical couplings where the membrane systems undergo phase transitions and have critical properties described by those models. However, physical membranes are of course self-avoiding, and it is known from the study of random walks with self-avoiding constraints (which are non-local in nature) that the scaling properties change drastically when the dimension d of target space $\mathbb{R}^d$ is less than four. In the case of surfaces one would expect this effect to be even more pronounced. We shall not discuss self-avoiding surfaces any further, since they cannot be associated with a local field theory.

In Chapter 2 we saw that the rigid random walk and the fermionic random walk are in some respects quite similar, and have the same critical exponents. Fermionic strings provide us with a strong field theoretical motivation to investigate random surface theories with extrinsic curvature terms in the action. They are defined in terms of a set of fermionic fields $\psi^\alpha(\xi)$ in addition to the d bosonic fields $X^\mu(\xi) \in \mathbb{R}^d$. Integrating out the fermionic variables in the path integral representation of the partition function, one is left with an effective bosonic string theory whose action in general contains non-local terms, as well as terms which are not real-valued. This action yields a weight factor for surfaces analogous to the one we used for the fermionic random walk in Chapter 2. Among the terms generated one finds an extrinsic curvature term and while we presently have no tools which allow us to deal with the non-real and the non-local terms in the effective action, the extrinsic curvature term fits nicely into the statistical mechanics approach to the study of random surfaces. The hope is that the critical behaviour of random surfaces with extrinsic curvature reflects important aspects of the critical behaviour of the full fermionic string theory in the same way as for random walks. One might be even more optimistic: the pathology of the random surfaces theories, i.e. the non-scaling of the string tension in dimensions $d \geq 2$, most likely reflects the tachyonic nature of the bosonic string. A generic fermionic string theory is also tachyonic and only specially designed fermionic string theories, the famous superstring theories, avoid the problem. If extrinsic curvature is able to cure the scaling problem of the string tension it seems to provide us with an example of a bosonic string theory without a tachyon problem, which is what we expect from the effective bosonic string theory originating from a superstring theory. While this is at present pure speculation, it is nevertheless a strong motivation for the study of random surface theories with an action depending on extrinsic curvature.

3.6.2 *Curvature-dependent action*

In order to choose an appropriate discretization of the extrinsic curvature we recall some facts from the classical theory of surfaces embedded in $\mathbb{R}^d$. Let $X(\xi_1, \xi_2)$ be a parametrization of a surface and let h_{ab} denote the induced metric (the first fundamental form of the surface)

$$h_{ab} = \partial_a X^\mu \partial_b X^\mu. \tag{3.173}$$

We let $K_{i;ab}$ denote the second fundamental form defined by

$$K_{i;ab} = -\partial_a n_i^\mu \partial_b X^\mu = n_i^\mu \partial_a \partial_b X^\mu, \tag{3.174}$$

where n_i^μ denotes $d - 2$ orthonormal vectors, normal to the surface, and $\partial_a \equiv \partial/\partial \xi^a$. The *mean curvature* (or *extrinsic curvature*) H and the scalar curvature R are given in terms of the trace and the determinant, respectively, of the second fundamental form by

$$(h^{ab} K_{i;ab})^2 = 4H^2, \qquad (h^{ab} K_{i;ab})^2 - h^{ad} h^{cb} K_{i;ab} K_{i;cd} = \frac{R}{2}. \tag{3.175}$$

Here h^{ab} denotes the matrix inverse to h_{ab}. The functions H and R can be expressed in terms of the principal curvatures κ_1 and κ_2 as

$$H = (\kappa_1 + \kappa_2)/2, \qquad R = 2\kappa_1\kappa_2.$$

While κ_1, κ_2 and H depend on the embedding X, it is well known that R depends only on the intrinsic geometry of the surface. Long thin tubes can have $R = 0$, while H is large for such objects. This suggests that extrinsic curvature terms such as H^2 might suppress branched polymer-like outgrowths more efficiently than R^2 terms.

The unit vectors $n_i^\mu(\xi)$ satisfy the equations

$$D_{a;ij}n_j^\mu = -K_{i;ab}h^{bc}\partial_c X^\mu, \qquad (3.176)$$

where

$$D_{a;ij} = \partial_a \delta_{ij} + n_i^\mu(\partial_a n_j^\mu) \qquad (3.177)$$

is the covariant derivative in the normal bundle of X. We can now express the square of the extrinsic curvature as

$$H^2 = \frac{1}{4}h^{ab}D_a n_i^\mu D_b n_i^\mu. \qquad (3.178)$$

In the previous sections we discussed two different (but presumably equivalent) ways of defining a path integral for bosonic strings. In one case the action is expressed exclusively in terms of the geometry of the surface X embedded in $\mathbb{R}^d$. A natural choice of action including extrinsic curvature is in this case given by

$$\tilde{S}_{NG}(X) = \int d^2\xi \sqrt{h}\,(\mu + \lambda H^2). \qquad (3.179)$$

This action refers only to the embedded surface X. In the second case an intrinsic metric g_{ab} on the underlying manifold enters as an independent dynamical variable. It is then natural to incorporate the extrinsic curvature term into the action by setting

$$\tilde{S}_P(X,g) = \int d^2\xi \sqrt{g}\, g^{ab}\left[\frac{1}{2}\partial_a X^\mu \partial_b X^\mu + \frac{1}{2}\mu g_{ab} + \lambda D_a n_i^\mu D_b n_i^\mu\right]. \qquad (3.180)$$

The actions given by Eqs. (3.179) and (3.180) are generalizations of the Nambu–Goto action and the Polyakov action, respectively.

The extrinsic curvature terms in Eqs. (3.179) and (3.180) are both reparametrization invariant and accord with the geometric point of view we have taken. In addition, they are both *scale invariant*, i.e. invariant under the scale transformation

$$X(\xi) \to \Omega X(\xi), \qquad (3.181)$$

where Ω is a constant.

The generalized Nambu–Goto action given by Eq. (3.179) can be discretized according to the same principles as we used for the pure area action. In particular, one can apply a hyper-cubic regularization. Here we concentrate on the action (3.180), and for reasons of simplicity we consider random surfaces in three dimensions only. Generalizations to higher dimensions are straightforward. For $d = 3$ there is only one normal vector n^μ and $D_a n$ reduces to $\partial_a n$. We discretize the space of intrinsic metrics by means of dynamical triangulations, as earlier. Let T be an abstract triangulation and X a piecewise linear surface in $\mathbb{R}^3$ based on T. A vertex i is mapped to $x_i \in \mathbb{R}^3$ by X and if (ijk) is a triangle in T then $(x_i x_j x_k)$ will define a triangle $\triangle$ in $\mathbb{R}^3$ with a unit normal $n_\triangle$ chosen in accordance with the orientation of X. For a given triangulation T, corresponding to an equivalence class of internal metrics, a discretized version (with lattice spacing $a = 1$) of the action (3.180) is given by

$$S_T(X) = \mu|T| + \frac{1}{2}\sum_{(ij)\in L(T)}(x_i - x_j)^2 + \lambda\sum_{(\triangle,\triangle')}(n_\triangle - n_{\triangle'})^2, \qquad (3.182)$$

where the last sum is over pairs of neighbouring triangles in the triangulation, i.e. pairs of triangles sharing a link. We call λ *the extrinsic curvature coupling*. It is convenient to introduce a special notation for the last two terms in (3.182):

$$S_T^{(G)}(X) = \frac{1}{2}\sum_{(ij)\in L(T)}(x_i - x_j)^2, \qquad (3.183)$$

$$S_T^{(E)}(X) = \sum_{(\triangle,\triangle')}(n_\triangle - n_{\triangle'})^2, \qquad (3.184)$$

where the labels G and E refer to "Gaussian" and "extrinsic", respectively. Note that in the first four sections of this chapter we denoted the Gaussian action by S_T.

Let $\theta_{\triangle,\triangle'}$ be the angle between the normals to the neighbouring triangles $\triangle$ and $\triangle'$, i.e.

$$\frac{1}{2}(n_\triangle - n_{\triangle'})^2 = 1 - \cos(\theta_{\triangle,\triangle'}) \qquad (3.185)$$

Recall now that in the case of random walks we found that the scaling behaviour was independent of the precise form of the function $\psi(\theta)$ entering the discretized extrinsic curvature action (2.103), θ being the angle between neighbouring links in the random walk, as long as the absolute minimum was at $\theta = 0$. We expect a similar universality to hold here. Accordingly, we define a more general form of the discretized extrinsic curvature action by

$$S_T^{(E)}(X) = \sum_{\ell\in L(T)}\psi(\theta_\ell), \qquad (3.186)$$

where ψ is a non-negative function on $[0,\pi]$ whose unique minimum is at $\theta = 0$, and where θ_l denotes the angle between the two neighbouring triangles sharing the link l. With this more general definition, the discretized extrinsic curvature action is still scale invariant. The form of the action (3.185) is that of a ferromagnetic Heisenberg spin system on the two-dimensional lattice provided by the triangulation T, where the spins are the normal vectors $n_\triangle$. We see that the situation is similar to the one encountered for the rigid random walk, where the scaling limit of a one-dimensional spin chain actually appeared. It is tempting to appeal to the Mermin–Wagner theorem in order to exclude the possibility of a phase with long-range order between the normals for finite values of λ. However, the theorem is not applicable in the present context and we shall find evidence in favour of a transition to a phase with long range order between the normals. The reasons that the Mermin–Wagner theorem does not apply are as follows. The normals are not independent unit vectors since they are normals to a surface in $\mathbb{R}^3$, i.e. they satisfy non-local constraints, as is clear from Eq. (3.176) in the case of smooth surfaces. In addition, the weight associated with a given spin configuration is complicated by the fact that we have to average over all embeddings with a fixed configuration of normals in order to obtain the effective spin action.

We shall now confine ourselves to analysing the simplest regularized partition function for rigid random surfaces:

$$Z(\mu,\lambda) = \sum_{T \in \mathcal{T}} \frac{1}{C_T} \int \prod_{i \in T \setminus \{i_0\}} dx_i \, \exp\left(-\mu|T| - S_T^{(G)}(X) - \lambda S_T^{(E)}(X)\right),$$

(3.187)

where $d = 3$, $S_T^{(G)}$ is the Gaussian action (3.183) and $S^{(E)}(X)$ is given by Eq. (3.184) or occasionally by the more general action (3.186). It is a theory with two coupling constants, μ and λ, as for the random walk with extrinsic curvature. It will be shown below that, qualitatively, the phase diagram looks identical to the diagram for the random walk except that in the two-dimensional system *there might be a phase transition for a finite value of λ*, as illustrated in Fig. 3.12. A detailed discussion of this possibility is the subject of the next subsection.

The loop functions are defined in the same way as for the pure Gaussian action and we denote them by $G_{\mu,\lambda}(\gamma_1,\ldots,\gamma_b)$. Let us first note the following simple result.

Lemma 3.11 *There is an open convex set $\mathcal{B}$ in $\mathbb{R}^2$ such that the loop functions $G_{\mu,\lambda}(\gamma_1,\ldots,\gamma_b)$ are finite for $(\mu,\lambda) \in \mathcal{B}$ and infinite outside the closure of $\mathcal{B}$.*

Proof Let $\mathcal{B}_0$ be the set of points (μ,λ) such that $G_{\mu,\lambda}(\gamma_1,\ldots,\gamma_b) < \infty$ for some fixed loops $\gamma_1,\ldots,\gamma_b$. The convexity of $\mathcal{B}_0$ is a consequence of Hölder's inequality: suppose that $(\mu_i,\lambda_i) \in \mathcal{B}_0$, $i = 1,2$, and let

$$(\mu,\lambda) = \varepsilon(\mu_1,\lambda_1) + (1-\varepsilon)(\mu_2,\lambda_2),$$

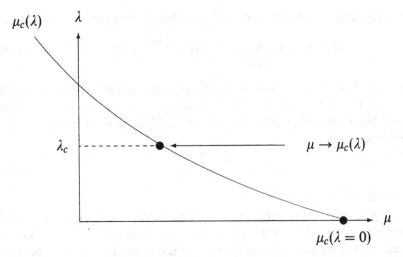

Fig. 3.12. The phase diagram for the model with partition function (3.187).

where $\varepsilon \in [0,1]$. Then

$$
\begin{aligned}
G_{\mu,\lambda}(\gamma_1,\ldots,\gamma_b) &= \sum_{T \in \mathscr{T}(n_1,\ldots,n_b)} \int \prod_{i \in T \setminus \partial T} dx_i \, e^{-S_T^{(G)} - \mu|T| - \lambda S_T^{(E)}} \\
&= \sum_{T \in \mathscr{T}(n_1,\ldots,n_b))} \int \prod_{i \in T \setminus \partial T} dx_i \\
&\quad \times \left(e^{-S_T^{(G)} - \mu_1|T| - \lambda_1 S_T^{(E)}} \right)^{\varepsilon} \left(e^{-S_T^{(G)} - \mu_2|T| - \lambda_2 S_T^{(E)}} \right)^{1-\varepsilon} \\
&\leq \left(G_{\mu_1,\lambda_1}(\gamma_1,\ldots,\gamma_b) \right)^{\varepsilon} \left(G_{\mu_2,\lambda_2}(\gamma_1,\ldots,\gamma_b) \right)^{1-\varepsilon},
\end{aligned}
$$

so that $(\mu,\lambda) \in \mathscr{B}_0$.

It is clear that the loop functions are monotonically decreasing in λ and that $(\mu,0) \in \mathscr{B}$ if $\mu > \mu_0$, where μ_0 it the critical coupling corresponding to the Gaussian model. It follows that $(\mu,\lambda) \in \mathscr{B}$ if $\mu > \mu_0$ and $\lambda \geq 0$. Thus, the interior of $\mathscr{B}_0$ is non-empty and we may define $\mathscr{B}$ as the interior of $\mathscr{B}_0$. Then $\mathscr{B}$ fulfils the claim of the theorem for fixed loops $\gamma_1,\ldots,\gamma_b$. The fact that $\mathscr{B}$ is independent of b and $\gamma_1,\ldots,\gamma_b$ follows by an argument analogous to the one given in the proof of Theorem 3.2.

By a slight extension of the proof of the bound (3.39) it follows that if we define

$$
N(\lambda) = \int_{S^{d-1}} d\Omega \, e^{-\lambda \psi(\theta)},
$$

where θ is the azimuthal angle on S^{d-1} and $d\Omega$ the uniform measure, then

$$G_{\mu,\lambda}(x) \leq C \sum_{T \in \mathcal{T}_1} \left(\pi^{d/2} N(\lambda) \right)^{N_v(T)} e^{-\mu|T|}. \tag{3.188}$$

It is clear that $N(\lambda) \to 0$ as $\lambda \to \infty$. If we denote the boundary of the set $\mathcal{B}$ by $\{(\mu_c(\lambda), \lambda) : \lambda \geq 0\}$ it follows that $\mu_c(\lambda) \to -\infty$ as $\lambda \to \infty$. This establishes the shape of $\mathcal{B}$ indicated in Fig. 3.12. If ψ is as in Eq. (3.185) then we have

$$\mu_c(\lambda) \sim \frac{d-2}{d} \log \frac{1}{\lambda}, \tag{3.189}$$

for λ large and $d > 2$ (see [25]).

As a consequence of Lemma 3.11 we can define the mass $m(\mu, \lambda)$ and the string tension $\tau(\mu, \lambda)$ in analogy with the Gaussian case. The existence will, however, not be discussed here. In [25] the following result is proved.

Theorem 3.12 *The mass and the string tension are positive in $\mathcal{B}$ and satisfy the upper bounds*

$$m(\mu, \lambda) \leq c_1 e^{2\mu/d}, \quad \tau(\mu, \lambda) \leq c_2 e^{2\mu/d}, \tag{3.190}$$

where c_1 and c_2 are constants.

We remark that the lower bound (3.99) implies that

$$\tau(\mu, \lambda) \geq 2e^{\frac{2}{d}\mu} \tag{3.191}$$

for all $\mu \geq \mu_c(0)$ and $\lambda \geq 0$.

From Eq. (3.189) and Eq. (3.190) it follows that there are trajectories (μ, λ) in $\mathcal{B}$ leading to $(-\infty, \infty)$ so that the mass and the string tension tend to zero as we move along this trajectory. It follows that $(-\infty, \infty)$ is a candidate point for the existence of a non-trivial scaling limit, but our bounds are not strong enough to allow us to conclude anything about the ratio

$$\frac{\tau(\mu, \lambda)}{m(\mu, \lambda)^2} \tag{3.192}$$

as $(\mu, \lambda) \to (-\infty, \infty)$. This ratio ought to be finite at a point where a non-trivial continuum limit exists.

3.6.3 *The crumpling transition*

It is consistent with our estimates that the mass and string tension scale to zero at a finite point $(\mu_0(\lambda_0), \lambda_0)$ on the boundary of $\mathcal{B}$. Unfortunately, there are no rigorous results pertaining to this hypothetical transition, but extensive computer simulations reveal that there exists a λ_0 where the geometry of the ensemble of surfaces undergoes a rapid change from rather "crumpled" surfaces for $\lambda < \lambda_0$ to much smoother surfaces for $\lambda > \lambda_0$. In addition the measurements of string tension and the mass gap in these numerical simulations are compatible with the existence of $v > 0$ such that

$$\tau(\lambda) \sim |\lambda - \lambda_0|^{2v}, \qquad m(\lambda) \sim |\lambda - \lambda_0|^v \qquad (3.193)$$

as $\lambda \uparrow \lambda_0$. In particular, this scaling behaviour is compatible with the existence of a continuum limit described by

$$\tau(\lambda) = \tau_{ph}\, a^2, \qquad m(\lambda) = m_{ph}\, a, \qquad a \sim |\lambda - \lambda_0|^v, \qquad (3.194)$$

where a is a cutoff, which is taken to zero in the scaling limit, and τ_{ph} and m_{ph} denote the corresponding physical string tension and mass.

In order to understand the numerical results and the reason why μ has disappeared from Eq. (3.193) and Eq. (3.194) it is useful to change the focus from the partition function (3.187), which we call the *grand canonical* partition function, to the *canonical* partition function $Z_N(\lambda)$ given by

$$Z_N(\lambda) = \sum_{T \in \mathcal{T}^{(N)}} \frac{1}{C_T} \int \prod_{i \in T \setminus \{i_0\}} dx_i\, e^{-S_T^{(G)}(X) - \lambda S_T^{(E)}(X)}, \qquad (3.195)$$

where $\mathcal{T}^{(N)}$ denotes the triangulations in $\mathcal{T}$ with N triangles. The functions $Z(\mu, \lambda)$ and $Z_N(\lambda)$ are related by

$$Z(\mu, \lambda) = \sum_{N=1}^{\infty} e^{-\mu N} Z_N(\lambda). \qquad (3.196)$$

Recall that we defined the string tension $\tau(\mu, \lambda)$ in the grand canonical ensemble, starting from the one-loop function $G_{\mu,\lambda}(\gamma(L, L; n))$, where $\gamma(L, L; n)$ is a square boundary loop in $\mathbb{R}^d$ (here $\mathbb{R}^3$) enclosing an area $A = L^2$ and with the number of boundary points $4n$ proportional to L. The function $G_{\mu,\lambda}(\gamma(L, L; n))$ can be regarded as the grand canonical partition function for random surfaces with the boundary fixed at the loop $\gamma(L, L, n)$. The *canonical* one-loop function $G_\lambda^N(\gamma)$ is defined in analogy with $Z_N(\lambda)$ by restricting the summation over surfaces to those that have a fixed number N of triangles. Associated with the partition functions $G_{\mu,\lambda}(\gamma(L, L; n))$ and $G_\lambda^N(\gamma(L, L; n))$ are the Gibbs free energy $\mathcal{G}(\mu, \lambda; A)$ and the Helmholtz free energy $\mathcal{F}(N, \lambda; A)$, defined by

$$G_{\mu,\lambda}(\gamma(L, L; n)) = e^{-\mathcal{G}(\mu, \lambda; A)}, \qquad G_\lambda^N(\gamma(L, L; n)) = e^{-\mathcal{F}(N, \lambda; A)}, \qquad (3.197)$$

where A is the area enclosed by $\gamma(L, L; n)$. The usual relation between $\mathcal{G}$ and $\mathcal{F}$ in the thermodynamic limit is

$$\mathcal{G}(\mu, \lambda; A) = \mu \bar{N} + \mathcal{F}(\bar{N}, \lambda; A) + o(\bar{N}), \qquad (3.198)$$

where

$$\bar{N} = \frac{\partial \mathcal{G}}{\partial \mu}. \tag{3.199}$$

Recall further that the string tension is defined as

$$\tau(\mu, \lambda) = \lim_{A \to \infty} \frac{\mathcal{G}(\mu, \lambda; A)}{A}. \tag{3.200}$$

We need to translate this relation to the canonical ensemble. It follows from Eqs. (3.198)–(3.200) that

$$A\tau(\mu, \lambda) = \mu\bar{N} + \mathcal{F}(\bar{N}, \lambda; A) + o(\bar{N}), \tag{3.201}$$

from which we conclude, using $\mu = -\partial \mathcal{F}/\partial N$, that

$$\tau(\mu, \lambda) = \frac{\partial \mathcal{F}(\bar{N}, \lambda; A)}{\partial A}, \qquad \bar{N} = A\frac{\partial \tau}{\partial \mu}, \tag{3.202}$$

up to terms associated with finite-size effects which are expected to vanish in the thermodynamic limit $N \to \infty$.

Let us now discuss the scaling of the string tension. For a given value of λ we have a critical value $\mu_0(\lambda)$ (see Fig. 3.12). Define $\Delta\mu = \mu - \mu_0(\lambda)$. As $\Delta\mu \to 0$ we expect, at least for small λ, that

$$\tau(\mu, \lambda) = \tau_0(\lambda) + d(\lambda)(\Delta\mu)^{2\nu(\lambda)} + o((\Delta\mu)^{2\nu(\lambda)}), \tag{3.203}$$

where $\nu(\lambda)$ is a critical exponent and $\tau_0(0) > 0$. From Eq. (3.202) we obtain

$$\bar{N} \sim \frac{(\Delta\mu)^{2\nu(\lambda)}A}{\Delta\mu}. \tag{3.204}$$

We define the relation between the *physical* area A_{ph} which is kept fixed as $\Delta\mu \to 0$ and the "bare area" A by

$$A_{ph} \sim \Delta\mu^{2\nu(\lambda)}A. \tag{3.205}$$

We then have that $\bar{N} \sim (\Delta\mu)^{-1}$ as $\Delta\mu \to 0$ and the relation between $\bar{N}$ and A is

$$\bar{N} \sim A^{1/2\nu(\lambda)} = A^{d_H/2}, \quad \text{i.e.} \quad d_H = \frac{1}{\nu(\lambda)}. \tag{3.206}$$

This discussion is parallel to our previous discussion of the Hausdorff dimension defined in terms of the two-point function and its relation to the critical exponent of the mass. If we assume that the two Hausdorff dimensions are identical, it follows that the exponent $\nu(\lambda)$ is to be identified with the critical exponent for the mass, i.e.

$$m(\mu, \lambda) \sim c(\lambda)(\Delta\mu)^{\nu(\lambda)}, \tag{3.207}$$

as one would indeed expect on dimensional grounds from Eq. (3.203).

If the bare mass scales to zero as in Eq. (3.207) we can define a continuum two-point function in terms of a *physical* length x_{ph} and a physical mass m_{ph} such that

$$m_{ph}(\lambda)x_{ph} = m(\mu, \lambda)x, \quad x = x_{ph}(\Delta\mu)^{-\nu(\lambda)}, \tag{3.208}$$

as discussed for the random walk in Chapter 2. The analogous proposal, i.e.

$$\tau_{ph}(\lambda)A_{ph} = \tau(\mu, \lambda)A, \tag{3.209}$$

for the one-loop function does not work as long as the string tension $\tau(\mu, \lambda)$ does not scale to zero for $\Delta\mu \to 0$, as discussed previously. In fact we obtain

$$\tau_{ph}(\lambda) \sim \frac{\tau_0(\lambda)}{\Delta\mu^{2\nu(\lambda)}} + d(\lambda), \tag{3.210}$$

which explicitly shows that $\tau_{ph}(\lambda)$ is infinite in the limit $\Delta\mu \to 0$ except perhaps at values of λ where $\tau_0(\lambda)$ vanishes.

Let us now *assume* that there exists a point λ_0 where $\tau_0(\lambda)$ vanishes such that

$$\tau_0(\lambda) \sim (\lambda_0 - \lambda)^\delta, \tag{3.211}$$

for some $\delta > 0$, as $\lambda \uparrow \lambda_0$. In this case it is natural to expect the two terms in Eq. (3.203) to be of the same order of magnitude, and since we assume that $d(\lambda_0) > 0$ the approach to the critical point $(\mu_0(\lambda_0), \lambda_0)$ along a curve of the form

$$\lambda_0 - \lambda = c(\Delta\mu)^\rho, \quad \text{where} \quad \delta\rho = 2\nu(\lambda_0), \tag{3.212}$$

will yield a *physical* string tension

$$\tau_{ph}(\lambda_0) = c + d(\lambda_0), \tag{3.213}$$

provided the curve defined by Eq. (3.212) lies inside $\mathscr{B}$.

If our assumptions are fulfilled then we can define a scaling limit where the ratio of the dimensionless string tension to the dimensionless mass[2] stays constant.

Let us now translate these considerations to the canonical ensemble. From Eq. (3.204) it follows that the quantity $r = A/\bar{N}$ can be expressed in terms $\Delta\mu$ and we can write

$$\tau(\mu, \lambda) = \tau_0(\lambda) + d(\lambda)r^{\frac{2\nu(\lambda)}{1-2\nu(\lambda)}} \equiv \tau_{can}(r, \lambda), \tag{3.214}$$

i.e. in the canonical ensemble r plays the role of $\Delta\mu$. If we can measure $\tau_{can}(r, \lambda)$ for various values of r we can, in principle, determine both $\tau_0(\lambda)$ and $\nu(\lambda)$. Rather surprisingly, it turns out to be easy to measure $\tau_{can}(r, \lambda)$ in a canonical framework by exploiting the scale invariance of the extrinsic curvature action functional $S_T^{(E)}$. Scaling the coordinates of the loop $\gamma(L, L; n)$ by β, taking the logarithm of $G_\lambda^N(\gamma(\beta L, \beta L; n))$, differentiating with respect to β and then putting $\beta = 1$ yields

$$\tau(\mu, \lambda) = \frac{\partial \mathscr{F}(\bar{N}, \lambda; A)}{\partial A} = \frac{\langle S_T^{(G)} \rangle_{\bar{N};A} - 3(\bar{N}/2 + 2n - 1)}{2A}, \qquad (3.215)$$

where $\langle \cdot \rangle_{N;A}$ is the expectation in the canonical ensemble used to define the string tension. Eq. (3.215) enables us to determine the string tension by computing the canonical expectation of the Gaussian action, i.e *it allows the determination of the quantity $\tau(\mu, \lambda)$ in terms of the expectation value of the Gaussian action.* Eq. (3.215) is very useful since it is much simpler to perform numerical simulations in the canonical ensemble than in the grand canonical ensemble, and since it is easy to measure the expectation value of the Gaussian action. In addition, one can use so-called finite-size scaling techniques in the canonical ensemble. Finite-size scaling is a powerful tool for determining the location of critical points and the values of critical exponents, see Chapter 5. Computer simulations indicate the existence of a critical λ_0 where τ_0 vanishes but this has not been established analytically.

The specific heat of the system with partition function $Z_N(\lambda)$ (see Eq. (3.195)) is defined as the second derivative of $\log Z_N(\lambda)$ with respect to the coupling constant multiplying the action, i.e.

$$C_N(\lambda) = \frac{3}{2} + \frac{d^2 \log Z_N(\lambda)}{d\lambda^2}, \qquad (3.216)$$

where the term 3/2 comes from the Gaussian part of the action. A peak in the specific heat is a good indicator for the presence of a phase transition point, provided its location converges to a fixed value as $N \to \infty$. In Fig. 3.13 we show the location $\lambda_0(N)$ of the peak of the specific heat as a function of $N^{-1/2}$. The position changes with N, but it seems to converge to a value $\lambda_0 = 1.50 \pm 0.03$. Associated with this peak, one observes the above-mentioned transition from surfaces which are on average rather crumpled to much more extended and smoother surfaces.

In Fig. 3.14 we have shown the measurements of $\tau_{can}(r, \lambda)$. It can be seen that the string tension has the qualitative features of Eq. (3.214) and the data are consistent with $\tau_0(\lambda) \to 0$ for $\lambda \to \lambda_0$. A fit to Eq. (3.214) gives $\nu(\lambda_0) = 0.4 \pm 0.02$. We can derive an expression analogous to Eq. (3.215) for the mass in terms of quantities measured in the appropriate canonical ensemble consisting of surfaces with two marked vertices separated by a distance $|x|$ in embedding space. Denoting the expectation in this ensemble

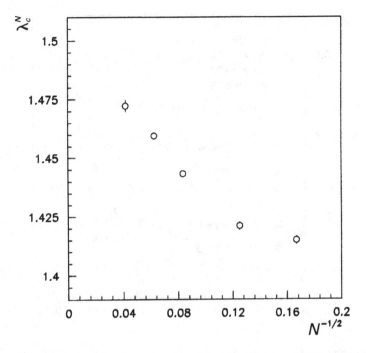

Fig. 3.13. The location of the maximum of the specific heat $\lambda_0(N)$ plotted against $N^{-1/2}$.

by $\langle \cdot \rangle_{N,|x|}$ we find by a scaling argument as given above that

$$m(\tilde{r}, \lambda) = \frac{\langle S_T^{(G)} \rangle_{\bar{N},|x|} - 3\bar{N}/2}{|x|}, \qquad (3.217)$$

where $\tilde{r} = |x|/\bar{N}$. In this way we obtain an independent measurement of the critical exponent $\nu(\lambda)$. The data yield $\nu(\lambda_0) = 0.42 \pm 0.02$.

Summing up, the numerical calculations suggest that there exists a critical point $\lambda_0 \approx 1.5$ where the bare string tension and the bare mass scale to zero. It is possible to fine-tune the approach to the critical point $(\mu_0(\lambda_0), \lambda_0)$, such that the ratio between the string tension $\tau(\mu, \lambda)$ and the mass squared $m(\mu, \lambda)^2$ stays constant. The associated phase transition will be referred to as the *crumpling transition*.

3.7 Crystalline surfaces

There is at present no proof of the existence of the crumpling transition described in the previous section. In fact, the sparse analytic insight we have into the dynamics of the system points in the opposite direction. Using the continuum action (3.179) one can study the expansion in $1/\lambda$

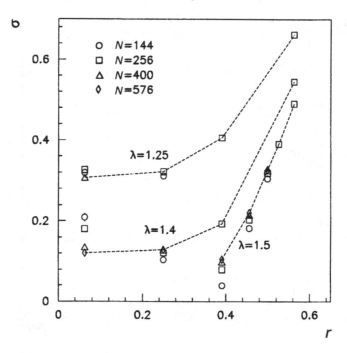

Fig. 3.14. The string tension measurement of $\tau_{can}(r, \lambda)$ plotted against r for various size systems.

around $1/\lambda = 0$. The β-function calculated to lowest order is negative, i.e. $1/\lambda = 0$ is an ultraviolet fixed point. If there is no additional fixed point for larger values of $1/\lambda$ we have an apparent contradiction with the computer simulations. If there is another fixed point it has to be an infrared fixed point. This disagrees with the computer experiments to be discussed below, according to which the phase transition point at $1/\lambda > 0$ is an ultraviolet fixed point. A way out of this dilemma is to realize that in the computer simulations one approaches the fixed point λ_0 from below, and it is very difficult numerically to move far into the region $\lambda > \lambda_0$ since there the normal–normal correlation is long ranged. This means that the discretized model for $\lambda < \lambda_0$ might have no relation to the formal continuum limit $1/\lambda \to 0$ of Eq. (3.179) investigated in the analytic one-loop calculations.

In order to understand how the above scenario may be realized, in the following we use the action (3.180), or the discretized version given by Eq. (3.182). In this model we have two dynamical fields g and X. Let us consider the partition function (3.187) for a fixed *intrinsic* volume $|T| = N$, i.e the canonical partition function (3.195). The metric degree of freedom can be eliminated by dropping the summation over the triangulations $\mathscr{T}^{(N)}$ and only retaining one approximately regular triangulation. If we consider

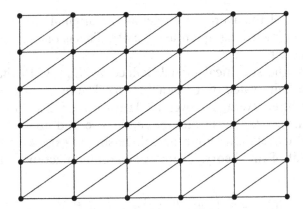

Fig. 3.15. A "regular" $N_1 \times N_2$ triangulation. The triangles are not chosen equilateral in order to emphasize the periodic nature of the triangulation.

triangulations of manifolds with the topology of a torus we can construct the $N \to \infty$ limit via a series of regular triangulations, as shown in Fig. 3.15, constructed from lattices Λ_L with periodic boundary conditions and with sides of lengths $N_1 = k_1 L$ and $N_2 = k_2 L$ and $N = k_1 k_2 L^2$ vertices, k_1 and k_2 being being fixed positive integers. The statistical system on Λ_L whose partition function is defined by

$$Z_L(\lambda) = \int \prod_{i \in \Lambda_L} dx_i \, \exp\left(-S_{\Lambda_L}^{(G)}(X) - \lambda S_{\Lambda_L}^{(E)}(X)\right) \delta\left(\sum_{i \in \Lambda_L} x_i\right) \qquad (3.218)$$

is called a *crystalline surface model* since it may serve as a model of membranes where the vertices of Λ_L are viewed as molecules which have fixed neighbours determined by the triangulation. On the other hand, the dynamically triangulated model may be viewed as a membrane model where the individual molecules do not have fixed neighbours, i.e. it may be viewed as a model of *fluid membranes*. Although interesting, we shall not pursue this point of view further here. Instead, we use the occasion to put forward an idea that will play an important role in the next chapter. It can be explained as follows. The partition function (3.218) represents a statistical system on a fixed lattice, whereas the partition function (3.195) represents the same model on a fluctuating lattice (with fixed topology). In the language of statistical mechanics the average over triangulations is *the annealed average** of the models defined on a single lattice. If the system defined on a single regular lattice has a critical point at which a continuum theory can be constructed and this critical point in some sense survives the averaging over triangulations, then we would like to interpret the resulting

---

* We shall discuss the concepts of annealed and quenched averages in Chapter 5.

continuum limit of the averaged model as the original continuum theory coupled to two-dimensional quantum gravity. This is one of the leitmotifs in the next chapter.

Assume that the statistical model defined by Eq. (3.218) has a phase transition at a finite value of λ when $L \to \infty$. The order of the transition might change after averaging over all lattices as in Eq. (3.195). If the transition in the crystalline surface ensemble is second order, there should be a divergent correlation length and there should exist a corresponding continuum limit of the theory. The naive continuum action of this system is

$$S(X) = \frac{1}{2} \int_M d^2\xi \, (\nabla X)^2 + \lambda \int_M d^2\xi \, (\nabla n)^2, \qquad (3.219)$$

where M is a torus if we use the same boundary conditions as in Eq. (3.218), and n is the unit normal to the surface X. We see that this is a massless Gaussian field theory perturbed by a non-polynomial interaction if we express the normals $n(\xi)$ in terms of the coordinates $X(\xi)$.

It is likely that taking the average over triangulations can only weaken the transition, so that a *necessary condition for the existence of a phase transition in the system defined by Eq. (3.180) is that the crystalline model as defined by Eq. (3.218) has a phase transition.* This motivates a closer study of the crystalline surface model.

3.7.1 *The kinematics of crumpling*

Let us begin by discussing the geometry of crystalline surfaces using the formal continuum action (3.219). We emphasize that the continuum formalism is only chosen for its calculational simplicity. The following discussion can also be carried out in the lattice framework. If $\lambda = 0$, then

$$X(\xi) = (x_1(\xi), x_2(\xi), x_3(\xi))$$

are three independent free massless fields. Even in this case the model has interesting features. Let us introduce an ultraviolet cutoff Λ and an infrared cutoff L^{-1} so the propagator of the massless Gaussian fields is given by

$$\langle x_\mu(\xi_1) x_\nu(\xi_2) \rangle = \delta_{\mu\nu} \int_{L^{-2}}^{\Lambda^2} \frac{du}{4\pi u} e^{-u(\xi_1 - \xi_2)^2},$$

where u is the proper time parameter. At each point on the surface there are two independent tangent vectors associated with the two-dimensional surface coordinate $\xi = (\xi^1, \xi^2)$:

$$t_\alpha = \frac{\partial X(\xi)}{\partial \xi^\alpha},$$

$\alpha = 1, 2$. A straightforward calculation yields, for $L \to \infty$, the tangent–tangent correlation function

$$\langle t_\alpha(\xi_1) \cdot t_\beta(\xi_2) \rangle = 3 v^\alpha v^\beta \delta_\Lambda(\xi_1 - \xi_2) \tag{3.220}$$
$$+ 3 \left(1 - e^{-\Lambda^2(\xi_1 - \xi_2)^2} \right) \frac{\delta_{\alpha\beta} - 2 v^\alpha v^\beta}{2\pi(\xi_1 - \xi_2)^2},$$

where v^α, $\alpha = 1, 2$, are the components of the two-dimensional unit vector

$$v = \frac{\xi_1 - \xi_2}{|\xi_1 - \xi_2|}$$

and

$$\delta_\Lambda(\eta) = \frac{\Lambda^2}{\pi} e^{-\eta^2 \Lambda^2}$$

denotes a regularized two-dimensional δ-function. We see that the tangent vectors can be *negatively correlated* along the surface, which is a characteristic for crumpled surfaces, as we discuss in more detail below. If $n(\xi)$ is the *unnormalized* vector field

$$n(\xi) = t_1(\xi) \times t_2(\xi).$$

orthogonal to the surface X we can use Eq. (3.220) to compute the correlation function

$$\langle n(\xi_1) \cdot n(\xi_2) \rangle = \frac{3\delta_\Lambda(\xi)}{\pi(\xi_1 - \xi_2)^2} - \frac{3 \left(1 - e^{-\Lambda^2(\xi_1 - \xi_2)^2} \right)^2}{2\pi^2(\xi_1 - \xi_2)^4}. \tag{3.221}$$

We note that the normal–normal correlation is *negative* at distances larger than $O(\Lambda^{-1})$. By continuity in λ (at least for finite L in Eq. (3.218)) there will also be some range $[0, \varepsilon]$ of λ where an analogous statement is true for the ensemble of crystalline surfaces defined by Eq. (3.218). This illustrates that although the partition function has some resemblance to the partition function of a ferromagnetic system, the normals playing the role of the spins, this analogy is misleading for small λ.

The random surface theory defined by Eq. (3.218), or in the continuum framework by the action (3.219), is an example of a *parametrized* random surface theory, contrary to the reparametrization invariant theories we have considered up to now. Previously, in this chapter, we worked with a Hausdorff dimension d_H defined by Eq. (3.109). It is convenient to give a slightly different definition here. The Hausdorff dimension relates the average extension of the surface in $\mathbb{R}^d$ (here $\mathbb{R}^3$) to the average area. Let us consider the part of the surface inside the disc $\{X(\xi) : |\xi| < r\}$. The average area is

$$\bar{A}(r) = \int_{|\xi| \le r} d^2\xi \, \langle A(\xi) \rangle, \tag{3.222}$$

where $A(\xi)$ denotes the area density:

$$A(\xi) = \sqrt{t_1(\xi)^2 t_2(\xi)^2 - (t_1(\xi) \cdot t_2(\xi))^2}. \qquad (3.223)$$

By rotational invariance $\langle (X(\xi) - X(0))^2 \rangle$ depends only on $|\xi|$, so we denote it by $\langle (X(r) - X(0))^2 \rangle$, $r = |\xi|$. This quantity is a measure of the mean-square size of the part of the surface corresponding to $|\xi| < r$. We define d_H by

$$\langle (X(r) - X(0))^2 \rangle \sim \bar{A}(r)^{2/d_H}. \qquad (3.224)$$

For any local operator $\mathcal{O}(\xi)$ we have

$$\left\langle \int_{|\xi| \le r} d^2\xi \, \mathcal{O}(\xi) \right\rangle \sim r^2, \qquad (3.225)$$

and hence,

$$\langle (X(r) - X(0))^2 \rangle \sim r^{4/d_H} \qquad (3.226)$$

as $r \to \infty$.

We can use the tangent–tangent correlation function to calculate the Hausdorff dimension d_H of a general ensemble of parametrized surfaces. We begin by noting that

$$X(\xi) - X(0) = \int_0^{|\xi|} ds \, t(s),$$

where

$$t(s) = \sum_{\alpha=1}^{2} t_\alpha(sv)v^\alpha \qquad (3.227)$$

and $v = \xi/|\xi|$. It follows that

$$\langle (X(r) - X(0))^2 \rangle = \int_0^r ds_1 \int_0^r ds_2 \, \langle t(s_1) \cdot t(s_2) \rangle. \qquad (3.228)$$

Using Eq. (3.220) to evaluate the integral in the Gaussian case $\lambda = 0$ we find

$$\langle (X(r) - X(0))^2 \rangle \sim \log(\Lambda r)$$

for large values of r. This well-known result could of course have been obtained more directly. In view of the application of Eq. (3.228) given below we find this derivation worth while.

Let us introduce the notation

$$g(s_1 - s_2) = \langle t(s_1) \cdot t(s_2) \rangle, \qquad (3.229)$$

where $t(s)$ is given by Eq. (3.227), and define

$$\Gamma(r) = \int_0^r ds \, g(s).$$

Note that

$$\frac{d}{dr}\langle(X(r) - X(0))^2\rangle = 2\Gamma(r).$$ (3.230)

First we consider the case where the function g is non-negative everywhere. If g is integrable, i.e. if $\lim_{r\to\infty}\Gamma(r) < \infty$, then it follows from Eq. (3.228) that the mean-square size of surfaces grows as r and, hence, $d_H = 4$. *In particular we see that in any theory of parametrized random surfaces with short-range correlations of tangents we have $d_H = 4$.* This result may be regarded as the simplest generalization of the universal result $d_H = 2$ for random paths whose tangents have short-range correlations as discussed in Chapter 2. We call the tangent–tangent correlations long-range if g is not integrable. In this case it is natural to expect

$$g(s) \sim s^{-\eta}$$

where $0 \le \eta < 1$. Then, by Eq. (3.228),

$$\langle(X(r) - X(0))^2\rangle \sim r^{2-\eta}$$

and

$$d_H = \frac{4}{2 - \eta}.$$

In particular, we see that $2 \le d_H \le 4$ if the correlation function g is positive. The Hausdorff dimension $d_H = 2$ can only occur if $\eta = 0$, i.e. if the tangents exhibit long-range order. This η should not be confused with the anomalous dimension of the two-point function introduced earlier.

Consider now the case where g is not positive everywhere. Since we assume, in accordance with (3.226), that $\langle(X(r) - X(0))^2\rangle$ is an increasing function of r, it follows from Eq. (3.230) that $\Gamma(s)$ is positive for all s and if $\Gamma(\infty) > 0$ then $d_H = 4$, as before. *We conclude that $\Gamma(\infty) = 0$ is a necessary condition for $d_H > 4$.* In the case of Gaussian surfaces considered in the previous subsection it can be checked by an explicit calculation that $\Gamma(\infty) = 0$, independent of the cutoff.

We now return to the system with action given by Eq. (3.219). For sufficiently small values of λ we expect it to have $d_H > 4$, as already mentioned. Clearly, g must be positive at sufficiently short distances and since it integrates to zero it must become negative at some value $s_0(\lambda)$. Beyond $s_0(\lambda)$ we expect g to decrease to a negative minimum and then increase monotonically to zero through negative values. The distance $s_0(\lambda)$ can be regarded as the correlation length or *persistence length* of the system. The Hausdorff dimension only depends on the asymptotic form of g so if

$$g(s) \sim -s^{-2+\varepsilon},$$ (3.231)

where $0 \leq \varepsilon < 1$ as $s \to \infty$, then

$$d_H = \frac{4}{\varepsilon},$$

as is easily inferred from Eq. (3.230). The value $\varepsilon = 0$ corresponds to the Gaussian surface and $\varepsilon \geq 1$ cannot occur in Eq. (3.231) since g must be integrable if it is negative at large distances.

We can now discuss how the theory can behave in the light of the above considerations. For sufficiently small λ we assume that the theory is governed by the Gaussian fixed point at $\lambda = 0$ and therefore the Hausdorff dimension is infinite and $\varepsilon = 0$. Let us indicate the dependence of the tangent–tangent correlation (3.229) on λ explicitly and write it as $g_\lambda(s)$. As λ increases from zero we expect the surface to flatten out locally and the persistence length $s_0(\lambda)$ to increase but for small λ the power in the asymptotic decay of g_λ should not change. There are now two possibilities:

(i) $s_0(\lambda) < \infty$ for all $\lambda < \infty$,

(ii) $s_0(\lambda) \uparrow \infty$ as $\lambda \uparrow \lambda_0$ for some finite λ_0.

In the first case there is no crumpling transition. In the second case $g_c = g_{\lambda_0} \geq 0$ and the Hausdorff dimension jumps to a value $d_H(\lambda_0) \in [2, 4]$ at the critical value of λ. In this case it is reasonable to assume that $g_c(s) \to 0$ as $s \to \infty$ but $g_\lambda(\infty) > 0$ for $\lambda > \lambda_0$. This means that there is a spontaneous global flattening of the surface at $\lambda > \lambda_0$ and its Hausdorff dimension is 2 for λ in this range. At the transition point the surface is in an intermediate state with a Hausdorff dimension between two and four.

Unfortunately, there are at present no analytic methods which allow us to decide whether case (i) or (ii) is realized for crystalline surfaces, but the system defined by Eq. (3.218) has been studied very carefully by computer simulations and they clearly favour case (ii). There seems to be a critical coupling value λ_0, such that $d_H = \infty$ for $\lambda < \lambda_0$ and $d_H = 2$ for $\lambda > \lambda_0$. The transition is most likely a second-order transition with the exponent of the specific heat being $\alpha \approx 0.5$. At λ_c the first computer simulations suggested $3.5 \leq d_H \leq 4$ but the latest simulations suggest $d_H \approx 2.7$. For $\lambda < \lambda_0$, i.e. in the *crumpled* phase, it is found numerically [220] that[*]

$$\langle x(\xi) \cdot x(0) \rangle \sim \int \frac{d^2 p}{(2\pi)^2} \frac{e^{i\xi \cdot p}}{p^2(p^2 + m(\lambda)^2)}, \tag{3.232}$$

---

[*] Strictly speaking, the comparison should be made with the discretized version of Eq. (3.232):

$$\langle x(\xi) \cdot x(0) \rangle \sim \sum_{p \in \tilde{\Lambda}_L \backslash \{0\}} \frac{e^{ip \cdot \xi}}{\Delta_T(p)(\Delta_T(p) + m(\lambda)^2)},$$

where $\Delta_T(p)$ is the lattice Laplacian in momentum space, $\tilde{\Lambda}_L$ is the first Brillouin zone in the lattice dual to Λ_L and ξ is a lattice vector.

where the effective mass $m(\lambda)$ scales to zero at λ_0 according to

$$m(\lambda) \sim (\lambda_0 - \lambda)^\nu,$$

with $\nu \approx 0.7$ [380]. Furthermore, the product $s_0(\lambda)m(\lambda)$ remains finite as $\lambda \uparrow \lambda_0$, confirming that there is only one length scale in the theory and this length scale can be identified with the persistence length for tangents (or normals).

Returning to the discussion given in the previous section we conclude that it is likely that the crumpling transition in the model defined by Eq. (3.195) represents the "annealed shadow" of the crumpling transition for crystalline surfaces.

3.7.2 *A lower bound on the size of crystalline surfaces*

The crumpling observed for the crystalline surfaces raises the question of how crumpled a statistical ensemble of parametrized surfaces can be. We close this section by establishing an analogue [246] of the Mermin–Wagner theorem [300] which shows that the mean-square extent of a parametrized random surface must grow at least logarithmically with the area of the surface for a large class of action functionals. In particular, this result excludes a hypothetical *supercrumpled* phase where the size of surfaces does not grow with L. One could imagine such a phase existing in a surface model with a sufficiently strong antiferromagnetic coupling between normal vectors.

We consider action functionals defined on a $L \times L$ regular hexagonal lattice Λ_L of the kind discussed earlier*. Let the action functional be of the form

$$S_L = \sum_{|\xi-\xi'|=1,\xi,\xi'\in\Lambda_L} U(x(\xi) - x(\xi')) \qquad (3.233)$$
$$+ \sum_{\xi\in\Lambda_L} W(x(\xi_1) - x(\xi),\ldots,x(\xi_6) - x(\xi)),$$

where ξ_i, $i = 1,\ldots,6$ are the nearest neighbours of ξ. In practice the potential U, which plays the role of the Gaussian, depends only on the norm of its argument, and W, which replaces the extrinsic curvature action, is invariant under cyclic permutations of its arguments. We need to assume that U and W are sufficiently smooth and the functions $|x(\xi)|^2$

---

* The detailed form of the regular lattice is of course of no importance. We have only chosen a specific one for reasons of notational simplicity.

and $|\nabla_{x(\xi)}\nabla_{x(\xi')}S_L|$ are integrable with respect to the measure

$$e^{-S_L}\delta(x_{cm})\prod_{\xi\in\Lambda_L} dx(\xi), \qquad x_{cm}\equiv\frac{1}{L}\sum_{\xi\in\Lambda_L} x(\xi).$$

We denote the corresponding partition function and expectation by Z_L and $\langle\cdot\rangle_L$, respectively. One can check that the quantity

$$R_L^2 = \left\langle L^{-2}\sum_{\xi\in\Lambda_L} x(\xi)^2 \right\rangle_L$$

has the same asymptotic behaviour as the mean-square extent discussed previously, see Eq. (3.226). With the above assumptions and definitions we have the following result.

Theorem 3.13 *There is a constant $C > 0$ such that*

$$R_L^2 \geq C\log L. \qquad (3.234)$$

Proof The inequality (3.234) follows from the Cauchy–Schwarz inequality

$$|\langle\psi\bar{\phi}\rangle_L|^2 \leq \langle|\psi|^2\rangle_L\langle|\phi|^2\rangle_L$$

with a judicious choice of functions ψ and ϕ. Define

$$\psi_j(p) = \sum_{\xi\in\Lambda_L} e^{-i\xi\cdot p}x_j(\xi)$$

and

$$\phi_j(p) = \sum_{\xi\in\Lambda_L} e^{-i\xi\cdot p}\frac{\partial S_L}{\partial x_j(\xi)},$$

where p is in the first Brillouin zone $\tilde{\Lambda}_L$. Then

$$\sum_{j=1}^{d}\sum_{p\in\tilde{\Lambda}_L}\langle|\psi_j(p)|^2\rangle_L = L^4 R_L^2. \qquad (3.235)$$

By partial integration we obtain

$$\langle\phi_j(p)\bar{\psi}_j(p)\rangle_L = Z_L^{-1}\sum_{\xi,\xi'\in\Lambda_L} e^{-ip\cdot(\xi-\xi')}\int\prod_{\zeta\in\Lambda_L} dx(\zeta)$$

$$\times\left(\delta(x_{cm})\delta_{\xi\xi'} + x_j(\xi)\frac{\partial}{\partial x_j(\xi')}\delta(x_{cm})\right)e^{-S_L}$$

$$= L^2 + Z_L^{-1}\sum_{\xi,\xi'\in\Lambda_L} e^{-ip\cdot(\xi-\xi')}\int\prod_{\zeta\in\Lambda_L} dx(\zeta)L^{-2}(x_{cm})_j\delta'(x_{cm})e^{-S_L}$$

$$= L^2(1-\delta_{p0})$$

by the translational invariance of S_L. Repeated indices j are not to be summed over in the above formula. By a similar argument we obtain

$$\left\langle |\phi_j(p)|^2 \right\rangle_L = \sum_{\xi,\xi' \in \Lambda_L} e^{ip \cdot (\xi - \xi')} \langle \frac{\partial^2 S_L}{\partial x_j(\xi) \partial x_j(\xi')} \rangle_L. \tag{3.236}$$

The expectation value given above is finite by our assumption about the integrability of the second derivative of the action. Since the action is of finite range, (3.236) is proportional to L, and it vanishes at $p = 0$ by translational invariance. Furthermore, the quantity given by Eq. (3.236) is an even bounded real analytic function of p so there is a constant $c > 0$ such that

$$\langle |\phi_j(p)|^2 \rangle_L \le cL^2 p^2. \tag{3.237}$$

It follows that

$$L^4(1 - \delta_{p0}) \le cL^2 p^2 \langle |\psi_j(p)|^2 \rangle_L. \tag{3.238}$$

Dividing by p^2 in (3.238), summing over p and j and using Eq. (3.235) yields

$$R_L^2 \ge \frac{d}{cL^2} \sum_{p \in \tilde{\Lambda}_L \setminus \{0\}} \frac{1}{p^2}. \tag{3.239}$$

The theorem now follows from the asymptotic behaviour of the right-hand side of Eq. (3.239) as $L \to \infty$.

3.8 Notes

Much of the content of Chapter 3 can be regarded as a discretization of bosonic string theory, even though the theory has applications to other subjects, as we describe below. The area action functional was introduced in [206, 308] and the old string theory was based on a canonical quantization of this action; see e.g. [344] for a comprehensive review. The Polyakov action was first discussed in [99, 100] and used for constructing a new type of string theory in [326]; see [15, 161, 162, 163] for a further discussion. For an account of string and superstring theory up to 1987, see [207]. The extrinsic curvature action for bosonic strings was first discussed by Polyakov in [327] and an analogous action was studied earlier in biophysics by Helfrich [227], as we describe below.

The systematic study of random surfaces in the spirit of this book started in the early 1980s. Much of the motivation came in the beginning from gauge theory, in particular lattice gauge theory, where the Wilson loop [388] expectation has a representation as a weighted sum over an ensemble of surfaces; see [68, 191, 107, 306, 179]. Related issues are considered in [11].

Early work in this field, in the context of lattice surfaces, was carried out by Weingarten, partly motivated by Euclidean quantum gravity, [374, 375], and continued in [165, 166, 266]. In this work it was realized that the sum over all lattice surfaces was ill-defined due to the superexponential growth of the number of surfaces with a given area. A rigorous approach to the subject was pursued in the series of papers [153, 152, 154, 155, 157].

This work culminated in the proof [154], given in Section 3.5, that branched polymers dominate the simplest models, see also [245] where the scaling limit was first constructed. For a discussion of the branched polymer picture in the context of continuum string theory see [115]. In [155] modifications of the models which might lead to a more sensible theory are suggested and reflection (Osterwalder–Schrader) positivity [317, 318] is established for a class of models. Improved entropy estimates for lattice surfaces are given in [365, 366]. Results about a related model of surfaces with fluctuating topology are obtained in [345]. A reformulation of the three-dimensional Ising model as a random surface theory is described in [238, 114]; see also [262, 316]. For reviews of some of this work see [187, 188, 149, 176].

Numerical simulations have always played an important guiding role in the study of random surfaces. See [352, 81] for some of the early work which concentrated on lattice surfaces. A more complete list of references to numerical results may be found in the notes to Chapter 5.

The study of dynamically triangulated surfaces began with three almost simultaneously presented papers in 1985 [23, 129, 271]. In [23] it was shown that the naive area action for dynamically triangulated surfaces leads in general to divergent partition functions due to the lack of short-distance cutoff in embedding space, and it was realized that the dynamically triangulated random surface model is the natural discretization of the Polyakov string theory. Further work on the phase diagram, $d \to \pm\infty$ limits and universality properties of the model is contained in [27, 24, 94, 93, 128], including the exact solution in zero dimensions which is based on the counting of ϕ^3-graphs in [96, 82]. In [130, 93] the exact solution of the model in -2 dimensions is given. The large d expansion is studied in [128]. The first numerical results on the DTRS-model were reported in [24, 94]. See [21] for large-scale simulations aimed at determining the susceptibility exponent γ. In [22] it was shown by the same method as we used in Section 3.4 that the string tension of the DTRS-model does not vanish at the critical point. For reviews of some of these results see [187, 149, 176].

The systematic study of discretized dynamically triangulated surface models with extrinsic curvature dependent action was initiated in [29] and continued in [25], where many of the results of Section 3.6 were first obtained. This work is to a large extent motivated by the fact that the string tension does not scale to zero at the critical point of the model with the extrinsic curvature action left out. Much of the information concerning this class of models has been obtained by numerical simulations, see [38, 19] and references therein. A recent review of this subject is given in [379]; for a further discussion, see below.

A *subdivision invariant* extrinsic curvature action for dynamically triangulated surfaces was suggested in [17]. In [158] it was shown that this action functional alone does not yield a convergent grand canonical partition function for any value of the coupling. Models with this action are studied numerically in [64, 63, 62]. Related work is found in [342].

Extrinsic curvature actions for random surfaces embedded in a hyper-cubic lattice are introduced in [157], where it is shown that the branched polymer picture may carry over to the model with extrinsic curvature action. A more thorough study of the model is contained in [26], where a real space renormalization group method is introduced for the study of lattice surfaces. The basic idea in this paper is to regard lattice surfaces as a tower of excitations living on an underlying flat surface. The renormalization group step then consists of shaving off one layer of excitations and renormalizing the couplings accordingly. It is shown that up to irrelevant terms the action is reproduced. In this paper the phase diagram of lattice surfaces with extrinsic curvature is mapped and candidates for critical points with scaling mass and string tension are identified. The scaling limit at these points has, however, not been controlled. A review of this work is given in [149].

As mentioned before, there is a large body of literature in membrane physics devoted to surfaces whose action (or energy) depends in an important fashion on the extrinsic curvature. Membranes (lipid bilayers, vesicles) whose energy is essentially area independent are a particularly interesting field of study and are believed to explain, at least partly, the shape of red blood cells [101, 139]. This work began with [227]; see [228, 229, 230] for more recent work in a similar vein. The papers [178, 284] and [346] study simple models for the shape of membranes. A review of membrane physics from the point of view of random surface theory may be found in the article by Leibler in [309].

The membranes that are of interest to biophysicists are usually what are called *fluid membranes*, meaning that the molecules that make up the membranes do not have a fixed relative position but can move freely inside the fluctuating surface. This corresponds to surfaces with dynamical triangulations. Membranes with fixed internal structure (corresponding to a fixed triangulation) are called crystalline membranes in Section 3.7 and those where the orientation of the molecules, but not necessarily their position, have a long-range order are called *hexatic membranes*. Crystalline and hexatic membranes have been studied intensively in recent years, some of this work being related to the theory of two-dimensional melting [252, 253, 310, 215]. For fluid membranes with hexatic order see [136]. For an overview of this subject, see [309] and the article by David in [212].

Random surfaces arise naturally in mathematical models of phase mixtures such as microemulsions [299, 357, 358, 138, 233, 381, 382]. For more information about this aspect of the theory of random surfaces, see [309, 339]. For an application of random surfaces with a Gaussian action to the study of macromolecules see [169].

We discussed the crumpling transition at length in Sections 3.6 and 3.7 in the case of fluid and crystalline surfaces, respectively, For crystalline surfaces the presence of the crumpling transition is well established in many slightly different models but we still have an incomplete analytical understanding of this transition. For early work on so called *tethered surfaces*, see [254, 255, 256]. Our discussion of the crystalline case follows [31]. Universality of the transition is discussed in [60]. For the most recent numerical results on the crystalline crumpling transition, see [380]. Models interpolating between crystalline and fluid surfaces are studied in [174, 175]. The crumpling transition for fluid surfaces is not well understood and perturbative calculations are at odds with the best numerical simulations. See [228, 180, 327, 276, 322, 321] for the perturbative calculations and [19] for the most recent numerical work.

The surfaces of interest to condensed matter physicists and biophysicists are usually self-avoiding. This is a constraint that is very difficult to deal with analytically; see the article by Duplantier in [309] for a review of the subject. For recent work on this problem which involves a novel renormalization group method see [135, 134].

An interesting model of self-avoiding lattice "tubes" has been studied in detail in [2, 1, 3]. In these papers techniques were developed which turned out to be useful for the mathematical analysis of the PLRS-model [245]. In [291] an approximate renormalization group method is applied to determine critical exponents for self-avoiding random surface models. In [292] the relation between self-avoiding random surface models and gauge theories are investigated; see also [290, 153]. Bounds on the number of random surfaces with a fixed area and genus are proved in [367]. Entropy bounds as well as bounds on the string tension of self-avoiding lattice surfaces are derived in [153]; see also [155], where it is argued that the upper critical dimension for self-avoiding surfaces is 8, meaning that self-avoidance is an irrelevant constraint for surfaces embedded in more than 8 dimensions. Monte Carlo simulations of lattice random surfaces with genus zero indicate that they are dominated by branched polymers in three dimensions and therefore also in higher dimensions [352, 202, 203].

It is natural to consider surfaces with fermionic degrees of freedom in the same way as we studied random walks with anticommuting degrees of freedom in the previous chapter. In addition there are at least two additional motivations from physics. First of all superstring theory includes explicitly fermionic degrees of freedom and if we wish to give a non-perturbative definition of such a theory using discretized random surfaces, it seems likely that one must include fermionic degrees of freedom explicitly. Secondly, as we have discussed in this chapter, it is necessary to include curvature-dependent terms in the action for random surfaces in order to obtain a well-behaved and non-trivial continuum limit. In the case of random paths we have seen that the influence of fermionic degrees of freedom is similar to the influence of extrinsic curvature terms; both serve to smoothen paths out and drive their Hausdorff dimension to 1. In the case of surfaces we know that narrow spikes must be suppressed and it is a natural working hypothesis that this can be done by including anticommuting degrees of freedom in the action. These two viewpoints are related by the fact that integrating out fermionic variables in superstring theory gives rise to an effective extrinsic curvature-dependent term in the action of the remaining bosonic string theory [383]. We finally mention that it has long been believed that the three-dimensional Ising model is related to a fermionic string theory in three dimensions but there is no convincing demonstration of such a relation, see however [238, 114, 262, 316].

There are, however, severe obstacles to carrying out the program of supersymmetrizing random surfaces. First of all it is well known that supersymmetric actions cannot be defined in a natural way on a discrete lattice [311] and, secondly, if we assign noncommuting weight factors to the triangles or links on a discrete surface, there is no unique way of ordering them as we had for the fermionic random walks where we could simply use path ordering. One suggestion for avoiding the first problem at least is to use the Green–Schwarz formulation of superstring theory where there is a space-time supersymmetry but no world-sheet supersymmetry. There is a proposal for the discretization of this theory [301] which is developed further in [53], where some numerical simulations are also undertaken. After integrating out the fermionic variables one obtains a complicated non-local interaction whose physical interpretation is not transparent.

4

Two-dimensional gravity

In the preceding chapters we focused on the properties of random walks and random surfaces in a Euclidean embedding space. We observed that embedded random walks and surfaces can be viewed as models of one- and two-dimensional quantum gravity coupled to scalar fields. The *pure quantum gravity* aspect of these theories is the main subject of this chapter. The regularization offered by dynamical triangulations works very well in this context and provides a soluble model of two-dimensional quantum gravity.

4.1 The continuum formalism

In this introductory section we discuss in a continuum framework the observables of two-dimensional quantum gravity using purely formal functional integrals. Most of the discussion applies to higher-dimensional quantum gravity as well.

Let M^h denote a closed, compact, connected and orientable surface of genus h and Euler characteristic $\chi(h) = 2 - 2h$. The partition function of two-dimensional Euclidean quantum gravity is formally given by

$$\mathscr{Z}(\Lambda, \kappa) = \sum_{h=0}^{\infty} \int \mathscr{D}[g] \, e^{-S(g;\Lambda,\kappa)}, \qquad (4.1)$$

where Λ denotes the cosmological constant, κ is the gravitational coupling constant and S is the continuum Einstein–Hilbert action defined by

$$S(g;\Lambda,\kappa) = \Lambda \int_{M^h} d^2\xi \sqrt{g} - \frac{1}{2\pi\kappa} \int_{M^h} d^2\xi \sqrt{g} \, R. \qquad (4.2)$$

In Eq. (4.1) the sum runs over all possible topologies of two-dimensional manifolds (i.e. over all genera h) and in Eq. (4.2) R denotes the scalar curvature of the metric g on the manifold M^h, and we have also taken the

149

liberty of denoting the determinant of the metric g by g. The functional integration is over all *diffeomorphism classes* $[g]$ of metrics on M^h; see Sections 2.2 and 3.1.

The curvature part of the Einstein–Hilbert action is a topological invariant according to the Gauss–Bonnet theorem and this allows us to write

$$\mathscr{Z}(\Lambda, \kappa) = \sum_{h=0}^{\infty} e^{\chi(h)/\kappa} \mathscr{Z}_h(\Lambda), \qquad (4.3)$$

where

$$\mathscr{Z}_h(\Lambda) = \int \mathscr{D}[g]\, e^{-S(g;\Lambda)}, \qquad (4.4)$$

and

$$S(g;\Lambda) = \Lambda V_g, \qquad (4.5)$$

where $V_g = \int d^2\xi \sqrt{g}$ is the volume of the universe for a given diffeomorphism class of metrics. In the remainder of this section we will, for simplicity, restrict our attention to manifolds homeomorphic to S^2 or S^2 with a number of holes unless explicitly stated otherwise. In this case we disregard the topological term in the action since it is a constant. However, the summation over topologies is a major issue in quantum gravity and will be discussed at length later in this chapter. The sphere S^2 with b boundary components will be denoted S_b^2 and we denote the partition function for the sphere, $\mathscr{Z}_0(\Lambda)$ in Eq. (4.4), by $\mathscr{Z}(\Lambda)$.

In the presence of a boundary it is natural to add to the action a boundary term

$$S(g;\Lambda, Z_1,...,Z_b) = \Lambda V_g + \sum_{i=1}^{b} Z_i L_{i,g}, \qquad (4.6)$$

where $L_{i,g}$ denotes the length of the ith boundary component with respect to the metric g. We refer to the Z_is as the cosmological constants of the boundary components. The partition function is in this case given by

$$W(\Lambda; Z_1,...,Z_b) = \int \mathscr{D}[g]\, e^{-S(g;\Lambda,Z_1,...,Z_b)}. \qquad (4.7)$$

Since the lengths of the boundary components are invariant under diffeomorphisms, it makes sense to fix them to values $L_1,...,L_b$ and define the *Hartle–Hawking wave functionals* [223] by

$$W(\Lambda; L_1,...,L_b) = \int \mathscr{D}[g]\, e^{-S(g;\Lambda)} \prod_{i=1}^{b} \delta(L_i - L_{i,g}), \qquad (4.8)$$

where $S(g; \Lambda)$ is given by Eq. (4.5). Since Eq. (4.7) is the Laplace transform of Eq. (4.8), i.e.

$$W(\Lambda; Z_1, ..., Z_b) = \int_0^\infty \prod_{i=1}^b dL_i e^{-Z_i L_i} W(\Lambda; L_1, ..., L_b), \qquad (4.9)$$

we denote them by the same symbol. We distinguish between the two by the names of the arguments.

It is sometimes convenient to consider wave functionals for fixed volume. They are defined by

$$W(V; L_1, ..., L_b) = \int \mathcal{D}[g] \, \delta(V - V_g) \prod_{i=1}^b \delta(L_i - L_{i,g}), \qquad (4.10)$$

so we can write

$$W(\Lambda; L_1, ..., L_b) = \int_0^\infty dV \, e^{-\Lambda V} W(V, L_1, ..., L_b). \qquad (4.11)$$

Here we have continued the same abuse of notation as introduced above and denote all the wave functionals by W. The special case $b = 0$ yields the partition functions

$$\mathcal{Z}(\Lambda) = \int_0^\infty dV \, e^{-\Lambda V} \mathcal{Z}(V), \qquad (4.12)$$

$$\mathcal{Z}(V) = \int \mathcal{D}[g] \, \delta(V - V_g). \qquad (4.13)$$

Next we define a reparametrization invariant two-point function by

$$G(R; \Lambda) = \int \mathcal{D}[g] \, e^{-\Lambda V_g} \int d^2\xi \, \sqrt{g} \int d^2\xi' \, \sqrt{g} \, \delta(d_g(\xi, \xi') - R). \qquad (4.14)$$

In this formula $d_g(\xi, \xi')$ denotes the *geodesic distance* between the points labelled by ξ and ξ' measured with respect to g. The two-point function can be viewed as the partition function for the ensemble of universes with two marked points separated by a fixed geodesic distance[*] R. Both the long-distance behaviour and the short-distance behaviour of the two-point function reflect the fractal structure of the surfaces that give the most important contribution to the loop functions of two-dimensional quantum gravity.

For the ensemble of manifolds with fixed volume V the two-point function is defined by

$$G(R; V) = \int \mathcal{D}[g] \, \delta(V_g - V) \int d^2\xi \, \sqrt{g} \int d^2\xi' \, \sqrt{g} \, \delta(d_g(\xi, \xi') - R) \qquad (4.15)$$

---

[*] We use the letter R to denote geodesic distance. The same letter is used to denote scalar curvature. No confusion should arise.

and the relation between $G(R;V)$ and $G(R;\Lambda)$ is again given by a Laplace transformation

$$G(R;\Lambda) = \int_0^\infty dV\, e^{-\Lambda V} G(R;V). \tag{4.16}$$

The function $G(R;V)$ has a simple geometrical interpretation for small R. For a given metric g the average

$$\langle n(R)\rangle_g = \frac{1}{V_g} \int d^2\xi \sqrt{g} \int d^2\xi' \sqrt{g}\, \delta(d_g(\xi,\xi')-R) \tag{4.17}$$

is reparametrization invariant and $\langle n(R)\rangle_g dR$ is the average volume of a spherical shell of thickness dR and radius R on the manifold with metric g and volume V_g. Taking the ensemble average of $\langle n(R)\rangle_g$ with fixed volume V we obtain

$$\langle n(R)\rangle = \frac{1}{\mathscr{Z}(V)} \int \mathscr{D}[g]\, \delta(V_g - V)\, \langle n(R)\rangle_g, \tag{4.18}$$

and the relation

$$G(R;V) = V\mathscr{Z}(V)\, \langle n(R)\rangle. \tag{4.19}$$

The *Hausdorff dimension*[*] d_h of the ensemble of manifolds with fixed volume is now defined by the asymptotic relation

$$\langle n(R)\rangle \sim R^{d_h-1}, \qquad R \ll V^{1/d_h}, \tag{4.20}$$

i.e. the small R dependence of $G(R;V)$ determines the Hausdorff dimension of this ensemble of manifolds and d_h is assumed not to depend on V.

Let us now study the relation between the *long-distance* behaviour of $G(R;\Lambda)$ and the fractal properties of surfaces. One can show by general considerations that $G(R;\Lambda)$ falls off exponentially for large distances (this will become clear when we have introduced a cutoff). Furthermore, we have the inequality

$$G(R_1 + R_2;\Lambda) \geq \text{const. } G(R_1;\Lambda)\, G(R_2;\Lambda), \tag{4.21}$$

which is established by methods analogous to the ones used in Section 3.4.4. The inequality (4.21) is basically an expression of the fact that by gluing together in a suitable fashion two manifolds of volumes V_1 and V_2, each of which has two marked points separated by geodesic distance R_1 and R_2, respectively, one obtains a manifold with volume $V_1 + V_2$ and distance $R_1 + R_2$ between two marked points.

---

[*] This notion of a Hausdorff dimension is an intrinsic Hausdorff dimension in the language of Section 3.4.5 since the manifolds are not embedded.

Since Eq. (4.21) implies that $-\log G(R; \Lambda)$ is a subadditive function up to a constant, i.e.

$$-\log G(R_1 + R_2; \Lambda) \le -\log G(R_1; \Lambda) - \log G(R_2; \Lambda) + \text{const.}, \quad (4.22)$$

it follows that

$$\lim_{R \to \infty} \frac{-\log G(R; \Lambda)}{R} = M(\Lambda) \quad (4.23)$$

exists. Moreover, as we shall see, the integrated two-point function is finite if $\Lambda > 0$ and the mass $M(\Lambda)$ is an increasing function of Λ. It follows that

$$M(\Lambda) > 0 \quad \text{if} \quad \Lambda > 0. \quad (4.24)$$

Let us *assume* that

$$M(\Lambda) = c\Lambda^\nu, \quad (4.25)$$

i.e. for $R \gg 1/M(\Lambda)$ we assume that

$$G(R; \Lambda) \sim R^\alpha \, e^{-c\Lambda^\nu R}, \quad (4.26)$$

where R^α is some unknown subleading correction to the exponential decay. Since $G(R; \Lambda)$ is the partition function corresponding to an ensemble of Riemannian manifolds with two marked points separated by a geodesic distance R, it follows from Eq. (4.14) that the average volume $\langle V \rangle_R$ of manifolds in this ensemble is

$$\langle V \rangle_R = -\frac{1}{G(R; \Lambda)} \frac{\partial G(R; \Lambda)}{\partial \Lambda}. \quad (4.27)$$

For $R \gg 1/M(\Lambda)$ this leads to

$$\langle V \rangle_R \sim \Lambda^{\nu-1} R, \quad (4.28)$$

which shows that for $R \gg 1/M(\Lambda)$ a typical universe has a volume that is proportional to R and is therefore just a long thin tube of length R. For $R \sim 1/M(\Lambda)$ the exponential decay ceases to be valid and inserting this relation in Eq. (4.28) yields

$$\langle V \rangle_R \sim R^{1/\nu} \quad \text{for} \quad R \sim 1/\Lambda^\nu. \quad (4.29)$$

By the definition of the Hausdorff dimension we expect

$$\langle V \rangle_R \sim R^{d_h}, \quad (4.30)$$

for $R \sim \Lambda^{-\nu}$ and we conclude that $d_h = 1/\nu$. It is not self-evident that this relation accords with Eq. (4.20), but we shall verify by an explicit calculation that this is the case.

The exponential decay given by Eq. (4.26) allows us, by Eq. (4.16), to calculate the large R behaviour of $G(R; V)$. From standard properties of the Laplace transform or by a saddle point calculation it follows that

$$- \log G(R; V) \sim (R/V^\nu)^{\frac{1}{1-\nu}} \qquad (4.31)$$

for $RV^{-\nu} \gg 1$. As we shall see the partition function $\mathscr{Z}(V)$ behaves like $V^{\gamma-3}$, where γ is an *entropy exponent*, similar to the exponent already encountered for random walks and random surfaces. Recalling Eq. (4.19) it follows that we can write

$$G(R; V) \sim V^{\gamma-1-\nu} F(R/V^\nu), \qquad (4.32)$$

where the *scaling function* $F(x)$ has the following properties:

$$F(x) \ \sim \ x^{d_h-1}, \qquad\qquad x \ll 1, \qquad (4.33)$$

$$F(x) \ \sim \ \exp\left(-c\, x^{\frac{1}{1-\nu}}\right) \qquad x \gg 1. \qquad (4.34)$$

In this way we see that the Hausdorff dimension d_h appears as the genuine fractal dimension *at all distances* if $\Lambda = 0$ and Eq. (4.29) agrees with Eq. (4.20).

Let us finally remark that the *short-distance behaviour of $G(R; \Lambda)$ is determined by that of $G(R; V)$*. From Eqs. (4.16) and (4.33)–(4.34) we obtain for $R \ll 1/\Lambda^\nu$

$$G(R; \Lambda) \sim R^{d_h-1} \int_{R^{d_h}}^\infty dV\ V^{\gamma-2} \sim R^{\gamma d_h-1}. \qquad (4.35)$$

It is clear from the above discussion that the functions $G(R; V)$ and $G(R; \Lambda)$ are ideal probes of the fractal structure of quantum gravity.

All the definitions of Hartle–Hawking wave functionals (loop functions) and invariant correlation functions can immediately be generalized to quantum gravity in higher dimensions. In the following sections we solve explicitly the model of pure gravity in two dimensions and calculate the wave functionals as well as the two-point function. Following the strategy used in the study of the DTRS-model we first define the functional integral over two-dimensional space-time by means of dynamical triangulations. In this case the action is very simple and the solution of the model at the discretized level reduces to the combinatorial problem of counting the number of elements in certain classes of triangulations. After the discretized models have been solved, one can construct a well-defined continuum limit.

4.2   The combinatorial solution

Let us consider two-dimensional quantum gravity without any matter fields. At the outset we restrict the topology of surfaces to be that of S^2 with a number of holes, deferring the treatment of higher-genus surfaces to later sections. As discussed in connection with the bosonic string we can view abstract triangulations of S_b^2 as defining a grid in the space of diffeomorphism classes of metrics on S_b^2. The precise identification is to some extent arbitrary as is clear from the very idea of imposing a reparametrization invariant cutoff of distances.

4.2.1 Regularization

Let T denote a triangulation of S_b^2. The regularized theory of gravity will be defined by replacing the action $S_g(\Lambda, Z_1, \ldots, Z_b)$ in Eq. (4.6) by

$$S_T(\mu, \lambda_1, \ldots, \lambda_b) = \mu N_t + \sum_{i=1}^{b} \lambda_i l_i, \qquad (4.36)$$

where N_t denotes the number of triangles in T and l_i is the number of links in the ith boundary component. The parameter μ is the bare cosmological constant and the λ_is are the bare cosmological constants of the boundary components. The integration over diffeomorphism classes of metrics in Eq. (4.7) becomes a summation over non-isomorphic triangulations. We define the loop functions (discretized versions of $W(\Lambda, Z_1, \ldots, Z_b)$) by summing over all triangulations of S_b^2:

$$w(\mu, \lambda_1, \ldots, \lambda_b) = \sum_{l_1, \ldots, l_b} \sum_{T \in \mathcal{T}(l_1, \ldots, l_b)} e^{-S_T(\mu, \lambda_1, \ldots, \lambda_b)}. \qquad (4.37)$$

Analogously, we define the partition function for closed surfaces by

$$\mathcal{Z}(\mu) = \sum_{T \in \mathcal{T}} \frac{1}{C_T} e^{-S_T(\mu)}, \qquad (4.38)$$

where C_T is the symmetry factor of T and $S_T(\mu) = \mu N_t$. We shall find it technically convenient to introduce classes of triangulations different from those discussed in Chapter 3. However, it is not important which one of the classes of triangulations is used here as we shall verify and discuss in more detail below. Since we consider surfaces of a fixed topology we have left out the curvature term in the action. It will be introduced later, when the restriction on topology is lifted.

Next we write down the regularized version of the Hartle–Hawking wave functionals $W(\Lambda, L_1, \ldots, L_b)$:

$$w(\mu, l_1, \ldots, l_b) = \sum_{T \in \mathcal{T}(l_1, \ldots, l_b)} e^{-S_T(\mu)} \qquad (4.39)$$

with an abuse of notation similar to the one in the previous section. This can also be written in the form

$$w(\mu, l_1, \ldots, l_b) = \sum_k e^{-\mu k} w_{k, l_1, \ldots, l_b}, \qquad (4.40)$$

where we have introduced the notation

$$w_{k,l_1,\ldots,l_b}$$

for the number of triangulations in $\mathcal{T}(l_1,\ldots,l_b)$ with k triangles.

The discretized analogues of the Laplace transformations which relate $W(\Lambda, Z_1, \ldots, Z_b)$ and $W(\Lambda, L_1, \ldots, L_b)$ or $W(V, L_1, \ldots, L_b)$ are

$$w(\mu, \lambda_1, \ldots, \lambda_b) = \sum_{l_1,\ldots,l_b} e^{-\sum_i \lambda_i l_i} w(\mu, l_1, \ldots, l_b) \qquad (4.41)$$

and Eq. (4.40). Similarly, we have for the partition functions

$$\mathcal{Z}(\mu) = \sum_k e^{-\mu k} \mathcal{Z}(k), \qquad (4.42)$$

$$\mathcal{Z}(k) = \sum_{T \in \mathcal{T}, N_t = k} \frac{1}{C_T}. \qquad (4.43)$$

It follows from the definitions (4.40)–(4.41) that $w(\mu, \lambda_1, \ldots, \lambda_b)$ is the generating function for the numbers $w_{k,l_1,\ldots,l_b}$, the arguments of the generating function being $e^{-\mu}$ and $e^{-\lambda_i}$. *In this way the evaluation of the loop functions of two-dimensional quantum gravity is reduced to the purely combinatorial problem of finding the number of non-isomorphic triangulations of S^2 or S_b^2 with a given number of triangles and boundary components of given lengths.*

We use the notation

$$w(g, z_1, \ldots, z_b) = \sum_{k,l_1,\ldots,l_b} w_{k,l_1,\ldots,l_b} g^k z_1^{-l_1-1} \cdots z_b^{-l_b-1} \qquad (4.44)$$

for the generating function with an extra factor $z_1^{-1} \ldots z_b^{-1}$, i.e. we make the identifications

$$g = e^{-\mu}, \qquad z_i = e^{\lambda_i}. \qquad (4.45)$$

The reason for this particular choice of variables in the generating function is motivated by its analytic structure, which will be revealed below.

In the following we consider two classes of triangulations. The first class we consider is the subclass of triangulations of type III, defined in Section 3.3, which satisfy the constraint that no two vertices in the same boundary component can be connected by an interior link. These are called *regular triangulations* in the present chapter and should not be confused with the regular triangulations discussed in Chapter 3, all of whose vertices had order 6.

The second class is obtained by enlarging the class of triangulations of type I by allowing degenerate boundaries. It may be defined as the class of complexes homeomorphic to the sphere with a number of holes that one obtains by successively gluing together a collection of triangles and

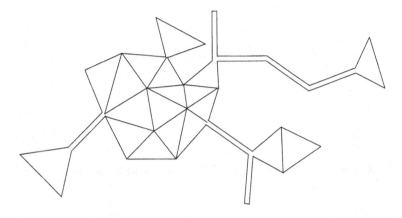

Fig. 4.1. A typical unrestricted "triangulation".

a collection of double-links which we consider as (infinitesimally narrow) strips, where links, as well as triangles, can be glued onto the boundary of a complex both at vertices and along links. Gluing a double-link along a link makes no change in the complex. An example of such a complex is shown in Fig. 4.1. We shall refer to complexes in this class as *unrestricted triangulations*.

The regular triangulations correspond most closely to our intuitive notion of a surface. Superficially, they look much nicer than the unrestricted ones and it is a strong indication of universality that the continuum limits (to be defined below) agree for the two classes of triangulations. This is a reflection of the fact that the degenerate structures present in the unrestricted triangulations appear on a slightly larger scale in the regular triangulations in the form of narrow strips consisting of triangles. We remark that among the classes of triangulations that we have considered the regular ones are the smallest while the unrestricted ones are the largest, so we expect the continuum limits of theories with triangulations of type I, II or III (defined in Section 3.3) to agree with the continuum limit obtained below.

Let $w(g,z)$ denote the generating function for the unrestricted triangulations with one boundary component and let $\tilde{w}(g,z)$ be the corresponding function for regular triangulations. Then we have

$$w(g,z) = \sum_{k=0}^{\infty} \sum_{l=0}^{\infty} w_{k,l}\, g^k\, z^{-(l+1)} \equiv \sum_{l=0}^{\infty} \frac{w_l(g)}{z^{l+1}} \qquad (4.46)$$

and

$$\tilde{w}(g,z) = \sum_{k=1}^{\infty} \sum_{l=3}^{\infty} \tilde{w}_{k,l}\, g^k z^{-(l+1)} \equiv \sum_{l=3}^{\infty} \frac{\tilde{w}_l(g)}{z^{l+1}}. \qquad (4.47)$$

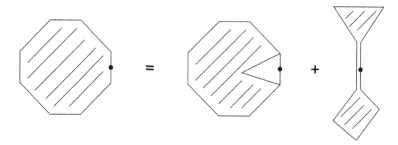

Fig. 4.2. Graphical representation of the loop equation for the class of unrestricted triangulations.

In the case of the unrestricted triangulations we have included the triangulation consisting of one point. It gives rise to the term $1/z$ and we have $w_0(g) = 1$. The function $w_1(g)$ starts with the term g, which corresponds to an unrestricted triangulation with a boundary consisting of one (closed) link with one vertex and containing one triangle. The coefficents $w_{k,1}$ in the expansion $w_{1,1}g + w_{3,1}g^3 + \cdots$ of $w_1(g)$ are the numbers of unrestricted triangulations with a boundary consisting of one link. In the case of regular triangulations the first term is g/z^4, corresponding to a single triangle. The terms proportional to $1/z^4$ make up a power series expansion of $\tilde{w}_3(g)$ whose coefficients are the numbers of regular triangulations with a boundary consisting of three links.

The coefficients of $w(g,z)$ and $\tilde{w}(g,z)$ fulfil recursion relations which have the simple graphical representations shown in Fig. 4.2 and Fig. 4.3. The diagrams indicate two operations that one can perform on a marked link on the boundary to produce a triangulation which has either fewer triangles or fewer boundary links. The first term on the right-hand side of Fig. 4.2 corresponds to the removal of a triangle. The second term corresponds to the removal of a double-link. Note that removing a triangle creates a new double-link if the triangle has two boundary links and note in addition that we count triangulations *with one marked link on each boundary component* and adopt the notation introduced above for the corresponding quantities.

In the case of regular triangulations the only operation is the removal of a triangle illustrated in Fig. 4.3. After a triangle has been removed some of the links might connect two boundary points without being boundary links. Cutting along these links yields a decomposition into regular triangulations T_i as shown in the diagram. The equations associated with the diagrams are

$$[w(g,z)]_{k,l} = [gzw(g,z)]_{k,l} + \left[\frac{1}{z}w^2(g,z)\right]_{k,l} \qquad (4.48)$$

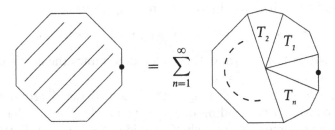

Fig. 4.3. Graphical representation of the loop equation for regular triangulations.

for the unrestricted triangulations and

$$[\tilde{w}(g,z)]_{k,l} = \sum_{n=1}^{\infty} \left[\frac{g}{z^2} (z^3 \tilde{w}(g,z))^n \right]_{k,l} = \left[\frac{g}{z^2} \frac{z^3 \tilde{w}(g,z)}{(1 - z^3 \tilde{w}(g,z))} \right]_{k,l} \quad (4.49)$$

for the regular triangulations. In these equations the subscripts k, l indicate the coefficient of g^k / z^{l+1}.

Let us explain the equations in some detail. The factor gz in Eq. (4.48) is present since the triangulation corresponding to the first term on the right-hand side of Fig. 4.2 has one triangle less and one boundary link more than the triangulation on the left-hand side. The function $w^2(g,z)$ in the last term in Eq. (4.48) arises from the two blobs connected by the double-link in Fig. 4.2 and the $1/z$ in front of $w^2(g,z)$ is inserted to make up for the decrease by two in the length of the boundary when removing the double link. The factor g on the right-hand side of Eq. (4.49) is present because the total number of triangles in the n regular triangulations shown on the right-hand side of Fig. 4.3 is $k - 1$. Finally, to explain the z-factors let us denote the length of the boundaries of the n regular triangulations of Fig. 4.3 by l_i, $i = 1, \ldots, n$, such that $\sum_i l_i = l - 1 + 2n$. In the term $\tilde{w}^n(z, g)$ the boundaries will contribute a factor

$$z^{-\sum_i (l_i + 1)} = z^{-(l-1+3n)}$$

whereas the corresponding factor on the left-hand side is $z^{-(l+1)}$. Thus, we have to multiply $w^n(z, g)$ by a prefactor z^{3n-2}.

As the reader may have discovered, Eqs. (4.48) and (4.49) are not correct for the smallest values of l. Consider, first, Fig. 4.2. The first term on the left-hand side of Eq. (4.48) (a single vertex) has no representation on the diagram. In order for Eq. (4.48) to be valid for $k = l = 0$ we have to add the term $1/z$ on the right-hand side of Eq. (4.48). Furthermore, it is clear from Fig. 4.2 that the first term on the right-hand side has at least two boundary links. Consequently, the term $gzw(g,z)$ on the right-hand side of Eq. (4.48) should be replaced by $gz(w(g,z) - 1/z - w_1(g)/z^2)$ such that all terms corresponding to triangulations with boundaries of length 0 and

1 are subtracted. It follows that the correct equation is

$$(z - gz^2)w(g,z) - 1 + g(w_1(g) + z) = w^2(g,z). \qquad (4.50)$$

In the next subsection we show that (4.50) determines the coefficients $w_{k,l}$, starting with $w_{0,0} = 1$, $w_{0,1} = 0$.

Next consider Fig. 4.3. The right-hand side of the diagram does not generate the term corresponding to a single triangle. Thus we have to add the term g/z^4 explicitly on the right-hand side of Eq. (4.49). Furthermore, the first term of the sum in Eq. (4.49), i.e. $gz\tilde{w}(z,g)$, is not correct since it follows from Fig. 4.3 that the triangulation T_1 for $n = 1$ has a boundary of at least length 4. This entails that we should subtract $g\tilde{w}_3(g)/z^3$ on the right-hand side of Eq. (4.49). After these modifications and a little algebra we arrive at the equation*

$$[z - gz^2 + g(1 - gz\tilde{w}_3(g))](z^3\tilde{w}(g,z)) - g[1 - gz\tilde{w}_3(g)] = z(z^3\tilde{w}(g,z))^2. \qquad (4.51)$$

Again, it is easy to show, by inserting the expansion (4.47) into the above equation that the coefficients $\tilde{w}_{k,l}$ are determined recursively by Eq. (4.51) and $\tilde{w}_{1,3} = 1$ and $\tilde{w}_{1,4} = 0$.

We will refer to Eqs. (4.50) and (4.51) as the *loop equations* corresponding to the two classes of triangulation. They are second-order equations in $w(g,z)$ and $\tilde{w}(g,z)$, respectively. As will be clear in the following this algebraic feature allows us to extract asymptotic formulas for the number of triangulations with k triangles in the limit $k \to \infty$. These formulas possess a number of universal properties which are independent of whether we use unrestricted or regular triangulations and only those universal features characterize the resulting continuum limit that we construct. In the following we shall, for technical convenience only, treat Eq. (4.50), corresponding to the case of unrestricted triangulations, but all results relevant for the construction of a continuum limit could as well have been derived starting from Eq. (4.51) corresponding to regular triangulations.

4.2.2 Counting planar graphs

Let us begin by solving Eq. (4.50) in the limit $g = 0$. In this case there are no internal triangles and the triangulations are in one to one correspondence with rooted branched polymers as defined in Section 2.5,

---

* This equation was derived in the seminal work of Tutte [363]. He used a generating functional $\psi(x,y) = \sum_{n,m=0}^{\infty} \psi_{n,m} x^n y^m$, where ψ denotes the number of non-isomorphic regular triangulations with a boundary of $3 + m$ links and n interior vertices. We have the relation $\psi(g^2, g/z) = z^4\tilde{w}(g,z)/g$ since $2n = k - l + 2$.

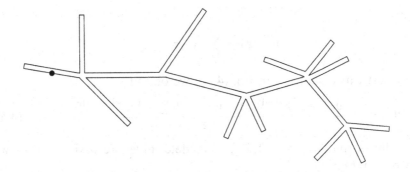

Fig. 4.4. Rooted branched polymers created by gluing of a boundary with one marked link.

the double-links corresponding to the links of the branched polymers and the root is the marked link, see Fig. 4.4. If $g = 0$ then Eq. (4.50) reads

$$w^2(z) - zw(z) + 1 = 0. \tag{4.52}$$

The above equation has two solutions but the one that corresponds to the counting problem has a Taylor expansion in z^{-1} whose first term is z^{-1} (recall that $w_{0,0} = 1$). This solution is given by

$$w(z) = \frac{1}{2}\left(z - \sqrt{z^2 - 4}\right). \tag{4.53}$$

Expanding in powers of $1/z$ yields

$$w(z) = \sum_{l=0}^{\infty} \frac{w_{2l}}{z^{2l+1}}, \tag{4.54}$$

where

$$w_{2l} = \frac{(2l)!}{(l+1)!\,l!} = \frac{1}{\sqrt{\pi}}\, l^{-3/2}\, 4^l\, (1 + O(1/l)), \tag{4.55}$$

and w_{2l} is the number of rooted polymers with l links. This result proves the inequality (2.196). Note that w_{2l} are the Catalan numbers, known from many combinatorial problems.

The generating function $w(z)$ is analytic in $\mathbb{C}$ with a cut on the real axis along the interval $[-2, 2]$. The endpoints of the cut determine the radius of convergence of $w(z)$ as a function of $1/z$ or, equivalently, the exponential growth of w_{2l}.

Let us now consider $w(g, z)$ as a function of g and write

$$w(g, z) = \sum_{k=0}^{\infty} w^{(k)}(z)g^k, \tag{4.56}$$

where

$$w^{(k)}(z) = \sum_{l=0}^{\infty} w_{k,l} z^{-l-1}.$$

If we insert this expansion in Eq. (4.50) we obtain

$$w^{(n)}(z) = \frac{-z\delta_{n,1} + z^2 w^{(n-1)}(z) - w_{n-1,1} + \sum_{k=1}^{n-1} w^{(k)}(z) w^{(n-k)}(z)}{\sqrt{z^2 - 4}}, \quad (4.57)$$

which shows that $w_{n,l}$, $l = 1, 2, \ldots$, are determined recursively from $w_{0,l}$ given by Eq. (4.55).

We can solve the second-order equation (4.50) and obtain

$$w(g, z) = \frac{1}{2}\left(V'(z) - \sqrt{(V'(z))^2 - 4Q(z)}\right), \quad (4.58)$$

where, anticipating generalizations, we have introduced the notation

$$V'(z) = z - gz^2, \quad Q(z) = 1 - gw_1(g) - gz. \quad (4.59)$$

The sign of the square root is determined as in Eq. (4.53) by the requirement that $w(g, z) = 1/z + O(1/z^2)$ for large z (since $w_{0,0} = 1$). If $g = 0$ then $V'(z)^2 - 4Q(z) = z^2 - 4$. For $g > 0$, on the other hand, $V'(z)^2 - 4Q(z)$ is a fourth-order polynomial of the form

$$V'(z)^2 - 4Q(z) = \{z - (2 + 2g) + O(g^2)\}$$
$$\times \{z + (2 - 2g) + O(g^2)\}\{gz - (1 - 2g^2) + O(g^3)\}^2 \quad (4.60)$$

in a neighbourhood of $g = 0$ since the analytic structure of $w(g, z)$ as a function of z cannot change discontinuously at $g = 0$. We can therefore write

$$V'(z)^2 - 4Q(z) = (z - c_+(g))(z - c_-(g))(c_2(g) - gz)^2,$$

and, by Eq. (4.58),

$$w(g, z) = \frac{1}{2}\left(z - gz^2 + (gz - c_2)\sqrt{(z - c_+)(z - c_-)}\right), \quad (4.61)$$

where c_-, c_+ and c_2 are functions of g, analytic in a neighbourhood of $g = 0$. We label the roots so that $c_- \le c_+$. They are uniquely determined by the requirement that $w(g, z) = 1/z + O(1/z^2)$, again originating from $w_{0,0} = 1$. This requirement gives three equations for the coefficients of z, z^0, z^{-1}. The first equation is

$$c_2(g) + \frac{1}{2}g(c_+(g) + c_-(g)) = 1 \quad (4.62)$$

with the initial conditions

$$c_2(0) = 1, \quad c_+(0) = 2, \quad c_-(0) = -2, \quad (4.63)$$

according to Eq. (4.53).

Rather than solving the loop equation explicitly let us discuss it in general terms. We can view g as the fugacity of triangles in the triangulations and $1/z$ as the fugacity of boundary links. The radius of convergence of the expansion of $w(z)$ in Eq. (4.54) in the case $g = 0$ can be viewed as the maximal allowed value of the fugacity. When z approaches this value the average length of a typical boundary (a branched polymer) will diverge. As g becomes positive, triangles begin to appear and a larger class of boundaries contributes. Consequently, the coeffients $w_l(g)$ increase, and therefore the minimal positive value allowed for z increases also. According to Eq. (4.61) this implies that $c_+(g)$ increases with g. Furthermore, $|c_-(g)| \leq c_+(g)$, since for $z < 0$ the boundaries with an odd number of links have negative weight, which can only increase the radius of convergence in $1/z$. Put differently: if the power series for w converges for some positive value of z, it also converges at $-z$ since all the coefficients are positive. These remarks are in agreement with Eq. (4.60), which shows that $c_+(g) = 2 + 2g + O(g^2)$ and $c_-(g) = -2 + 2g + O(g^2)$. Considering the expansion (4.46) we see that an exponential growth of $w_{k,l}$ in k, independent of l, would imply that the power series of the functions $w_l(g)$ all had the same radius of convergence g_0.

If we look at the expansion of (4.61) in terms of $1/z$ we see that the coefficients $w_l(g)$ are polynomials in $c_+(g)$, $c_-(g)$ and $c_2(g)$ so $w_l(g)$ can only have a finite radius of convergence if the coefficients themselves have a finite radius of convergence. In order to determine the singular behaviour of $c_+(g)$, $c_-(g)$ and $c_2(g)$ without detailed calculations let us reformulate Eq. (4.50) in a way which reveals the analytic structure imposed by the loop equation and the requirement that $w(g,z) = 1/z + O(1/z^2)$. With $V'(z)$ given by Eq. (4.59) we can rewrite Eq. (4.50) as

$$\oint_C \frac{d\omega}{2\pi i} \frac{V'(\omega)}{z - \omega} w(g,\omega) = w^2(g,z), \qquad (4.64)$$

where the contour C encloses counterclockwise the cut of $w(g,z)$ but not z, see Fig. 4.5. If we use the expansion (4.46) of $w(g,z)$ in the integral (4.64) and deform the contour to infinity we pick up a pole term at z, which is the first term on the left-hand side of Eq. (4.50), while the residue at infinity precisely reproduces the rest of the left-hand side of Eq. (4.50).

Let us for a moment consider Eq. (4.64) for a general polynomial $V'(z)$ of the form

$$V'(z) = z - g(t_1 + t_2 z + t_3 z^2 + \cdots + t_n z^{n-1}), \qquad (4.65)$$

and generalize Eq. (4.61) to

$$w(g,z) = \frac{1}{2} \left(V'(z) - M(z)\sqrt{(z - c_+(g))(z - c_-(g))} \right), \qquad (4.66)$$

where $M(z)$ is a polynomial of a degree which is one less than that of $V'(z)$.

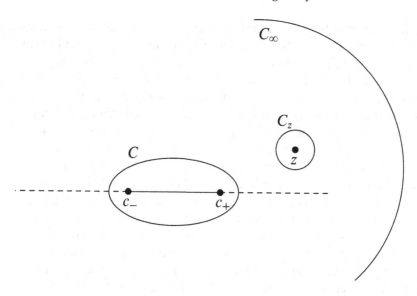

Fig. 4.5. The integration contour C and the cut from c_- to c_+. When deforming the contour to infinity we obtain two contributions: one from the circle C_∞ and one from the circle C_z around the pole at z.

In the sequel we shall call the function V the *potential*. The polynomial M is uniquely determined by the requirement that $w(g,z)$ falls off at infinity as before, i.e. $w(g,z) = 1/z + O(1/z^2)$. This follows from the relation

$$M(z) = \oint_{C_\infty} \frac{d\omega}{2\pi i} \frac{M(\omega)}{\omega - z} = \oint_{C_\infty} \frac{d\omega}{2\pi i} \frac{V'(\omega)}{\omega - z} \frac{1}{\sqrt{(\omega - c_+)(\omega - c_-)}}, \qquad (4.67)$$

where C_∞ is a curve with winding number 1 with respect to the point z and enclosing the cut $[c_-, c_+]$. The first equality sign above is valid for any polynomial. The second equality is obtained by expressing $M(z)$ in terms of $V'(z)$ and $w(g,z)$ and using $w(g,z) = O(1/z)$. Eq. (4.67) determines the coefficients of the polynomial $M(z)$ in terms of g, $c_+(g)$ and $c_-(g)$. Let us next show how c_+ and c_- are determined as functions of g. If we multiply Eq. (4.67) by $\sqrt{(z - c_+)(z - c_-)}$ and deform the loop C_∞ to that of C (see Fig. 4.5), then we pick up $V'(z)$ from the residue at z, so by Eq. (4.66) we obtain

$$w(g,z) = \frac{1}{2} \oint_C \frac{d\omega}{2\pi i} \frac{V'(\omega)}{z - \omega} \frac{\sqrt{(z - c_+)(z - c_-)}}{\sqrt{(\omega - c_+)(\omega - c_-)}}. \qquad (4.68)$$

If we expand the left-hand side of Eq. (4.68) in powers of $1/z$ and demand that $w(g,z) = 1/z + O\left(1/z^2\right)$, then we obtain

$$M_{-1}(g, c_-, c_+) = 2, \qquad M_0(g, c_-, c_+) = 0, \qquad (4.69)$$

where for later convenience we have introduced the notation

$$M_k(g, c_-, c_+) = \oint_C \frac{d\omega}{2\pi i} \frac{V'(\omega)}{(\omega - c_+)^{k+1/2}(\omega - c_-)^{1/2}}, \quad (4.70)$$

$$J_k(g, c_-, c_+) = \oint_C \frac{d\omega}{2\pi i} \frac{V'(\omega)}{(\omega - c_+)^{1/2}(\omega - c_-)^{k+1/2}}. \quad (4.71)$$

Eqs. (4.69) are two algebraic equations which allow us to determine $c_+(g)$ and $c_-(g)$.

4.2.3 Generalization

The formalism developed above allows a generalization of the counting problem for triangulations since it is independent of the detailed form of the polynomial $V'(z)$. In order to see this let us assume that the constants t_i in Eq. (4.65) fulfil $t_1, \ldots, t_{n-1} \geq 0$ and $t_n > 0$. For $k > n$ we can move the integration contour C in Eq. (4.70) to infinity and obtain $M_k = 0$. As a consequence of the positivity of the coefficients t_i we have

$$M_k(g, c_-, c_+) < 0 \quad \text{for} \quad 1 < k \leq n, \quad (4.72)$$

for $g > 0$, while we have

$$M_1(g, c_-, c_+) > 0 \quad (4.73)$$

in a neighbourhood of $g = 0$. The solution $w(g, z)$ of Eq. (4.64) which falls off as $1/z$ is given explicitly by Eq. (4.68) and we can express it in terms of the *moments* M_k:

$$w(g, z) = \frac{1}{2} \left\{ V'(z, g) - \left(\sum_{k=1}^{n-1} M_k(z - c_+)^{k-1} \right) \sqrt{(z - c_+)(z - c_-)} \right\}. \quad (4.74)$$

We now associate an enumeration problem with the polynomial given by Eq. (4.65). Deforming the contour C in Eq. (4.64) to infinity we pick up a pole term $V'(z)w(g, z)$ at z, and at infinity we obtain a polynomial in z:

$$-Q(z) = \oint_{C_\infty} \frac{d\omega}{2\pi i} \frac{V'(\omega)}{z - \omega} w(g, \omega), \quad (4.75)$$

as is seen by expanding the integrand in powers of $1/\omega$. The polynomial $Q(z)$ is of order $n - 2$, given by

$$Q(z) = 1 - g \sum_{j=2}^{n} t_j \sum_{l=0}^{j-2} z^l w_{j-2-l}(g), \quad w_0(g) = 1, \quad (4.76)$$

and the Eq. (4.64), satisfied by $w(z, g)$, takes the form

$$w(g, z)^2 = V'(z)w(g, z) - Q(z). \quad (4.77)$$

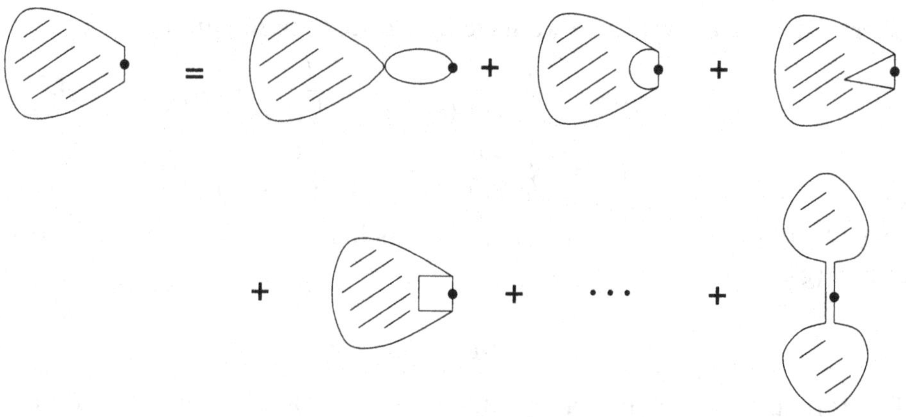

Fig. 4.6.   The graphical representation of the general loop equation.

Writing Eq. (4.77) as

$$w(g,z) = g\left(t_1\frac{1}{z} + t_2 + t_3 z + \cdots + t_n z^{n-2}\right) w(g,z)$$
$$+ \frac{1}{z}Q(z) + \frac{1}{z}w^2(g,z), \tag{4.78}$$

we can give it the combinatorial interpretation we are aiming for which is to count planar complexes made up of polygons with an arbitrary number $j \le n$ of sides, including "one-sided" and "two-sided" polygons. In addition there are double-links as in the unrestricted triangulations. If we attribute a weight gt_j to each j-sided polygon and a weight z to each boundary link, then Eq. (4.78) is analogous to Eq. (4.50). The graphical representation of the equation is shown in Fig. 4.6. The subtraction of the polynomial $Q(z)$ in Eq. (4.77) reflects the fact that the term with a j-sided polygon in Fig. 4.6 must have a boundary of length at least $j-1$ for $j > 1$. The constant term 1 in Q corresponds to the complex consisting of a single vertex.

In general, we are not interested in graphs with one-sided and two-sided polygons and for the determination of $w(g,z)$ from the loop equation we could have chosen $t_1 = t_2 = 0$ and left out the corresponding graphs in Fig. 4.6. However, it is convenient to keep the variables t_1 and t_2 different from zero in the intermediate calculations if we want to use $w(g,z)$ to determine the generating functions for planar graphs with more than one boundary. For this purpose let us change notation and introduce $g_j = gt_j$ as new variables as well as the notation $\underline{g}$ for the set of coupling constants $g_1, g_2, g_3, \ldots$ and $\underline{t}$ for the set of constants $t_1, t_2, t_3, \ldots$.

Then

$$V'(z) = z - \sum_{j=1}^{n} g_j z^{j-1}. \qquad (4.79)$$

and

$$w(\underline{g}, z) = \sum_{l, k_1, \dots, k_n} w_{\{k_j\}, l} \, z^{-(l+1)} \prod_{j=1}^{n} g_j^{k_j}, \qquad (4.80)$$

where $w_{\{k_j\}, l}$ is the number of planar graphs with k_j j-sided polygons, $j = 1, \dots, n$, and a boundary of length l.

From $w(\underline{g}, z)$ we can derive the generating function for planar graphs with two boundary components by applying the *loop insertion operator*

$$\frac{d}{dV(z)} = \sum_{j=1}^{\infty} \frac{j}{z^{j+1}} \frac{d}{dg_j}. \qquad (4.81)$$

One should think of this operator as acting on formal power series in an arbitrary number of variables g_j. The action of $d/dV(z_2)$ on $w(\underline{g}, z_1)$ has in each term of the power series the effect of reducing the power k_j of a specific coupling constant g_j by one and adding a factor jk_j/z_2^{j+1}. The geometrical interpretation is that a j-sided polygon is removed, leaving a marked boundary of length j to which the new indeterminate z_2 is associated. The factor k_j is due to the possibility to make the replacement at any of the k_j j-sided polygons present in the planar graph, while j is the number of possibilities to choose the marked link on the new boundary component. The generating function for planar graphs with b boundary components can therefore be expressed as

$$w(\underline{g}, z_1, \dots, z_b) = \frac{d}{dV(z_b)} \cdots \frac{d}{dV(z_2)} w(\underline{g}, z_1). \qquad (4.82)$$

Below we shall alternate between the notations

$$w(g, z_1, \dots, z_b) \quad \text{and} \quad w(\underline{g}, z_1, \dots, z_b).$$

The first is appropriate when $\underline{t}$ is kept fixed and $\underline{g} = \underline{g}t$, the second when we consider more general changes of $\underline{g}$, i.e. when we apply the loop insertion operator. If no misunderstanding is possible we occasionally drop the reference to the coupling constants and write $w(z_1, \dots, z_b)$.

4.2.4 *An easy example*

The problem of counting *planar* surfaces constructed from even-sided polygons corresponds to the case when the potential V is even, so its derivative V' is odd. In order to illustrate the formalism developed above we consider this case *with the additional restriction imposed that the length of each boundary component has to be even* if the triangulation has more than one boundary component. It turns out that this additional requirement simplifies drastically the problem and explicit formulas can be obtained.

If the planar surface has only one boundary component and is constructed from even-sided polygons the length of the boundary is even, so the function $w(\underline{g}, z)$ is odd in z and $-c_- = c_+$ which we shall denote by c. Therefore, by Eq. (4.66) and Eq. (4.67),

$$w(\underline{g}, z) = \frac{1}{2} \left(V'(z) - M(\underline{g}, z)\sqrt{z^2 - c^2} \right), \qquad (4.83)$$

where

$$M(\underline{g}, z) = \oint_{C_\infty} \frac{d\omega}{2\pi i} \frac{\omega V'(\omega)}{(\omega^2 - z^2)\sqrt{\omega^2 - c^2}}. \qquad (4.84)$$

Expanding $1/(\omega^2 - z^2)$ in Eq. (4.84) in powers of $1/(\omega^2 - c^2)$,

$$\frac{1}{\omega^2 - z^2} = \frac{1}{\omega^2 - c^2} \sum_{n=0}^{\infty} \left(\frac{z^2 - c^2}{\omega^2 - c^2} \right)^n,$$

we obtain

$$M(\underline{g}, z) = \sum_{k=1}^{\infty} \tilde{M}_k(\underline{g}, c^2)(z^2 - c^2)^{k-1},$$

where

$$\tilde{M}_k(\underline{g}, c^2) = \oint_C \frac{d\omega}{2\pi i} \frac{\omega V'(\omega)}{(\omega^2 - c^2)^{k+1/2}}, \qquad (4.85)$$

and the position c of the endpoint of the cut is determined by an analogue of the first equation in (4.69):

$$\tilde{M}_0(\underline{g}, c^2) = 2. \qquad (4.86)$$

The calculations leading to this result are similar to the ones in the general model.

While the loop insertion operator seems rather formal, it is very useful and its use leads to a remarkable simplification. In the present context it can be written

$$\frac{d}{dV(z)} = \sum_j \frac{2j}{z^{2j+1}} \frac{d}{dg_{2j}} = \frac{\partial}{\partial V(z)} + \frac{dc^2}{dV(z)} \frac{\partial}{\partial c^2}, \qquad (4.87)$$

where

$$\frac{\partial}{\partial V(z)} = \sum_j \frac{2j}{z^{2j+1}} \frac{\partial}{\partial g_{2j}}. \qquad (4.88)$$

Here we think of the loop insertion operator as acting on functions which depend on the g_js and c. From this definition it follows that

$$\frac{\partial V'(\omega)}{\partial V(z)} = \frac{-2\omega z}{(z^2 - \omega^2)^2}, \qquad \frac{\partial \tilde{M}_k}{\partial V(z)} = \frac{\partial}{\partial z} \frac{z}{(z^2 - c^2)^{k+1/2}}. \qquad (4.89)$$

Definition (4.85) implies that

$$\frac{\partial \tilde{M}_k}{\partial c^2} = \left(k + \frac{1}{2}\right) \tilde{M}_{k+1}, \qquad (4.90)$$

and from the constraint (4.86) we obtain

$$\frac{dc^2}{dV(z)} = \frac{2}{\tilde{M}_1(\underline{g}, c(\underline{g}))} \frac{c^2}{(z^2 - c^2)^{3/2}}. \qquad (4.91)$$

The loop insertion operator (4.87) can now be expressed as

$$\frac{d}{dV(z)} = \frac{\partial}{\partial V(z)} + \frac{1}{\tilde{M}_1(\underline{g}, c(\underline{g}))} \frac{2c^2}{(z^2 - c^2)^{3/2}} \frac{\partial}{\partial c^2}, \qquad (4.92)$$

and using Eqs. (4.89) and (4.90) it is straightforward to apply it to $w(\underline{g}, z)$ given by Eq. (4.83). We find

$$w(\underline{g}, z_1, z_2) = \frac{1}{2(z_1^2 - z_2^2)^2} \left[z_2^2 \sqrt{\frac{z_1^2 - c^2}{z_2^2 - c^2}} + z_1^2 \sqrt{\frac{z_2^2 - c^2}{z_1^2 - c^2}} - 2z_1 z_2 \right], \qquad (4.93)$$

$$w(\underline{g}, z_1, z_2, z_3) = \frac{c^4}{2\tilde{M}_1(c^2)} \left(\frac{1}{(z_1^2 - c^2)(z_2^2 - c^2)(z_3^2 - c^2)} \right)^{3/2} \qquad (4.94)$$

and

$$w(\underline{g}, z_1, \ldots, z_b) = \left(\frac{2}{\tilde{M}_1(c^2)} \frac{d}{dc^2} \right)^{b-3} \frac{1}{2c^2 \tilde{M}_1(c^2)} \prod_{k=1}^{b} \frac{c^2}{(z_k^2 - c^2)^{3/2}}. \qquad (4.95)$$

While Eq. (4.93) and Eq. (4.94) are obtained by a direct calculation, Eq. (4.95) is most easily proved by induction, starting from Eq. (4.94).

Let us now study the singularities of the above functions. For this purpose we fix as above the variables t_j and write $\underline{g} = g\underline{t}$. Then c is a function of the variable g. As remarked above the only possibility for a singular behaviour of $w(g, z)$ as a function of g is due to the singular behaviour of $c(g)$. From Eq. (4.94) and Eq. (4.95) we see that $w(g, z_1, \ldots, z_b)$ is singular when $\tilde{M}_1 = 0$ and from Eq. (4.91) it is clear that $c(g)$ is singular only at $\tilde{M}_1 = 0$. Summarizing:

Theorem 4.1 *The generating functions $w(g, z_1, \ldots, z_b)$ are analytic in a neighbourhood of $g = 0$. They become singular in the variable g when $\tilde{M}_1(g, c^2(g)) = 0$. They are analytic in the complex variables z_i except at a cut along $[-c, c]$.*

Let us denote the value of g where $\tilde{M}_1 = 0$ by g_0, the value of c corresponding to g_0 by c_0 and define

$$\Delta g = g_0 - g, \quad \Delta c^2 = c_0^2 - c^2(g), \quad \tilde{M}_2^0 = \tilde{M}_2(g_0, c_0^2). \quad (4.96)$$

From Eq. (4.90) we obtain, using that $\tilde{M}_1(g_0, c_0^2) = 0$,

$$\tilde{M}_1(g, c^2(g)) = -\frac{3}{2}\tilde{M}_2^0 \Delta c^2 + O(\Delta g). \quad (4.97)$$

By Eq. (4.86) we can write

$$2 = \tilde{M}_0(g_0, c_0^2) = \tilde{M}_0(g + \Delta g, c^2 + \Delta c^2)$$

and by Taylor expanding $\tilde{M}_0(g + \Delta g, c^2 + \Delta c^2)$ and using Eq. (4.90) and Eq. (4.97) we obtain

$$\left(\Delta c^2\right)^2 = -\frac{4(c_0^2 - 4)}{3\tilde{M}_2^0}\frac{\Delta g}{g_0} + o(\Delta g), \quad (4.98)$$

and

$$\tilde{M}_1(g, c^2(g)) = \sqrt{3|\tilde{M}_2^0|(c_0^2 - 4)}\sqrt{\frac{\Delta g}{g_0}} + o(\sqrt{\Delta g}). \quad (4.99)$$

The singularity structure of $w(g, z_1, \ldots, z_b)$ for $z_i > c$ fixed and g close to g_0 now follows directly from Eq. (4.95):

$$w(g, z_1, \ldots, z_b) \sim \frac{1}{(\Delta g)^{b-5/2}}\prod_{i=1}^{b}\frac{1}{(z_i^2 - c^2)^{3/2}}\left(1 + O(\sqrt{\Delta g})\right), \quad (4.100)$$

since the most singular contribution arises when all $b - 3$ differentiations with respect to c^2 act on $\tilde{M}_1(c)$. Since

$$\frac{1}{(g_0 - g)^{b-5/2}} = g_0^{5/2-b}\sum_{k=0}^{\infty}\binom{b - 7/2 + k}{k}\left(\frac{g}{g_0}\right)^k, \quad (4.101)$$

we can determine the asymptotic behaviour of $w_{k,l_1,\ldots,l_b}$, i.e. the coefficient of the product $g^k z_1^{-l_1-1}\cdots z_b^{-l_b-1}$ in $w(g, z_1, \ldots, z_b)$, for fixed l_j and $k \to \infty$. If we take one $t_j = 1$ and the rest of the t_js equal to zero, then $w_{k,l_1,\ldots,l_b}$, with $g = g_0$, is the number of planar surfaces that can be made of k j-sided polygons, such that the boundary components have the lengths $l_1, \ldots, l_b$. Explicitly, we find in this case

$$w_{k,l_1,\ldots,l_b} \sim \left(\frac{1}{g_0(\underline{t})}\right)^k k^{b-7/2} \quad (4.102)$$

for k large, and we have indicated the dependence of g_0 on the t_js. The value $g_0(\underline{t})$ will of course vary with the weights $\underline{t}$, but, as we shall see below, the precise value of g_0 plays no role in the theory of quantum gravity. It is only the subleading power correction which has an interpretation in terms of continuum physics and we have now seen that this power is independent of the class of polygons one chooses for constructing planar surfaces.

From Eq. (4.95) we can likewise obtain the asymptotic form of $w_{k,l_1,...,l_b}$ as the lengths of the boundaries l_i go to infinity. The situation is similar to the one encountered in Eqs. (4.52) and (4.54) for branched polymers (pure boundaries). The generating function is singular as $z \to c(g)$. Expanding in powers of $c^2(g)/z^2$ gives the numbers $w_{k,l_1,...,l_b}$ in the limit $l_i \to \infty$ with k, the number of polygons, fixed. In general we are interested in the limit where k and the l_is go simultaneously to infinity. This limit will be studied below.

It is worth emphasizing the structure of Eq. (4.95):

Theorem 4.2 *For any odd polynomial $V'(z)$ of the form given by Eq. (4.79), the generating function $w(\underline{g}, z_1, z_2)$ depends only on z_1, z_2 and c^2 and is given by Eq. (4.93), while $w(\underline{g}, z_1, ..., z_b)$, $b > 2$, is a function of $\tilde{M}_k$, $k = 1, 2, ..., b - 2$, c^2 and $z_i^2 - c^2$, explicitly given by Eqs. (4.94) and (4.95). Hence, for $b > 3$ it has the form*

$$
w(\underline{g}, z_1, ..., z_b) = \sum_{m=0}^{b-3} \sum_{\alpha_l, n_i} C_m(\alpha_1, ..., \alpha_k; n_1, ..., n_b)
$$
$$
\times \frac{\tilde{M}_{\alpha_1} \cdots \tilde{M}_{\alpha_k}}{c^{2(m+1)} \tilde{M}_1^{b-2+k}} \prod_{i=1}^{b} \frac{c^2}{(z_i^2 - c^2)^{n_i+3/2}}, \quad (4.103)
$$

where the coefficients $C_m(\alpha_1, ..., \alpha_k; n_1, ..., n_b)$ are rational numbers independent of $V'(z)$. The indices in the sum above satisfy $2 \leq \alpha_l \leq b - 2$ and $0 \leq n_i \leq b - 3$ and obey the constraint

$$
b - 3 = m + \sum_{i=1}^{b} n_i + \sum_{l=1}^{k} (\alpha_l - 1). \quad (4.104)
$$

4.2.5 The general model

Having seen the solution in the special case of the previous section it is now straightforward to lift the restriction that the polynomial $V'(z)$ be odd and each boundary component of even length. The general structure does not change but $c_+(g) > -c_-(g)$ if $V'(z)$ is not odd. In addition we will have to work with the two sets of "moments" M_k and J_k instead of the single set of moments $\tilde{M}_k$ used in the previous subsection. For fixed non-negative values of the weights t_j it is still true that the critical value g_0 of g is determined by

$$
M_1(g_0, c_-(g_0), c_+(g_0)) = 0. \quad (4.105)
$$

The machinery we use is similar to the one developed for an odd polynomial $V'(z)$. Let us list the corresponding formulas:

$$\frac{\partial M_k}{\partial c_+} = \left(k + \frac{1}{2}\right) M_{k+1}, \qquad \frac{\partial J_k}{\partial c_-} = \left(k + \frac{1}{2}\right) J_{k+1}, \qquad (4.106)$$

$$M_1(g, c(g)) = -\frac{3}{2} M_2(g_0)\Delta c_+ + O(\Delta g, \Delta c_-) + o(\Delta c_+), \qquad (4.107)$$

and

$$(\Delta c_+)^2 = -\frac{4(c_+^2 - 4 - \frac{1}{4}(c_+ + c_-)^2)}{3(c_+ - c_-)M_2}\frac{\Delta g}{g} + o(\Delta g), \qquad (4.108)$$

$$\Delta c_- = \frac{4 - c_-^2 + \frac{1}{4}(c_+ + c_-)^2}{(c_+ - c_-)J_1}\frac{\Delta g}{g} + o(\Delta g), \qquad (4.109)$$

where $\Delta g = g_0 - g$ and $\Delta c_\pm = c_\pm(g_0) - c_\pm(g)$. The loop insertion operator can now be written as

$$\frac{d}{dV(z)} = \frac{\partial}{\partial V(z)} + \frac{dc_+}{dV(z)}\frac{\partial}{\partial c_+} + \frac{dc_-}{dV(z)}\frac{\partial}{\partial c_-}, \qquad (4.110)$$

where

$$\frac{\partial}{\partial V(z)} = \sum_j \frac{j}{z^{j+1}}\frac{\partial}{\partial g_j}, \qquad (4.111)$$

as before. It follows that

$$\frac{\partial V'(\omega)}{\partial V(z)} = \frac{\partial}{\partial z}\left(\frac{1}{z - \omega}\right), \qquad (4.112)$$

$$\frac{\partial M_k}{\partial V(z)} = \frac{\partial}{\partial z}\left(\frac{1}{(z - c_+)^{k+\frac{1}{2}}(z - c_-)^{\frac{1}{2}}}\right), \qquad (4.113)$$

$$\frac{\partial J_k}{\partial V(z)} = \frac{\partial}{\partial z}\left(\frac{1}{(z - c_+)^{\frac{1}{2}}(z - c_-)^{k+\frac{1}{2}}}\right), \qquad (4.114)$$

and Eq. (4.69) leads to

$$\frac{dc_+}{dV(z)} = \frac{1}{M_1}\frac{1}{(z - c_+)^{\frac{3}{2}}(z - c_-)^{\frac{1}{2}}} \qquad (4.115)$$

$$\frac{dc_-}{dV(z)} = \frac{1}{J_1}\frac{1}{(z - c_+)^{\frac{1}{2}}(z - c_-)^{\frac{3}{2}}}. \qquad (4.116)$$

Eqs. (4.112) and (4.106) are all we need to apply the loop insertion operator to $w(\underline{g}, z)$ when written in the form of Eq. (4.66). The result of the first insertion is

$$w(\underline{g}, z_1, z_2) \tag{4.117}$$

$$= \frac{1}{2(z_1 - z_2)^2} \left(\frac{z_1 z_2 - \frac{1}{2}(z_1 + z_2)(c_+ + c_-) + c_+ c_-}{\sqrt{[(z_1 - c_+)(z_1 - c_-)][(z_2 - c_+)(z_2 - c_-)]}} - 1 \right).$$

The structure is very similar to that of Eq. (4.93). In particular, the two-loop function $w(\underline{g}, z_1, z_2)$ is a function of c_-, c_+, z_1 and z_2 only. Note that Eq. (4.117) does not reduce to Eq. (4.93) when $c_+ = -c_-$. The reason is that even though the total number of the boundary links has to be even in this case, each boundary component can have an odd number of links. This is not possible in the simplified model considered in the previous subsection.

Repeated application of the loop insertion operator produces results of the same structure as in Eq. (4.94) and Eq. (4.95), although it appears to be difficult to write the b-loop function as a simple closed expression like Eq. (4.95), valid for arbitrary $b \geq 3$. It is, however, easy to prove that $w(\underline{g}, z_1, \ldots, z_b)$ depends only on $c_\pm$, $M_1, \ldots, M_{b-2}$ and $J_1, \ldots, J_{b-2}$ and it can be expressed as a sum of terms of the form

$$\frac{M_{\alpha_1} \cdots M_{\alpha_k} J_{\beta_1} \cdots J_{\beta_l}}{M_1^\alpha J_1^\beta d^\gamma} \prod_{i=1}^b \frac{1}{(z_i - c_+)^{n_i + 3/2}(z_i - c_-)^{m_i + 3/2}}, \tag{4.118}$$

where $\alpha_i, \beta_i \geq 2$, and α, β and γ are integers which depend on the numbers $\alpha_i, \beta_i, n_i, m_i$ and

$$d = c_+ - c_-. \tag{4.119}$$

The coefficients of the terms given by Eq. (4.118) are again rational numbers independent of the polynomial $V'(z)$.

Needless to say, the asymptotic form of $w_{k, l_1, \ldots, l_b}$ for large k or large $l_1, \ldots, l_b$ is identical to the one we found in the last subsection for the reduced model.

4.3 Counting higher-genus surfaces

4.3.1 The loop equation for genus $h > 0$

Up to now we have confined our discussion to triangulations with spherical topology. The extension to higher genera is surprisingly easy as the loop equations can be generalized. The graphical equation shown in Fig. 4.6 for the one-loop function is still valid, except that one more term has to be added on the right-hand side, corresponding to the case when the removal of a double link does not decompose the triangulation into two disjoint parts but corresponds to the cutting of a handle. This

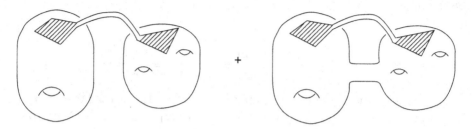

Fig. 4.7. An illustration of the two situations which can occur when removing a double-link from a surface of genus $h > 0$. In the first situation the surface consists of two subsurfaces of genus h_1 and h_2, $h_1 + h_2 = h$, only connected via the double link. In the second situation the removal of the double-link will not separate the surface in two, but instead lower h to $h - 1$. Note that we have chosen, for clarity of presentation, to keep the surface itself unshaded, contrary to the convention used in Figs. 4.2 and 4.6.

operation lowers the genus by one and increases the number of boundary components to two. Under this operation the boundary component of length l containing the marked link splits in two parts of length l_1 and l_2, where $l = l_1 + l_2 - 2$, see Fig. 4.7. In order to write down the algebraic equation corresponding to the modified Fig. 4.6 we generalize the generating functions for triangulations discussed above and sum over triangulations of all genera. Let us first consider the generalization of Eq. (4.46). We denote the sum over triangulations of fixed genus h by $w_h(g,z)$ and introduce a new variable, which will be written $1/N^2$ for later convenience, and the generalized generating function

$$w(g,z) = \sum_{h=0}^{\infty} \frac{1}{N^{2h}} w_h(g,z). \tag{4.120}$$

This is a purely formal power series. In fact we shall see that it is always divergent. With this notation the generalization of the loop equation (4.78) becomes an infinite system of coupled equations, one for each power $1/N^{2h}$ of the indeterminate $1/N^2$:

$$w_h(g,z) = g(t_1 \frac{1}{z} + t_2 + t_3 z + \cdots + t_n z^{n-2}) w_h(g,z) + \frac{1}{z} Q_h(z) \tag{4.121}$$

$$+ \frac{1}{z} \sum_{h_1+h_2=h} w_{h_1}(g,z) w_{h_2}(g,z) + \frac{1}{z} w_{h-1}(g,z,z),$$

where $Q_h(z)$ is defined as for $h = 0$, see Eq. (4.76), but summing over triangulations of genus h. Eq. (4.121) can also be written as (dropping explicit reference to g)

$$\int_C \frac{d\omega}{2\pi i} \frac{V'(\omega) - 2w_0(\omega)}{z - \omega} w_h(\omega) = \sum_{h'=1}^{h-1} w_{h'}(z)w_{h-h'}(z) + w_{h-1}(z,z), \quad (4.122)$$

or, even more compactly, as

$$\int_C \frac{d\omega}{2\pi i} \frac{V'(\omega)}{z - \omega} w(\omega) = w^2(z) + \frac{1}{N^2} \frac{dw(z)}{dV(z)}. \quad (4.123)$$

We shall refer to both of these equations as the *generalized loop equation* or just the *loop equation*, and the simpler equation (4.64) as the *spherical loop equation* or the *genus zero loop equation* if there is a danger of misunderstanding. In the previous subsection we determined $w_0(z)$ and $w_0(z_1, \ldots, z_b)$ and it can be seen that Eq. (4.122) is an iterative equation for $w_h(z)$ since the right-hand side only refers to genera less than h and for a fixed genus one can calculate the two-loop function from the one-loop function using the loop insertion operator.

Let us define an operator $\hat{K}$, acting on functions of a complex variable, by

$$\begin{aligned}
(\hat{K}f)(z) &= \oint_C \frac{d\omega}{2\pi i} \frac{V'(\omega) - 2w_0(\omega)}{z - \omega} f(\omega), \quad &(4.124)\\
&= \oint_C \frac{d\omega}{2\pi i} \frac{M(\omega)\sqrt{(\omega - c_+)(\omega - c_-)}}{z - \omega} f(\omega),
\end{aligned}$$

Then the loop equation can be written

$$\hat{K}w_h(z) = \sum_{h'=1}^{h-1} w_{h'}(z)w_{h-h'}(z) + \frac{dw_{h-1}(z)}{dV(z)}, \quad (4.125)$$

for $h > 0$.

4.3.2 *Solution of the loop equation for $h > 0$*

In order to solve Eq. (4.125) genus by genus we introduce the following notation:

$$\Phi^{(0)}(z) = \frac{1}{\sqrt{(z - c_+)(z - c_-)}}, \quad (4.126)$$

$$\Phi_+^{(n)}(z) = \frac{\Phi^{(0)}(z)}{(z - c_+)^n}, \quad \Phi_-^{(n)}(z) = \frac{\Phi^{(0)}(z)}{(z - c_-)^n}, \quad (4.127)$$

$$\Psi_+^{(n)}(z) = \frac{1}{M_1}\left(\Phi_+^{(n)}(z) - \sum_{k=1}^{n-1} \Psi_+^{(k)}(z)M_{n-k+1}\right), \quad (4.128)$$

$$\Psi_-^{(n)}(z) = \frac{1}{J_1}\left(\Phi_-^{(n)}(z) - \sum_{k=1}^{n-1} \Psi_-^{(k)}(z) J_{n-k+1}\right). \qquad (4.129)$$

From the definitions of the moments M_k and J_k we have

$$M_k = \oint_C \frac{d\omega}{2\pi i} V'(\omega)\Phi_+^{(k)}(\omega), \quad J_k = \oint_C \frac{d\omega}{2\pi i} V'(\omega)\Phi_-^{(k)}(\omega). \qquad (4.130)$$

and it is easy to show that

$$\hat{K}\Psi_+^{(n)}(z) = \frac{1}{(z-c_+)^n}, \quad \hat{K}\Psi_-^{(n)}(z) = \frac{1}{(z-c_-)^n}. \qquad (4.131)$$

Let us now consider the *ansatz*

$$w_h(z) = \sum_{n=1}^{3h-1}\left(A_h^{(n)}\Psi_+^{(n)}(z) + B_h^{(n)}\Psi_-^{(n)}(z)\right), \quad h>0, \qquad (4.132)$$

where $A_h^{(n)}$ and $B_h^{(n)}$ are independent of z and depend on g only through the moments M_k, J_k and d. Alternatively, this *ansatz* can be written as

$$w_h(z) = \frac{1}{\sqrt{(z-c_+)(z-c_-)}} \sum_{n=1}^{3h-1}\left(\frac{a_h^{(n)}}{(z-c_+)^n} + \frac{b_h^{(n)}}{(z-c_-)^n}\right), \qquad (4.133)$$

where $a_h^{(n)}$ and $b_h^{(n)}$ are independent of z and depend on g in a similar way to $A_h^{(n)}$ and $B_h^{(n)}$.

We now justify the claim that the *ansatz* (4.132), or, equivalently, the *ansatz* (4.133), provides a solution to the generalized loop equation. First consider the right-hand side of Eq. (4.125). Acting with $\hat{K}$ on both sides of Eq. (4.132) gives, according to Eq. (4.131),

$$\hat{K}w_h(z) = \sum_{n=1}^{3h-1}\left(\frac{A_h^{(n)}}{(z-c_+)^n} + \frac{B_h^{(n)}}{(z-c_-)^n}\right). \qquad (4.134)$$

With the *ansatz* (4.133) it is clear that the product $w_{h'}w_{h-h'}$ can be written as a linear combination of products of the form

$$\frac{1}{(z-c_+)^n}\frac{1}{(z-c_-)^m} = \sum_{i=1}^{n} \frac{\alpha_i(d)}{(z-c_+)^i} + \sum_{j=1}^{m} \frac{\beta_j(d)}{(z-c_-)^j}, \qquad (4.135)$$

where $d = c_+ - c_-$ (see Eq. (4.119)) and the coefficients α_i, β_j are integer multiples of powers of $1/d$. It follows that $w_{h'}w_{h-h'}$ is of the form indicated on the right-hand side of Eq. (4.134). Next we discuss the action of the loop insertion operator on the term $w_{h-1}(z)$, assumed to be given by Eq. (4.132). We have already seen in Eqs. (4.110)–(4.112) that we can write the loop insertion operator as

$$\frac{d}{dV(z)} = \frac{\partial}{\partial V(z)} + \Psi_+^{(1)}(z)\frac{\partial}{\partial c_+} + \Psi_-^{(1)}(z)\frac{\partial}{\partial c_-}, \qquad (4.136)$$

where $\partial/\partial V(z)$ only acts on M_k and J_k, and its action is given by Eq. (4.113) and Eq. (4.114). It is now easy to check that the loop insertion operator produces negative powers of $(z - c_+)$ and $(z - c_-)$ when applied to $\Psi_+^{(n)}(z)$ and $\Psi_-^{(n)}(z)$, the lowest power being $-(n+3)$ in both cases. This opens up the possibility that one can restrict the dependence of the $A_h^{(n)}$ and $B_h^{(n)}$ on M_k and J_k in such a way that the highest value of n in the *ansatz* (4.132) grows only by three when h is increased by one (see Eq. (4.141) for the explicit form of $A_h^{(n)}$ and $B_h^{(n)}$). Below we show that $n = 1, 2$ for $h = 1$ and in this way one can prove by induction that the range of the summation index n can be restricted to be from 1 to $3h - 1$ in Eq. (4.132). We omit the trivial, but cumbersome, details of these arguments. We have thus made plausible that the terms on the right-hand side of Eq. (4.125) can be matched with a suitable choice of coefficients $A_h^{(n)}$ and $B_h^{(n)}$ and hence that the *ansatz* (4.132) is valid.

It is of interest to uncover the structure of the coefficients $A_h^{(n)}$ and $B_h^{(n)}$ in Eq. (4.132). For $h = 1$ we have

$$\sum_{n=1}^{2} \left(\frac{A_1^{(n)}}{(z - c_+)^n} + \frac{B_1^{(n)}}{(z - c_-)^n} \right) = w_0(z, z). \qquad (4.137)$$

The function $w_0(z, z)$ is given by Eq. (4.117):

$$w_0(z, z) = \frac{d^2}{16(z - c_+)^2(z - c_-)^2}, \qquad (4.138)$$

and we conclude that

$$A_1^{(2)} = B_1^{(2)} = \frac{1}{16}, \quad A_1^{(1)} = -B_1^{(1)} = -\frac{1}{8d}, \qquad (4.139)$$

so

$$w_1(g, z) = \frac{1}{16\sqrt{(z - c_+)(z - c_-)}} \left\{ \frac{1}{M_1(z - c_+)^2} + \frac{1}{J_1(z - c_-)^2} \right. \qquad (4.140)$$
$$\left. - \left(\frac{M_2}{M_1} + \frac{2}{d} \right) \frac{1}{M_1(z - c_+)} - \left(\frac{J_2}{J_1} - \frac{2}{d} \right) \frac{1}{J_1(z - c_-)} \right\}.$$

If the middle term in the loop insertion operator (4.136) acts on M_k it increases the index k by one and introduces an additional negative power of M_1, while the two other terms cannot increase the index of M_k, but they can produce lower values of k as well as negative powers of M_1 and d. Similar remarks apply to J_k and J_1 and the last term in Eq. (4.136). In this way it can be shown by induction that the coefficient $A_h^{(n)}$ can be written in the form

$$A_h^{(n)} = \sum_{\alpha_i, \beta_j, \alpha, \beta} C_h^{(n)}(\alpha_1, \ldots, \alpha_k; \beta_1, \ldots, \beta_l; \alpha, \beta, \gamma) \frac{M_{\alpha_1} \cdots M_{\alpha_k} J_{\beta_1} \cdots J_{\beta_l}}{M_1^\alpha J_1^\beta d^\gamma},$$
$$(4.141)$$

where the coefficients $C_h^{(n)}(\alpha_1,\ldots,\alpha_k;\beta_1,\ldots,\beta_l;\alpha,\beta,\gamma)$ are rational numbers *independent* of the polynomial V'. The indices α_i,β_j, $i=1,\ldots,k$, $j=1,\ldots,l$, are integers in the interval $[2,3h-n]$ which, together with the integers α,β,γ, satisfy the following constraints: $l+k \le 3h-n-1$, $k \le \alpha$, $l \le \beta$, $h-1 \le \gamma \le 4h-2-n$ and $\alpha+\beta = k+l+2h-2$. Furthermore,

$$\gamma = 4h-2-n-\sum_{i=1}^{k}(\alpha_i-1)-\sum_{j=1}^{l}(\beta_j-1).$$

There is an expression for $B_h^{(n)}$ similar to Eq. (4.141). In fact, one can obtain the formula for $B_h^{(n)}$ from that for $A_h^{(n)}$ by interchanging the moments M_k and J_k and reversing the sign of d. Note that the first coefficients satisfy

$$A_h^{(1)}(\{M_k\},\{J_k\},d) = -B_h^{(1)}(\{M_k\},\{J_k\},d) \tag{4.142}$$

as a consequence of the fact that the functions $\Psi_{\pm}^{(1)}(z)$ only appear on the right-hand side of the loop equation (4.125) as a result of the decomposition

$$\frac{1}{(z-c_+)(z-c_-)} = \frac{1}{d}\Big(\frac{1}{z-c_+}-\frac{1}{z-c_-}\Big).$$

The difference in sign between the two terms inside the bracket on the right-hand side of this equation implies Eq. (4.142).

Finally, we can use the loop insertion operator to construct

$$w_h(z_1,\ldots,z_b) = \frac{d}{dV(z_b)}\cdots\frac{d}{dV(z_2)}w_h(z_1), \tag{4.143}$$

which is seen to depend only on M_k,J_k for $k \le 3h-2+b$.

4.3.3   The generating function $\mathscr{Z}_h$ for closed triangulations

For $h>0$ there exists a function $\mathscr{Z}_h$ of M_k,J_k and d which satisfies

$$w_h(z) = \frac{d}{dV(z)}\mathscr{Z}_h. \tag{4.144}$$

Let us first prove that w_h can be written as a derivative with respect to $d/dV(z)$. We have

$$\frac{dw_h(z_1)}{dV(z_2)} = \frac{dw_h(z_2)}{dV(z_1)}. \tag{4.145}$$

This equation simply expresses that $w_h(z_1,z_2) = w_h(z_2,z_1)$, which is satisfied as a consequence of the combinatorial interpretation of $w_h(z_1,z_2)$. Eq. (4.145) can be viewed as an integrability condition with respect to the derivative $d/dV(z)$ in the same way as $\partial_y f_1(x,y) = \partial_x f_2(x,y)$ implies the existence of a function $F(x,y)$ such that $\partial_x F = f_1$ and $\partial_y F = f_2$. We conclude that there exists a function $\mathscr{Z}_h$ such that Eq. (4.144) is satisfied.

Next, let us prove that $\mathscr{Z}_h$ is only a function of M_k, J_k and d for $h > 0$, a fact which is not a direct consequence of Eq. (4.144) and not true for $h = 0$. The coefficients $A_h^{(n)}$ and $B_h^{(n)}$ in the expansion of $w_h(z)$ (see Eq. (4.132)) are functions of the moments M_k and J_k and d, as shown in Eq. (4.141). Using Eqs. (4.106)–(4.116) allows us to conclude that the functions

$$\tilde{\Phi}_+^{(n)}(z) = \Phi_+^{(n)}(z) - \frac{M_n}{M_1}\Phi_+^{(1)}(z)$$

$$\tilde{\Phi}_-^{(n)}(z) = \Phi_-^{(n)}(z) - \frac{J_n}{J_1}\Phi_-^{(1)}(z)$$

for $n \geq 1$ can be expressed as

$$\left(n+\frac{1}{2}\right)\tilde{\Phi}_+^{(n+1)}(z) = -\frac{dM_n}{dV(z)} + \sum_{k=0}^{n-1}\frac{\tilde{\Phi}_+^{(n-k)}(z) + M_{n-k}\frac{d(d)}{dV(z)}}{2(-d)^{k+1}},$$

$$\left(n+\frac{1}{2}\right)\tilde{\Phi}_-^{(n+1)}(z) = -\frac{dJ_n}{dV(z)} + \sum_{k=0}^{n-1}\frac{\tilde{\Phi}_-^{(n-k)}(z) + J_{n-k}\frac{d(-d)}{dV(z)}}{2d^{k+1}},$$

where $\tilde{\Phi}_\pm^{(1)} = 0$ and $\tilde{\Phi}_\pm^{(0)} \equiv 0$.

From the definitions (4.128) and (4.129) it follows that $\Psi_\pm^{(n)}(z)$ can be expressed in terms of the functions $\tilde{\Phi}_\pm^{(m)}(z)$, $m = 2,\ldots,n$, for $n > 1$. Hence, the functions $\Psi_\pm^{(n)}(z)$ depend only on M_k, J_k and d and derivatives with respect to $d/dV(z)$ of these quantities for $n > 1$. For $n = 1$ we obtain from Eq. (4.142) the term

$$A_h^{(1)}\Psi_+^{(1)}(z) + B_h^{(1)}\Psi_-^{(1)}(z) = A_h^{(1)}\frac{d}{dV(z)}d. \qquad (4.146)$$

From these remarks it follows that $w_h(z)$ can be written as a sum of functions of the form given by Eq. (4.141) times derivatives $d/dV(z)$ of M_k, J_k and d. Since we have argued above that a function $\mathscr{Z}_h$ exists such that $w_h(z) = d\mathscr{Z}_h/dV(z)$, it becomes a matter of inspection to arrange $\mathscr{Z}_h$ as a function of the moments $\{M_k\}$, $\{J_k\}$ and d. As an example we can write Eq. (4.140) as

$$w_1(z) = \frac{1}{16}\left(\frac{1}{M_1}\tilde{\Phi}_+^{(2)}(z) + \frac{1}{J_1}\tilde{\Phi}_-^{(2)}(z) - \frac{2}{d}\frac{d}{dV(z)}d\right)$$

$$= -\frac{1}{24M_1}\frac{dM_1}{dV(z)} - \frac{1}{24J_1}\frac{dJ_1}{dV(z)} - \frac{1}{6d}\frac{d}{dV(z)}d, \qquad (4.147)$$

which implies that

$$\mathscr{Z}_1 = \frac{1}{24}\log\frac{1}{M_1 J_1 d^4}. \qquad (4.148)$$

Similarly, for $h = 2$, some algebra leads to

$$
\begin{aligned}
\mathscr{L}_2 &= -\left(\frac{181}{480J_1{}^2} + \frac{181}{480M_1{}^2} + \frac{5}{16J_1M_1}\right)\frac{1}{d^4} \\
&\quad + \left(\frac{181J_2}{480J_1{}^3} - \frac{181M_2}{480M_1{}^3} + \frac{3J_2}{64J_1{}^2M_1} - \frac{3M_2}{64J_1M_1{}^2}\right)\frac{1}{d^3} \\
&\quad - \left(\frac{11J_2{}^2}{40J_1{}^4} + \frac{11M_2{}^2}{40M_1{}^4} - \frac{43M_3}{192M_1{}^3} - \frac{43J_3}{192J_1{}^3} - \frac{J_2M_2}{64J_1{}^2M_1{}^2}\right)\frac{1}{d^2} \\
&\quad + \left(\frac{21J_2{}^3}{160J_1{}^5} - \frac{21M_2{}^3}{160M_1{}^5} - \frac{29J_2J_3}{128J_1{}^4d} + \frac{29M_2M_3}{128M_1{}^4}\right. \\
&\quad \left. + \frac{35J_4}{384J_1{}^3} - \frac{35M_4}{384M_1{}^3}\right)\frac{1}{d}.
\end{aligned}
\tag{4.149}
$$

$\mathscr{L}_h$ is of course only detemined up to a constant by Eq. (4.144). While the expression for $\mathscr{L}_2$ might look complicated it should be emphasized that it is valid for *any* polynomial $V'(z)$.

The power series expansion of $\mathscr{L}_h$ in $\underline{g}$ can be interpreted as the generating function for *closed* triangulations of genus h with a given number of *j*-sided polygons, $j = 1, 2, \ldots, n$. We summarize the above results as follows [20].

Theorem 4.3 *The generating function $\mathscr{L}_h$, $h \geq 1$, depends only on the $2(3h-2)$ parameters M_k, J_k, $k = 1, 2, \ldots, 3h-2$ and $d = c_+ - c_-$. The function $\mathscr{L}_h$ is given by Eq. (4.148) for $h = 1$ and*

$$
\mathscr{L}_h = \sum_{\alpha_i,\beta_j,\alpha,\beta} \langle \alpha_1,\ldots,\alpha_k; \beta_1,\ldots,\beta_l | \alpha,\beta,\gamma \rangle_h \frac{M_{\alpha_1}\cdots M_{\alpha_k} J_{\beta_1}\cdots J_{\beta_l}}{M_1^{\alpha} J_1^{\beta} d^{\gamma}}
\tag{4.150}
$$

for $h > 1$. The coefficients $\langle \alpha_1,\ldots,\alpha_k; \beta_1,\ldots,\beta_l | \alpha,\beta,\gamma \rangle_h$ are rational numbers independent of the particular form of the polynomial $V'(z)$. The integer indices $k, l, \alpha_i, \beta_j, \alpha, \beta$ and γ in Eq. (4.150) satisfy the relations

$$
0 \leq k \leq \alpha, \quad 0 \leq l \leq \beta, \quad \alpha + \beta = 2h - 2 + k + l,
\tag{4.151}
$$

and

$$
\sum_{i=1}^{k}(\alpha_i - 1) + \sum_{j=1}^{l}(\beta_j - 1) + \gamma = 4h - 4.
\tag{4.152}
$$

In addition the generating function $w_h(z_1,\ldots,z_b)$ can be written as

$$
w_h(z_1,\ldots,z_b) = \sum_{\alpha_i,\beta_j,\alpha,\beta,n_i,m_i} C_h(\alpha_1,\ldots,\alpha_k; \beta_1,\ldots,\beta_l; \alpha,\beta,\gamma; n_i,m_i)
$$

$$
\frac{M_{\alpha_1}\cdots M_{\alpha_k} J_{\beta_1}\cdots J_{\beta_l}}{M_1^{\alpha} J_1^{\beta} d^{\gamma}} \prod_{i=1}^{b} \frac{1}{(z_i - c_+)^{n_i+\frac{1}{2}}(z_i - c_-)^{m_i+\frac{1}{2}}},
\tag{4.153}
$$

where the coefficients $C_h(\alpha_1, \ldots, \alpha_k; \beta_1, \ldots, \beta_l; \alpha, \beta, \gamma; n_i, m_i)$ are related to the coefficients which appear in Eq. (4.141). The coefficients are again rational numbers independent of $V'(z)$. Relations similar to Eq. (4.151) and Eq. (4.152) can be derived for the indices α_i, β_j which enter Eq. (4.153). In particular, we note [20] that *the generating functions* $w_h(z_1, \ldots, z_b)$, $h \geq 1$, *depend on the parameters* M_1, J_1, *as well as on the* $2(3h - 3 + b)$ *parameters* M_k, J_k, $k \in [2, 3h - 2 + b]$. The complex dimension of the moduli space of Riemann surfaces of genus h and with b punctures is $3h - 3 + b$ and in Section 4.8.4 it will become clear why the dimension of this space coincides with the number of different M_ks which can appear in the numerator in Eq. (4.153).

4.3.4   *The number of triangulations of genus h*

Let us now reintroduce the coupling constants $\underline{g}$ into our notation and write $\underline{g} = g\underline{t}$. As in the case of genus 0 the singularity of $w_h(g, z_1, \ldots, z_b)$ in g (with fixed $\underline{t}$) is determined by the equation $M_1(g, c_-(g), c_+(g)) = 0$, while the singular behaviour in z_i occurs as $z_i \to c_+(g)$. In order to count the number of triangulations with boundary components of fixed length we identify the most singular term in Δg, as in the case of $h = 0$, which is given by the highest inverse power of M_1 in Eq. (4.153). This power can be determined from Eq. (4.141) and the explicit form of the functions $\Psi_+^{(n)}(z)$, but it might be more instructive to use the loop equation. It is easy to see from the structure of the loop equation and the loop insertion operator that the most singular term in $w_h(g, z)$ is the one involving the powers $(z - c_+)^{-3/2}(z - c_-)^{-1/2}$. On the right-hand side of the loop equation (4.125) these powers are transformed to $(z - c_+)^{-3}(z - c_-)^{-1}$. From Eqs. (4.127)–(4.132) it follows that consistency of the loop equation for the lowest powers $(z - c_+)^{-3/2}$ requires that the operator $\hat{K}$ multiplies these singular terms in $w_h(g, z)$ with M_1^3. Finally, the loop insertion operation will increase the singularity of $w_{h-1}(g, z)$ by a factor M_1^{-2}, as already observed in the genus zero case. The conclusion is that the singular behaviour is increased by a factor M_1^{-5} in going from $w_{h-1}(g, z)$ to $w_h(g, z)$. For $h = 1$ we have seen by an explicit calculation that $w_1(g, z) \sim M_1^{-2}$ and it follows that $w_h(g, z) \sim M_1^{-5(h-1)-2}$.

We can construct the multi-loop function $w_h(g, z_1, \ldots, z_b)$ by applying the loop insertion operator $b - 1$ times to $w_h(g, z)$. Each time, the singularity is increased by a factor M_1^{-2}. Since $M_1 \sim \sqrt{\Delta g}$ it follows that

$$w_h(g, z_1, \ldots, z_b) \sim \left(\frac{1}{\sqrt{g_0 - g}} \right)^{5(h-1)+2b}, \qquad (4.154)$$

as $g \to g_0$. This implies that the generating function $w_h(g, l_1, \ldots, l_b)$ for the number of triangulations, $w^{(h)}_{k,l_1,...,l_b}$, of genus h, constructed from k triangles with b boundary components of length $l_1,...,l_b$, has a singularity as $g \to g_0$ that is independent of the length of the boundary components and is given by

$$w_h(g, l_1, ..., l_b) \sim \left(\frac{1}{\sqrt{g_0 - g}} \right)^{5(h-1)+2b}. \tag{4.155}$$

Finally, we obtain from Eq. (4.155) the asymptotic behaviour of $w^{(h)}_{k,l_1,...,l_b}$ as $k \to \infty$:

$$w^{(h)}_{k,l_1,...,l_b} \sim \left(\frac{1}{g_0} \right)^k k^{\frac{5}{2}(h-1)+b-1}, \tag{4.156}$$

a result first derived for arbitrary h in [76].

4.4   The continuum limit

We now show how continuum physics is related to the asymptotic behaviour of $w_h(k, l_1, ..., l_b)$ for $k \to \infty$ *and* $l_1, ..., l_b \to \infty$ in a specific way, and we use the results for the generating functions $w_h(\underline{g}, z_1, ..., z_b)$ derived in the previous sections to study this limit.

4.4.1   *Renormalization of the cosmological constant*

Before discussing details it is useful to clarify how we expect the continuum wave functionals $W(\Lambda, Z_1, ..., Z_b)$ to renormalize. Since the cosmological constants Λ and Z_i have dimensions $1/a^2$ and $1/a$, respectively, a being the length of the lattice cutoff, it is natural to expect that they are subject to an additive renormalization

$$\Lambda_0 = \frac{\mu_0}{a^2} + \Lambda, \quad Z_{i,0} = \frac{\lambda_{i,0}}{a} + Z_i, \tag{4.157}$$

where Λ_0 and $Z_{i,0}$ are the *bare* cosmological coupling constants; see Eq. (4.6). Since our regularization is represented in terms of discretized two-dimensional manifolds, the bare cosmological constants should be related to the dimensionless coupling constants μ, λ_i by

$$\Lambda_0 = \frac{\mu}{a^2}, \quad Z_{i,0} = \frac{\lambda_i}{a}, \tag{4.158}$$

so that Eq. (4.157) can be written

$$\mu - \mu_0 = a^2 \Lambda, \quad \lambda_i - \lambda_{i,0} = a Z_i. \tag{4.159}$$

In the following we assume for simplicity that all the $\lambda_{i,0}$s are equal to λ_0. As in the case of random walks and random embedded surfaces, discussed in previous chapters, we identify the constants μ_0 and λ_0 with the critical couplings g_0 and $c_+(g_0)$ via the relations

$$\frac{1}{g_0} = e^{\mu_0}, \qquad c_+(g_0) = e^{\lambda_0}. \qquad (4.160)$$

Recalling the relation (4.45) between μ, g and z, λ, it follows that the $a \to 0$ limit of the functions $w(\mu, \lambda_1, \ldots, \lambda_b)$ is determined by their singular behaviour at g_0. The renormalizations (4.157) have the effect of cancelling the exponential entropy factor for the triangulations, see Eq. (4.156). Note that since we have the same exponential factors for all genera, we expect the renormalization of the cosmological constants to be independent of genus.

We begin by studying the continuum limit for planar surfaces. Afterwards we take higher genera into account, reintroduce the gravitational coupling constant κ and discuss its renormalization. This will lead us to the so-called *double scaling limit*.

4.4.2 *Continuum results for genus zero*

We are interested in a limit of the discretized models where the length a of the links goes to zero while the number of triangles k and the lengths of the boundary components l_i go to infinity in such a way that

$$V = ka^2 \quad \text{and} \quad L_i = l_i a \qquad (4.161)$$

remain finite. The asymptotic behaviour of $w_{k,l_1,\ldots,l_b}$ is given by Eq. (4.102) if the $l_1, \ldots, l_b$ remain bounded. In this case the leading term is of the form

$$e^{\mu_0 V/a^2} (V/a^2)^\beta,$$

where β is a critical exponent. If the boundary lengths $l_1, \ldots, l_b$ diverge according to Eq. (4.161) we expect a corresponding factor

$$e^{\lambda_0 L_i/a} (L_i/a)^\alpha,$$

where α is another critical exponent. This form of the entropy was encountered for branched polymers in Eq. (4.55). We can therefore express the expected asymptotic behaviour of the coefficients $w_{k,l_1,\ldots,l_b}$ as

$$w_{k,l_1,\ldots,l_b} \sim e^{\frac{\mu_0}{a^2} V} e^{\lambda_0 \sum_i l_i} a^{-\alpha b - 2\beta} W(V, L_1, \ldots, L_b), \qquad (4.162)$$

as $a \to 0$, with V and L_i defined by Eq. (4.161) fixed. The factor $a^{-\alpha b - 2\beta}$ may be thought of as a wave-function renormalization. As we saw in Chapter 2 such factors are present even for the free relativistic particle in $\mathbb{R}^d$, which can be viewed as one-dimensional gravity coupled to d scalar fields. Thus, we expect a wave-function renormalization for each boundary loop and an additional renormalization associated with the volume of surfaces. It is natural to assume the linear b dependence indicated in Eq. (4.162). The wave-function renormalization in Eq. (4.162) has the same form as the wave-function renormalization encounted for the b-point function of branched polymers studied in Section 2.5, see Eq. (2.214).

From Eq. (4.162) we deduce that the scaling behaviour of the discretized wave functional $w(\mu, \lambda_1, \ldots, \lambda_b)$ is given by

$$
\begin{aligned}
w(\mu, \lambda_1, \ldots, \lambda_b) &= \sum_{k, l_1, \ldots, l_b} e^{-\mu k}\, e^{-\sum_i \lambda_i l_i}\, w_{k, l_1, \ldots, l_b} \\
&\sim \frac{1}{a^{\alpha b + 2\beta}} \sum_{k, l_1, \ldots, l_b} e^{-(\mu - \mu_0)k}\, e^{-\sum_i (\lambda_i - \lambda_0)l_i}\, W(V, L_1, \ldots, L_b) \\
&\sim \frac{1}{a^{(\alpha+1)b + (2\beta+2)}}\, W(\Lambda, Z_1, \ldots, Z_b),
\end{aligned}
\tag{4.163}
$$

where we have used the relation

$$
W(\Lambda, Z_1, \ldots, Z_b) = \int_0^\infty dV \prod_{i=1}^b dL_i\, e^{-\Lambda V - \sum_i Z_i L_i}\, W(V, L_1, \ldots, L_b).
$$

Our next goal is to show that the expression (4.95) which we have found for $w(g, z_1, \ldots, z_b)$ allows us to take a limit as suggested by Eqs. (4.159) and (4.163). In terms of the variables g, z_i we have

$$
g = g_0(1 - \Lambda a^2), \qquad z_i = z_0(1 + a Z_i),
\tag{4.164}
$$

where we have introduced the notation

$$
z_0 = c_+(g_0) = e^{\lambda_0}
\tag{4.165}
$$

for the critical value $c_+(g_0)$ of z corresponding to the largest allowed value of g. From Eq. (4.164) it follows that $\Delta g = \Lambda a^2$ and applying Eq. (4.107) we have

$$
M_1(g) = -\frac{3}{2} M_2(g_0)\Delta c_+ + o(\Delta c_+),
\tag{4.166}
$$

$$
\Delta c_+ = \tilde{\alpha}\sqrt{\frac{\Delta g}{g_0}} = \tilde{\alpha}\, a\sqrt{\Lambda},
\tag{4.167}
$$

where $\tilde{\alpha}$ is a constant which depends on g_0.

Inserting Eq. (4.164) and Eq. (4.166) in the expression (4.95) for the

b-loop function[*] we obtain, for $b \geq 3$,

$$w(g, z_1, \ldots, z_b) \sim \frac{1}{a^{7b/2-5}} \left(-\frac{d}{d\Lambda}\right)^{b-3} \left[\frac{1}{\sqrt{\Lambda}} \prod_{i=1}^{b} \frac{1}{(Z_i + \sqrt{\Lambda})^{3/2}}\right], \quad (4.168)$$

after rescaling the variables Z_i suitably. In other words, for $b \geq 3$, we can define the continuum functions $W(\Lambda, Z_1, \ldots, Z_b)$ as anticipated in Eq. (4.163) by

$$W(\Lambda, Z_1, \ldots, Z_b) = \lim_{a \to 0} a^{7b/2-5} w(g, z_1, \ldots, z_b) \qquad (4.169)$$

$$= C\left(-\frac{d}{d\Lambda}\right)^{b-3} \left[\frac{1}{\sqrt{\Lambda}} \prod_{i=1}^{b} \frac{1}{(Z_i + \sqrt{\Lambda})^{3/2}}\right],$$

where C is a constant. Calculating the inverse Laplace transforms of the functions $W(\Lambda, Z_1, \ldots, Z_b)$ with respect to Z_i and/or Λ yields (up to constant multiples)

$$W(\Lambda, L_1, \ldots, L_b) = \left(-\frac{d}{d\Lambda}\right)^{b-3} \left[\frac{1}{\sqrt{\Lambda}} \prod_{i=1}^{b} \sqrt{L_i} e^{-\sqrt{\Lambda} L_i}\right],$$

$$W(V, L_1, \ldots, L_b) = V^{b-7/2} \sqrt{L_1 \cdots L_b} \, e^{-(L_1 + \cdots + L_b)^2/(4V)}.$$

The continuum limits of the one- and two-loop functions are obtained similarly by inserting Eq. (4.164) and Eq. (4.166) into Eq. (4.83) and Eq. (4.93):

$$W(\Lambda, Z_1, Z_2) = \frac{1}{(Z_1 - Z_2)^2} \left(\frac{\frac{1}{2}(Z_1 + Z_2) + \sqrt{\Lambda}}{\sqrt{(Z_1 + \sqrt{\Lambda})(Z_2 + \sqrt{\Lambda})}} - 1\right), \quad (4.170)$$

$$W(\Lambda, Z) = (Z - \frac{1}{2}\sqrt{\Lambda})\sqrt{Z + \sqrt{\Lambda}}, \qquad (4.171)$$

where in Eq. (4.171) we have dropped the analytic part of $w(\mu, z)$. Likewise, it is again possible to calculate the inverse Laplace transforms of these functions, e.g.

$$W(\Lambda, L) = L^{-5/2}(1 + \sqrt{\Lambda} L)e^{-\sqrt{\Lambda} L}, \qquad (4.172)$$

up to a constant multiple. This completes the calculation of the Hartle–Hawking wave functionals in the genus 0 sector of two-dimensional quantum gravity.

---

[*] Strictly speaking, Eq. (4.95) is only valid for the restricted class of graphs considered in Section 4.2.4, but a closer study of the loop insertion operator in the scaling limit shows that the results in this limit are the same for general polynomials and arbitrary planar graphs.

We remark that if we multiply the right-hand side of Eq. (4.169) by $Z_b^{3/2}$ and take the limit $Z_b \to \infty$, then we obtain the derivative of $W(\Lambda, Z_1, \ldots, Z_{b-1})$ with respect to Λ. The reason is that in the limit $Z_b \to \infty$ only triangulations with boundary loops of length $L_b = 0$ survive. It follows that this boundary component is a "puncture". This puncture can be anywhere on the surface and thus has an entropy factor proportional to the volume V. A factor of V is also obtained by differentiating with respect to the cosmological constant Λ. It follows that we can calculate the one- and two-loop functions given by Eqs. (4.170) and (4.171) by integration from the three-loop function which is given by Eq. (4.169).

We can apply these remarks to the partition function $\mathscr{Z}(\mu)$ itself, i.e. the case where $b = 0$. Recalling that $\mathscr{Z}(k)$ denotes the number of closed planar triangulations with k triangles (see Eq. (4.43)), we have, as in Eq. (4.163),

$$
\begin{aligned}
\mathscr{Z}(\mu) &= \sum_{k=1}^{\infty} e^{-\mu k} \mathscr{Z}(k) \sim \sum_{k=1}^{\infty} e^{-(\mu-\mu_0)k} k^{\gamma-3} (1 + O(1/k)) \\
&\sim (\mu - \mu_0)^{2-\gamma} = (\mu - \mu_0)^{5/2} \sim (a^2 \Lambda)^{5/2},
\end{aligned}
\tag{4.173}
$$

where we have used Eq. (4.102) in the case where $b = 0$, and Eq. (4.164). The last line in Eq. (4.173) is only correct up to less singular terms in $\mu - \mu_0$, but we see that it agrees with the formal expression (4.168) in the case where $b = 0$ if we integrate three times with respect to Λ according to the recipe mentioned in the previous paragraph. The exponent γ is called either the *entropy exponent* or the *susceptibility exponent*, since it has an interpretation similar to the susceptibility exponents discussed in previous chapters. Defining the susceptibility χ as the second derivative of the partition function with respect to μ we have

$$
\chi(\mu) = \mathscr{Z}''(\mu) \sim \frac{1}{(\mu - \mu_0)^{\gamma}}.
\tag{4.174}
$$

Eq. (4.173) shows that $\gamma = -1/2$.

4.4.3 Continuum results for higher-genus surfaces

The results derived for the higher-genus generating functions allow us to deduce the singular behaviour of $w_h(z)$ as $z - c_+(g) \to 0$. We use the same scaling *ansatz* as in the previous subsection, i.e.

$$
z_i - c_+(g) \sim a(Z_i + \sqrt{\Lambda}) \quad \text{and} \quad M_1 \sim a\sqrt{\Lambda},
$$

as $a \to 0$. In discussing the scaling properties of the Hartle–Hawking wave functionals we saw that the loop insertion operator increases the singularity at $a = 0$ by a factor $a^{-7/2}$. The operator $\hat{K}(z)$ decreases the singularity of $w_h(z)$ by a factor $a^{3/2}$, as is clear from Eq. (4.131) since

$\Psi_+^{(n)}(z)$ behaves as $a^{-n-3/2}$ for a small. This implies that the singularity is increased by a factor a^{-5} for each time the genus h is increased by one, and since the singular behaviour of $w_1(g,z)$ is $a^{-7/2}$, see Eq. (4.168), we find that

$$w_h(g,z) \sim a^{3/2-5h} W_h(\Lambda, Z), \qquad (4.175)$$

for $h > 0$. For example, the genus one continuum limit can be constructed explicitly from Eq. (4.140) and we find, using Eq. (4.107), that

$$W_1(\Lambda, Z) \sim \frac{1}{\sqrt{\Lambda}(Z + \sqrt{\Lambda})^{5/2}} + \frac{2}{3\Lambda(Z + \sqrt{\Lambda})^{3/2}}. \qquad (4.176)$$

By taking inverse Laplace transforms it is easy to calculate $W_1(\Lambda, L)$ and $W_1(V, L)$.

Applying the loop insertion operator to Eq. (4.140) yields

$$w_h(g, z_1, \ldots, z_b) \sim \frac{1}{a^{7b/2+(5h-5)}} W_h(\Lambda, Z_1, \ldots, Z_b). \qquad (4.177)$$

In particular, taking $b = 0$ leads to an expression for the singular part of the partition function:

$$\mathscr{Z}_h(g) \sim \frac{\tau_h}{(a^2\Lambda)^{5(h-1)/2}}, \qquad (4.178)$$

where the constants τ_h are determined from Eq. (4.150). More precisely, the term which dominates the scaling limit of Eq. (4.150) is

$$\langle \alpha_1 = 2, \ldots, \alpha_{3h-3} = 2; \, | \, 5h - 5, 0, h - 1 \rangle_h \frac{M_2^{3h-3}}{M_1^{5h-5} d^{h-1}}. \qquad (4.179)$$

Up to a rescaling of the cosmological constant we can identify the coefficient τ_h in Eq. (4.178) with the rational number $\langle \cdot \rangle_h$ in Eq. (4.179). Eq. (4.178) shows that the susceptibility exponent for closed surfaces of genus h is

$$\gamma_h = 2 + \frac{5}{2}(h - 1). \qquad (4.180)$$

From Eq. (4.177) it follows that the wave-function renormalization of the genus h multi-loop functions is identical, up to a factor a^{-5h}, to the one found in the genus zero case, given by Eq. (4.169). We now show how this factor can formally be absorbed into a renormalization of the gravitational coupling constant κ.

In analogy with Eq. (4.120) it is natural to introduce the generating function

$$\mathscr{Z}(g) = N^2 \sum_{h=0}^{\infty} \frac{1}{N^{2h}} \mathscr{Z}_h(g) \qquad (4.181)$$

for closed triangulations of arbitrary topology. We have

$$w(z) = \frac{1}{N^2} \frac{d\mathscr{Z}}{dV(z)}. \tag{4.182}$$

We remark that while for $h \geq 1$ it is straightforward to construct $\mathscr{Z}_h$ from w_h, which is uniquely determined by the loop equations, we have not derived an expression for $\mathscr{Z}_0$. Like $w_0(z)$ it depends explicitly on the coupling constants g, and the trivial term $1/z$ in $w_0(z)$ cannot be represented by the right-hand side of Eq. (4.182). In the scaling limit, where these non-universal terms in $w_0(z)$ play no role, we can use Eq. (4.178) and write

$$\mathscr{Z}(g) = \sum_{h=0}^{\infty} \tau_h \left(a^{5/2} N \right)^{2-2h} \Lambda^{5(1-h)/2}. \tag{4.183}$$

We want to compare $\mathscr{Z}(g)$ in Eq. (4.183) to the full partition function for two-dimensional gravity, including surfaces of arbitrary topology. Reintroducing the curvature term in the two-dimensional gravitational action, the discretized form of the continuum action (4.2) is

$$S_T(\mu, \kappa) = \mu N_T - \frac{\chi}{\kappa}, \tag{4.184}$$

where $\chi = 2 - 2h$ is the Euler characteristic. The partition function is then given by

$$\mathscr{Z}(\mu, \kappa) = \sum_{h=0}^{\infty} \sum_{T \in \mathscr{T}(h)} \frac{1}{C_T} e^{-S_T(\mu, \kappa)}. \tag{4.185}$$

In the scaling limit we obtain from Eq. (4.178)

$$\mathscr{Z}(\mu, \kappa) = \sum_{h=0}^{\infty} \tau_h e^{\frac{2-2h}{\kappa}} a^{5(1-h)} \Lambda^{\frac{5(1-h)}{2}}. \tag{4.186}$$

Comparing this equation with Eq. (4.183) it can be seen that the indeterminate N in Eq. (4.183) can be identified with $e^{1/\kappa}$, i.e.

$$\frac{1}{\kappa} = \log N. \tag{4.187}$$

The factor a^{-5} present for each genus can be absorbed in a renormalization of the gravitational coupling constant

$$\frac{1}{\kappa_{ren}} = \frac{1}{\kappa} - \frac{5}{2} \log \frac{1}{a}, \tag{4.188}$$

where κ_{ren} denotes the renormalized gravitational coupling. If we express Eq. (4.188) in terms of the variables a and N we obtain

$$N a^{5/2} = e^{1/\kappa_{ren}}. \tag{4.189}$$

The limit $N \to \infty$ and $a \to 0$, with κ_{ren} fixed, is called the *double scaling limit*. It follows that the partition function $\mathscr{Z}$, given by Eq. (4.183), can be written as

$$\mathscr{Z} = \sum_{h=0}^{\infty} \tau_h \left(\frac{G}{\Lambda^{\frac{5}{4}}} \right)^{2h-2} \tag{4.190}$$

where

$$G = e^{-1/\kappa_{ren}} = \frac{1}{Na^{5/2}}. \tag{4.191}$$

Since two-dimensional gravity can be viewed as string theory in a zero-dimensional target space, G can be interpreted as the renormalized *string coupling constant*. Clearly, $\mathscr{Z}$ is only a function of *one variable*

$$t = \Lambda \, G^{-4/5}. \tag{4.192}$$

In writing Eq. (4.190) we have glossed over the fact that the genus zero and genus one cases are slightly different. The genus one partition function $\mathscr{Z}_1$ is logarithmically divergent in the scaling limit, see Eq. (4.148). We have not constructed $\mathscr{Z}_0$ since the one-loop correlator $w_0(\underline{g}, z)$ contains, as already mentioned, non-universal parts, and the same will be true when we integrate the loop insertion operator to obtain $\mathscr{Z}_0$. We shall discuss how to circumvent this problem below.

The above considerations are formal in the sense that the sum (4.185) is divergent for all values of the couplings μ and κ. The entropy factor for all triangulations is too large for the partition function $\mathscr{Z}(\mu, \kappa)$ to converge, no matter how we choose the coupling constants. The number of non-isomorphic triangulations with k triangles grows as $(ck)!$, $c > 0$, since we can glue k triangles together with almost no restrictions if the topology is not fixed. On the other hand, $\chi > -k$ for a triangulation constructed from k triangles and the action grows at most exponentially with k. It follows that the sum over all triangulations is not Borel summable since all terms are positive. If we ignore these problems and view Eq. (4.185) as a perturbative expansion in topology, we are lead to the double scaling limit if we want to consider surfaces with fluctuating topology. Later we shall discuss various attempts to blow life into Eq. (4.185).

4.5 Multi-critical models

4.5.1 General considerations

We have implicitly assumed that the signs of the coupling constants (coefficients) g_i in the polynomial $V'(z)$ given by Eq. (4.79) are positive. In this regime the second moment M_2 is negative, and, as we have seen, the critical behaviour is governed by the vanishing of the first moment M_1.

We now propose to lift the constraint on the coupling constants so that the weights of some polygons become negative. As already mentioned, the formulae derived from the loop equations, when written in terms of the moments, are valid for *any* potential. It is then possible to obtain different scaling limits since one can fine-tune the coupling constants g_j to values g_{0j} such that

$$M_k(\underline{g}_0, c_+(\underline{g}_0), c_-(\underline{g}_0)) = 0, \quad k < m, \qquad (4.193)$$
$$M_m(\underline{g}_0, c_+(\underline{g}_0), c_-(\underline{g}_0)) \neq 0. \qquad (4.194)$$

The hyper-surface in coupling constant space determined by Eqs. (4.193) and (4.194) is called the *mth multi-critical hyper-surface* and pure gravity corresponds to the case $m = 2$. We now study the critical behaviour for arbitrary m and approach the critical surface transversally as in Eq. (4.164)

$$\underline{g} = \underline{g}_0(1 - \Delta g). \qquad (4.195)$$

Let $\Delta c_+ = c_+(\underline{g}_0) - c_+(\underline{g})$ denote the corresponding change in the critical point for the variables z_i. From Eqs. (4.193) and (4.194) it follows that

$$(\Delta c_+)^m \sim \Delta g, \qquad \Delta c_- \sim \Delta g; \qquad (4.196)$$

see Eq. (4.108) and (4.109). As in Eq. (4.164) we set

$$z_i = z_0(1 + aZ_i) \qquad (4.197)$$

since the cutoff a is interpreted as the length of links in the triangulations. We therefore demand $\Delta c_+ \sim a$ and consequently $\Delta g \sim a^m \sim \Delta c_-$ and we can write

$$\underline{g} = \underline{g}_0(1 - (\Lambda a^2)^{m/2}), \quad c_+(\underline{g}) = z_0(1 - \sqrt{\Lambda}a), \qquad (4.198)$$

where Λ is the renormalized cosmological constant. Repeating the arguments used for pure gravity we find that Eq. (4.100) is now replaced by

$$w_0(\underline{g}, z_1, \dots, z_b) \sim \frac{1}{(\Lambda a^2)^{\frac{1}{2}bm - (m + \frac{1}{2})}} \qquad (4.199)$$

for z_i away from z_0. If we let z_i approach its singular value according to Eq. (4.197) we obtain an additional singular factor such that Eq. (4.168) is replaced by

$$w_0(\underline{g}, z_1, \dots, z_b) \sim \frac{1}{a^{b(m + \frac{3}{2}) - (2m + 1)}} \qquad (4.200)$$

in the case of genus 0. The continuum loop function is, as before, given by the coefficient of the most singular term in w_0, and generalizing immediately to the higher genus functions $w_h(\underline{g}, z_1, \ldots, z_b)$ we obtain the extension of Eq. (4.177) to $m > 2$:

$$w_h(\underline{g}, z_1, \ldots, z_b) \sim \frac{W_h(\Lambda, Z_1, \ldots, Z_b)}{a^{b(m+\frac{3}{2})+(h-1)(2m+1)}}. \qquad (4.201)$$

Comparing Eq. (4.199) for $b = 0$ to the partition function in pure gravity, whose singular behaviour is given by Eq. (4.173), we conclude that for general m the critical exponent of the susceptibility is given by

$$\gamma = -m + \frac{3}{2}, \qquad (4.202)$$

and similarly in the case of higher genus we obtain from Eq. (4.201)

$$\gamma_h = 2 + (h - 1)\left(m + \frac{1}{2}\right), \qquad (4.203)$$

which generalizes Eq. (4.180).

4.5.2 The dimer model

It is possible to give a statistical mechanical interpretation of the new critical points found above and characterized by an integer $m \geq 2$. They correspond to so-called $(2, 2m - 1)$-conformal field theories coupled to two-dimensional quantum gravity. Let us outline how this identification comes about in the simplest case of $m = 3$, which can be realized in terms of a cubic polynomial $V'(z)$:

$$V'(z) = z - g_3 z^2 - g_4 z^3. \qquad (4.204)$$

The corresponding surfaces are obtained by gluing together triangles and squares. We can view the squares as being composed of two triangles. However, in order to ensure the correct counting of triangulations those triangles have to be distinguished from the triangles associated with the variable g_3. We can do this by "colouring" a diagonal in each of the squares. In this way the original counting problem becomes identical to counting triangulations with "dimers" inserted on some of the links, regarding the coloured diagonal as a dimer. The only restriction on the dimer configurations is that there can be at most one dimer on the boundary of each triangle. The problem of counting all possible triangulations with weight g_3 for each triangle and all dimer configurations with weigth ξ for each dimer and the restriction mentioned, is equivalent to the problem of counting triangulations made up of triangles and squares with weights g_3 and $g_4 = 2g_3^2\xi$, respectively. If we view the counting problem in terms of dual graphs, i.e. ϕ^3 graphs, the dimers will still be placed on links, and the restriction on dimer configurations is that no dimers can meet at vertices. This is called a *hard dimer problem*, see Fig.

192 Two-dimensional gravity

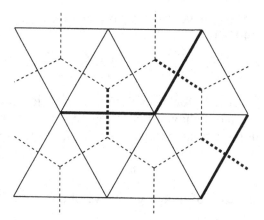

Fig. 4.8. Dimers, shown as thick links, on a triangulation and on the dual ϕ^3 graph.

4.8. It is clear that in order to have multi-criticality it is necessary that $g_4 < 0$, i.e. the fugacity ξ of the dimers has to be negative.

For simplicity we restrict our attention to surfaces of spherical topology. We can write the generating function for hard dimers on the ensemble of random planar ϕ^3 graphs as

$$Z(g,\xi) = \sum_{G \in \mathscr{G}(\phi^3)} g^{V_G} \sum_{\psi} \xi^{L_\psi}, \tag{4.205}$$

where $\mathscr{G}(\phi^3)$ denotes the class of planar ϕ^3 graphs, V_G is the number of vertices in G, the sum $\sum_\psi$ runs over all hard dimer configurations on the graph G and the number of links in G covered by dimers is denoted by L_ψ.

We next relate the hard dimer model described above to an Ising-type model on a random triangulation. Consider first a finite regular sublattice of $\mathbb{Z}^2$ with Ising spins $\sigma = \pm 1$ placed on the vertices. The partition function of an Ising model on such a lattice is given by

$$Z(\beta,H) = \sum_{\{\sigma_i\}} e^{\beta \sum_{(ij)} \sigma_i\sigma_j + H \sum_i \sigma_i}, \tag{4.206}$$

where the first sum in the exponent is over all links (pairs of nearest neighbour vertices), the second sum runs over all vertices on the lattice and the sum $\sum_{\{\sigma_i\}}$ is over all spin configurations.

Using the identity

$$e^{\sigma H} = \cosh H + \sigma \sinh H, \quad \sigma = \pm 1, \tag{4.207}$$

we can write down the standard high-temperature expansion of the model. If V and L denote the number of vertices and links, respectively, on the lattice, we have

$$Z(\beta, H) = \cosh^V H \cosh^L \beta \sum_{\{\sigma_i\}} \prod_i (1 + \sigma_i \tanh H) \prod_{(ij)} (1 + \sigma_i \sigma_j \tanh \beta).$$
(4.208)

Expanding the products we see that terms where σ_i appears an odd number of times vanish when we sum over all spin configurations. At high temperatures, i.e. for small $\beta = T^{-1}$, Eq. (4.208) may therefore be written as

$$\begin{aligned} Z(\beta, H) &= (2 \cosh H)^V \cosh^L \beta \left(1 + \tanh^2 H(\theta(1)\beta + O(\beta^2)) \right. \\ &\quad \left. + \tan^4 H(\theta(2)\beta^2 + O(\beta^4)) + \cdots \right), \end{aligned}$$
(4.209)

where $\theta(n)$ denotes the number of ways one can put down n hard dimers on the lattice. Setting

$$\xi = \beta \tanh^2 H,$$
(4.210)

it follows that

$$\tilde{Z}(\xi) = \lim_{H \to i\frac{\pi}{2},\ \beta \to 0} (2 \cosh H)^{-V} Z(\beta, H),$$
(4.211)

with $\xi = \beta \tanh^2 H$ fixed, is the partition function of an ensemble of hard dimers on the lattice, *with negative fugacity* ξ.

The Ising model on a regular lattice without an external magnetic field has a second-order transition between a magnetized and a disordered phase at a critical temperature T_c. By a well-known result from Lee and Yang [396] the Ising model in a *purely imaginary* magnetic field has a phase transition for fixed $T > T_c$ as a function of the imaginary magnetic field $H = i\tilde{H}$. Denoting the critical value of the magnetic field H by $H_c(T)$, where $\tilde{H}_c(T) \geq 0$, the "magnetization" is of the form

$$M = \frac{1}{V} \frac{\partial \log Z}{\partial \tilde{H}} \sim (\tilde{H} - \tilde{H}_c(T))^{\sigma_0},$$
(4.212)

as $\tilde{H} \downarrow \tilde{H}_c(T)$ and σ_0 is a critical exponent. For $T < T_c$ the critical field $\tilde{H}_c(T)$ vanishes and for $T \to \infty$ we have $\tilde{H}_c(T) \to \pi/2$. The transition at $H_c(T)$ is called the *Lee–Yang edge singularity*. It is shown in [177] and [111] that a scaling limit at $\tilde{H} = \pi/2$ can be identified with the so-called minimal (2,5) conformal field theory. This non-unitary conformal field theory is characterized by two numbers: a *central charge* $c = -22/5$ and a *conformal weight* $\Delta^{(0)} = -1/5$ associated with a scalar field. The exponent σ_0 is related to $\Delta^{(0)}$ by a general scaling relation (in two dimensions) $\sigma_0 = \Delta^{(0)}/(1 - \Delta^{(0)})$. This relation yields the value $\sigma_0 = -1/6$.

As a consequence of the mapping between the Ising model in an imaginary magnetic field and the hard dimer model, it follows that the latter has a phase transition at a certain negative value ξ_c of the fugacity, such that

$$\frac{1}{V}\left(\frac{d\log Z}{d\xi}\right) \sim (\xi - \xi_0)^{\sigma_0}, \qquad (4.213)$$

as $\xi \to \xi_0$, and the corresponding scaling limit is the non-unitary conformal field theory described above.

The Ising model can be generalized in a straightforward way to a random lattice. We consider this in some detail in Section 4.9. Summing over random lattices defines the coupling to quantum gravity, as we have discussed above. Random planar ϕ^3 graphs with Ising spins on the vertices are equivalent to random triangulations with Ising spins located at the center of the triangles. We adopt the latter point of view in the following. Taking for granted that the mapping between the Ising model and the dimer model on a regular lattice extends to arbitrary lattices we conclude that the Ising model coupled to quantum gravity reduces in the limit of high temperature and imaginary magnetic field $H = i\pi/2$ to the dimer model on random graphs.

Let us now compare the critical exponents of the systems before and after coupling to quantum gravity. First, note that the partition function of the Ising model coupled to gravity can be written as

$$Z(\mu, \beta, H) = \sum_{N=1}^{\infty} e^{-\mu N} Z_N(\beta, H) = \sum_{N=1}^{\infty} e^{-\mu N + F_N(\beta, H)}, \qquad (4.214)$$

where $F_N(\beta, H) = \log Z_N(\beta, H)$ is the free energy in a fixed volume N, obtained by summing over all planar triangulations with N triangles and all spin configurations on such triangulations. From Eq. (4.214) it follows that the critical cosmological constant $\mu_0(\beta, H)$ is given by

$$\mu_0(\beta, H) = \lim_{N \to \infty} \frac{F_N(\beta, H)}{N}, \qquad (4.215)$$

and can be interpreted as the free energy per unit volume. Similarly, in the limit (4.211) the partition function (4.214) is replaced by the dimer partition function (4.205), and the critical point $\mu_0(\xi) = \log g_0(\xi)$ equals the free energy per unit volume of the ensemble of hard dimers. We anticipate that the critical coupling ξ_0 in Eq. (4.213) survives the coupling to quantum gravity (not necessarily at the same value of ξ) and that the critical behaviour given by Eq. (4.213) is replaced by

$$\frac{d\log g_0(\xi)}{d\xi} \sim (\xi - \xi_0)^{\sigma}, \qquad (4.216)$$

where the critical exponent σ can be interpreted as a dressed version of σ_0 due to the coupling to gravity. Eq. (4.216) can indeed be established and σ determined by the machinery developed above. The detailed calculations are tedious so we only present an outline of the argument. In principle the one-loop function $w(\underline{g}, z)$ can be expressed in terms of the coupling constants g_j. The equations $M_{-1}(\underline{g}) = 2$ and $M_0(\underline{g}) = 0$ determine the cut $[c_-(\underline{g}), c_+(\underline{g})]$, while the equation $M_1(\underline{g}) = 0$ yields a relation between g_3 and g_4 which allows us to express the critical condition $g_3 = g_0$ in terms of ξ, recalling that $g_4 = 2g_3^2\xi$. The critical value ξ_0 is determined by $M_2(\underline{g}) = 0$ and in a neighbourhood of ξ_0 one finds the relation

$$\frac{dg_0(\xi)}{d\xi} \sim (\xi - \xi_0)^{1/2}. \tag{4.217}$$

Eq. (4.217) implies that $\sigma = \frac{1}{2}$ and we conclude that the coupling to gravity shifts the critical exponent $\sigma_0 = -\frac{1}{6}$ to $\sigma = \frac{1}{2}$.

4.5.3 Connection with conformal field theory

The considerations above make it plausible that the scaling limit of the multicritical $m = 3$ theory can be identified with a non-unitary continuum scalar field theory. This theory is one in a series of conformal field theories, called minimal conformal theories, which are described in the fundamental work of Belavin, Polyakov and Zamolodchikov [73]. They are labelled by two positive integers p and q with no common divisor and $2 \leq p < q$. The number

$$c = 1 - 6\frac{(p-q)^2}{pq}$$

is called the *central charge* of the model and can be given a physical interpretation, see [112, 113] for reviews. For our purposes it suffices to note that the theory is unitary if and only if $q = p + 1$.

It is possible to give formal continuum arguments, generalizing the discussion above, showing that the scaling limit at points on the mth critical hyper-surface can be identified with the minimal $(2, 2m-1)$ conformal field theory coupled to quantum gravity. These theories are all non-unitary, except in the case $m = 2$, which corresponds to pure quantum gravity.

Let us take a closer look at the content of the (p, q) conformal theory in order to interpret the singular behaviour expressed by Eq. (4.199) in continuum terms. There exist so-called *primary fields* $\phi_{r,s}$ labelled by positive integers r, s satisfying $1 \leq r \leq q - 1$ and $1 \leq s \leq p - 1$, with associated scaling dimensions $\Delta_{r,s}^{(0)}$ such that when the model is realized on a surface M equipped with a metric g whose volume is V, then the expectation value of the operator $\int_M d^2\xi \sqrt{g(\xi)} \phi_{r,s}(\xi)$ behaves as

$$\left\langle \int_M d^2\xi \sqrt{g(\xi)} \phi_{r,s}(\xi) \right\rangle_g \sim V^{1-\Delta_{r,s}^{(0)}} \tag{4.218}$$

for large V. When the theory is coupled to gravity, expectation values also involve an integration over diffeomorphism classes of metrics with fixed volume V and the gravitationally dressed scaling dimension $\Delta_{r,s}$ is defined by

$$\left\langle \int_M d^2\xi \sqrt{g(\xi)}\, \phi_{r,s}(\xi) \right\rangle \sim V^{1-\Delta_{r,s}} \qquad (4.219)$$

as $V \to \infty$. In [73] it is shown that

$$1 - \Delta_{r,s}^{(0)} = \frac{(p+q)^2 - (pr-qs)^2}{4pq}, \qquad (4.220)$$

while the methods of [277, 131, 143] lead to

$$1 - \Delta_{r,s} = \frac{p + q - |pr - qs|}{2p}. \qquad (4.221)$$

The strongest divergence in Eq. (4.219) is obtained for the field $\phi_{min} = \phi_{r,s}$ with the smallest scaling dimension Δ_{min}. For the $(2, 2m-1)$ theories it can easily be seen that

$$1 - \Delta_{min} = \frac{m}{2}$$

so $\Delta_{min} < 0$ for $m > 2$. Integrating Eq. (4.219) over volume with a cosmological constant Λ we find for the $(2, 2m-1)$ model

$$\left\langle \int_M d^2\xi \sqrt{g(\xi)}\, \phi_{min}(\xi) \right\rangle \sim \Lambda^{-\frac{m}{2}}.$$

The factor $(a^2\Lambda)^{-\frac{m}{2}}$ in Eq. (4.199) associated with each of the b punctures on the surface can be viewed as an insertion of the operator ϕ_{min}. Naively one would have expected a factor $(a^2\Lambda)^{-1}$ for each puncture since integration over the position of the puncture contributes a factor V. Put differently, each puncture corresponds in continuum language to an insertion of the identity operator $\phi_{1,1}$ with a vanishing scaling dimension and so contributes a factor V in Eq. (4.219). The presence in the theory of operators with negative scaling dimensions invalidates this naive expectation.

Another piece of evidence in favour of the identification of the scaling limits described above is the agreement between the formula (4.202) for the susceptibility exponent γ and the formula

$$\gamma = 1 - \frac{q}{p}, \qquad (4.222)$$

which may be derived by continuum arguments [277].

It is beyond the scope of this book to derive Eqs. (4.221) and (4.222), which are special cases of the following formulas which express γ in terms of the central charge c and relate the scaling dimension $\Delta^{(0)}$ of a given primary field to the corresponding dressed scaling dimension Δ in the presence of fluctuating geometry:

$$\gamma = \frac{c - 1 - \sqrt{(c-1)(c-25)}}{12}, \tag{4.223}$$

$$\Delta = \frac{\sqrt{1 - c + 24\Delta^{(0)}} - \sqrt{1-c}}{\sqrt{25 - c} - \sqrt{1-c}}. \tag{4.224}$$

These relations are valid for any conformal field theory on a surface of genus zero and were first obtained in [277] and generalized to arbitrary genus in [131, 143].

The scaling limits of the discrete random surface models are perhaps best described in the language of the renormalization group. In the infinite-dimensional space of coupling constants $\underline{g}$, which parametrize the set of models, there are certain critical hyper-surfaces of finite co-dimension given by the equations

$$M_1(\underline{g}) = 0, \ldots, M_{m-1}(\underline{g}) = 0, \qquad M_m(\underline{g}) \neq 0. \tag{4.225}$$

Let $\underline{g}_0$ be a point on the critical surface. Let us approach this point along a curve $\underline{g}(a)$, $\underline{g}(a) \to \underline{g}_0$ as $a \to 0$, such that

$$M_k(\underline{g}(a)) = \mu_k a^{m-k} + O(a^{m-k+1}), \quad 1 \le k < m, \tag{4.226}$$
$$M_k(\underline{g}(a)) = O(1), \quad k \ge m. \tag{4.227}$$

This scaling limit is "natural" since the generic change of M_k for $k \le m$, while moving away from the mth critical surface according to Eq. (4.196), is given by

$$M_k(\underline{g}(a)) = (k + \frac{1}{2})M_{k+1}(\underline{g}_0)\Delta c_+ + \cdots \tag{4.228}$$

$$+ \frac{(k + \frac{1}{2}) \cdots (m + \frac{1}{2})}{(m-k)!} M_m(\underline{g}_0)(\Delta c_+)^{m-k} + O(\Delta g, \Delta c_-),$$

where a is the lattice spacing and $\Delta c_+ \sim a$ and we have used Eq. (4.106). We note that more general variations exist where the various $\Delta g_j \neq O(a^m)$, but where Eqs. (4.226) and (4.227) are still satisfied. We consider a class of such variations in the next section.

Inserting Eqs. (4.226) and (4.227) into the genus expansion of the partition function and using powers of the indeterminate $1/N^2$ to label terms in the expansion, it follows that the $a \to 0$ limit is dominated by

$$\mathscr{Z} = \sum_{h=0}^{\infty} \left(\frac{1}{N^2 a^{2m+1}} \right)^{h-1} \mathscr{Z}_h, \tag{4.229}$$

where for $h > 1$

$$\mathscr{L}_h = \sum_{\alpha_i > 1} \langle \alpha_1 \cdots \alpha_k \rangle_h \frac{\mu_{\alpha_1} \cdots \mu_{\alpha_k}}{\mu_1^\alpha d^\gamma}, \qquad (4.230)$$

with μ_α as in Eq. (4.226) and we have introduced the shorthand notation

$$\langle \alpha_1 \cdots \alpha_k \rangle_h = \langle \alpha_1, \cdots, \alpha_k ; |\alpha = 2h - 2 + k, \beta = 0, \gamma = h - 1 \rangle_h \qquad (4.231)$$

for the rational coefficients given by Eq. (4.150), with $l = 0$, i.e. $\sum_{i=1}^k \alpha_i = 3h - 3 + k$; see Eq. (4.152). In Section 8 of this chapter we show that the values of these coefficients coincide with certain topological invariants defined on the moduli space of Riemann surfaces. From Eqs. (4.148) and (4.149) the first two terms are given by

$$\mathscr{L}_1 = -\frac{1}{24} \log\left(\mu_1 d^4\right), \qquad (4.232)$$

$$\mathscr{L}_2 = \frac{-21}{160} \frac{\mu_2^3}{\mu_1^5 d} + \frac{29}{128} \frac{\mu_2 \mu_3}{\mu_1^4 d} + \frac{-35}{384} \frac{\mu_4}{\mu_1^3 d}. \qquad (4.233)$$

Explicit expressions up to $\mathscr{L}_4$ have been calculated [20, 168]. The result for $h = 3$ reads

$$\begin{aligned}
\mathscr{L}_3 =\ & \frac{2205\mu_2^6}{256d^2\mu_1^{10}} - \frac{8685\mu_2^4\mu_3}{256d^2\mu_1^9} + \frac{15375\mu_2^2\mu_3^2}{512d^2\mu_1^8} + \frac{5565\mu_2^3\mu_4}{256d^2\mu_1^8} \\[2mm]
& - \frac{5605\mu_2\mu_3\mu_4}{256d^2\mu_1^7} - \frac{72875\mu_3^3}{21504d^2\mu_1^7} - \frac{3213\mu_2^2\mu_5}{256d^2\mu_1^7} \\[2mm]
& + \frac{2515\mu_3\mu_5}{512d^2\mu_1^6} + \frac{21245\mu_4^2}{9216d^2\mu_1^6} + \frac{5929\mu_2\mu_6}{1024d^2\mu_1^6} - \frac{5005\mu_7}{3072d^2\mu_1^5}.
\end{aligned} \qquad (4.234)$$

We emphasize that Eqs. (4.232)–(4.234) are valid for *any* multi-critical models.

We have thus arrived at the same type of expression as obtained for pure gravity in Eqs. (4.183) and (4.190), provided we introduce a generalized *string coupling constant*

$$G = \frac{1}{N a^{m+1/2}}. \qquad (4.235)$$

This definition agrees with Eq. (4.191) for $m = 2$.

At the critical surface of a statistical theory we usually have infinite-range correlations. Associated with this hyper-surface there is a universality class of massive continuum theories, whose masses are determined by the different ways one has of approaching the critical surface. Usually one demands that the critical hyper-surface has a finite co-dimension, in order that only a finite number of coupling constants need to be fine-tuned in approaching the critical surface. This is satisfied for the multi-critical

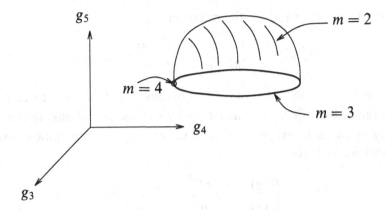

Fig. 4.9. A multi-critical surface of finite co-dimension in the infinite-dimensional coupling constant space, here symbolized by the couplings g_3, g_4 and g_5.

models. The continuum theories associated with the multi-critical hyper-surfaces (4.225) can, as already explained, be identified with conformal $(2, 2m - 1)$ theories coupled to quantum gravity, while the massive scaling limits are characterized by the additional parameters $\mu_2, \mu_3, \ldots, \mu_{m-1}$.

4.6   The continuum loop equation

Above we have presented the loop equation as a combinatorial identity, and used it to derive the asymptotic behaviour of the number of graphs on a surface of genus h. We observed that the moments M_k appear as natural variables, especially when it comes to a discussion of the scaling limits. Below we introduce yet another set of coupling constants T_k which are closely related to the moments M_k. In terms of these coupling constants the loop insertion operator has a transparent scaling behaviour in the neighbourhood of any given multi-critical point. Furthermore, the use of these coupling constants allows us to derive a loop equation for the continuum loop functions. This continuum loop equation is valid at *all* multi-critical points and in this way we achieve a unified description of the critical theories considered so far. We show that the continuum loop equation is equivalent to an integrable hierarchy of non-linear differential equations, the so called Korteweg–de Vries hierarchy (KdV–hierarchy) for the continuum loop functions.

In order to define the coupling constants T_k let us recall that the approach to a given multi-critical point $\underline{g}_0$ of order m is characterized by $M_k(\underline{g}) \sim \mu_k a^{m-k}$ for $1 \leq k \leq m$. We now choose a parametrization of the approach to the given multi-critical point which makes the KdV-structure transparent. Let us introduce the notation

$$M_k^0(\underline{g}) = \oint \frac{d\omega}{2\pi i} \frac{V'(\omega)}{(\omega - c_+^0)^{k+\frac{1}{2}}(\omega - c_-^0)^{\frac{1}{2}}}. \qquad (4.236)$$

Here $c_\pm^0$ are the values of c_+ and c_- corresponding to the critical point $\underline{g}_0$. The functions $M_k^0(\underline{g})$ are *linear* functions in the deviations $\Delta \underline{g}$ from $\underline{g}_0$ and they allow us to express $\Delta \underline{g}$ in terms of $\{M_k^0\}$. Let us now consider deviations $\Delta \underline{g}$ such that

$$M_k^0(\underline{g}) \sim a^{m-k}, \quad k < m, \qquad (4.237)$$

$$M_k^0(\underline{g}) \sim O(1), \quad k \geq m. \qquad (4.238)$$

It can easily be seen that Eqs. (4.237)–(4.238) imply that $M_k(\underline{g}) \sim a^{m-k}$ for $k \leq m$. We finally impose the further condition that the deviations $\Delta \underline{g}$ away from $\underline{g}_0$ are such that $c_-(\underline{g}) = c_-(\underline{g}_0)$. This restriction is in no way essential and is only imposed to simplify the following discussion. The general case can be treated by replacing c_-^0 in Eq. (4.236) by $c_-(\underline{g})$. We define the coupling constants T_k by

$$T_k(\underline{g}) = \frac{\sqrt{d_0}}{k + \frac{1}{2}} a^{k-m} M_k^0(\underline{g}), \quad k \geq 0, \qquad (4.239)$$

where m refers to the order of a given multi-critical point $\underline{g}_0$ and where $d_0 = c_+^0 - c_-^0$. Strictly speaking, both M_k^0 and T_k should be considered as functions of $\underline{g}_0$ and $\Delta \underline{g}$ rather than $\underline{g}$ but the shorthand notation given above should lead to no misunderstanding. The coupling constants T_k are designed such that they, like the μ_ks, have a finite value when $\Delta \underline{g}$ goes to zero according to Eqs. (4.237)–(4.238). However, contrary to the μ_ks they can be used to obtain a simple expression for the loop insertion operator for $\underline{g}$ in the neighbourhood of $\underline{g}_0$ (see Eqs. (4.244) and (4.245) below).

If we parametrize the distance $c_+^0 - c_+(\underline{g}) = \sqrt{\Lambda} a$ as in Eq. (4.167), Λ being the cosmological constant and a the lattice spacing, we can expand $M_k(\underline{g})$ in powers of $c_+(\underline{g}) - c_+^0$ and obtain

$$M_k = \frac{a^{m-k}}{\sqrt{d_0}} \sum_{l=k}^{\infty} \left(l + \frac{1}{2}\right) T_l \, (-1)^{l-k} \Lambda^{\frac{l-k}{2}} c_l^{(k)}, \quad k \geq 0, \qquad (4.240)$$

where

$$c_l^{(k)} = \frac{\Gamma(l + \frac{1}{2})}{\Gamma(k + \frac{1}{2})\,(l - k)!} \qquad (4.241)$$

are the Taylor coefficients of $(1-x)^{-k-\frac{1}{2}}$. Note that the equation $M_0(\underline{g}) = 0$

determines Λ as an implicit function of the T_ks:

$$\sum_{l=0}^{\infty} (-1)^l \left(l + \frac{1}{2}\right) c_l\, T_l\, \Lambda^{\frac{l}{2}} = 0, \qquad c_l = c_l^{(0)}. \tag{4.242}$$

For $k > m$ we have $M_k = O(1)$ (or zero for k larger than the order of the polynomial $V'(z)$), while $T_k = O(a^{k-m})$.

We can express the loop insertion operator as (see Eq. (4.113))

$$\frac{d}{dV(z)} = \sum_{k=0}^{\infty} \frac{dM_k^0}{dV(z)} \frac{d}{dM_k^0} \tag{4.243}$$

$$= \sum_{k=0}^{\infty} \frac{\partial}{\partial z} \left[\frac{1}{(z - c_+^0)^{k+\frac{1}{2}}(z - c_-^0)^{\frac{1}{2}}} \right] \frac{\sqrt{d_0}\, a^{k-m}}{k + \frac{1}{2}} \frac{d}{dT_k}.$$

In taking the scaling limit we set $z = c_+^0 + aZ$ (see Eq. (4.197)), so

$$\frac{d}{dV(z)} = \frac{1}{a^{m+\frac{3}{2}}} \left(\frac{d}{dV_{cnt}(Z)} + O(a) \right), \tag{4.244}$$

where

$$\frac{d}{dV_{cnt}(Z)} = -\sum_{k=0}^{\infty} \frac{1}{Z^{k+3/2}} \frac{d}{dT_k}. \tag{4.245}$$

We use the label *cnt* to indicate continuum quantities, introduce the notation $\underline{T} = (T_0, T_1, \ldots)$ analogous to the notation used for the coupling constants g_j and define

$$V_{cnt}(\underline{T}, Z) = \sum_{k=0}^{\infty} Z^{k+\frac{1}{2}}\, T_k, \qquad V'_{cnt}(\underline{T}, Z) = \sum_{k=0}^{\infty} \left(k + \frac{1}{2}\right) Z^{k-\frac{1}{2}}\, T_k. \tag{4.246}$$

We now recall the expressions given by Eq. (4.201) for the multi-loop functions $w_h(\underline{g}, z_1, \ldots, z_b)$ and $W_h(\underline{T}, Z_1, \ldots, Z_b)$. Except for $h = 0$ and $b = 1, 2$ we have

$$w_h(\underline{g}, z_1, \ldots, z_b) \sim \frac{W_h(\underline{T}, Z_1, \ldots, Z_b) + O(a)}{a^{b(m+\frac{3}{2})+(h-1)(2m+1)}}. \tag{4.247}$$

The notation is slightly more general than in (4.201) since Λ is replaced by the coupling constants $\underline{T}$. The reason is that we consider a more general approach (4.237)-(4.238) to the critical surface than represented by (4.198).

For $h > 0$ or $b > 2$ we define the continuum loop function by

$$W_h^{cnt}(\underline{T}, Z_1, \ldots, Z_b) = W_h(\underline{T}, Z_1, \ldots, Z_b). \tag{4.248}$$

This definition depends on the choice of a specific multi-critical point $\underline{g}_0$ but, as we will show below, the equation which determines the loop function $W_h^{cnt}(\underline{T}, Z_1, \ldots, Z_b)$ has no reference to $\underline{g}_0$ nor to the specific multi-critical order m of $\underline{g}_0$. A characteristic property of these multi-loop functions is that they have the following expansion for large Z_is

$$W_h^{cnt}(\underline{T}, Z_1, \ldots, Z_b) = \sum_{n_i \geq 0} \frac{W_h^{cnt}(\underline{T}, n_1, \ldots, n_b)}{Z_1^{n_1 + 3/2} \cdots Z_b^{n_b + 3/2}}, \qquad (4.249)$$

where the coefficients $W_h^{cnt}(\underline{T}, n_1, \ldots, n_b)$ depend on the coupling constants $\underline{T}$.

The expansion (4.249) fails for the spherical two-loop function, and Eq. (4.247), as well as the expansion (4.249), fails for the spherical one-loop function. Let us now *define* the continuum loop functions $W_0^{cnt}(\underline{T}, Z)$ and $W_0^{cnt}(\underline{T}, Z_1, Z_2)$ as the parts of $W_0(\underline{T}, Z)$ and $W_0(\underline{T}, Z_1, Z_2)$ which scale according to Eq. (4.247) and allow the expansion (4.249). These functions will fall off at least as fast as $Z_i^{-3/2}$. We find

$$w_0(\underline{g}, z) - \frac{1}{2} V'(\underline{g}, z) = a^{m - \frac{1}{2}} \left(W_0^{cnt}(\underline{T}, Z) - \frac{1}{2} V'_{cnt}(Z) + O(a) \right) \qquad (4.250)$$

$$w_0(\underline{g}, z_1, z_2) + \frac{1}{2} \frac{1}{(z_1 - z_2)^2} \qquad (4.251)$$

$$= \frac{1}{a^2} \left(W_0^{cnt}(\underline{T}, Z_1, Z_2) + \frac{\frac{1}{2}(Z_1 + Z_2)}{2(Z_1 - Z_2)^2 \sqrt{Z_1 Z_2}} + O(a) \right).$$

In particular,

$$w_0(\underline{g}, z, z) = \frac{1}{a^2} \left(W_0^{cnt}(\underline{T}, Z, Z) + \frac{1}{16 Z^2} + O(a) \right). \qquad (4.252)$$

With these definitions we can express the continuum multi-loop functions by the use of the continuum loop insertion operator:

$$W_h^{cnt}(\underline{T}, Z_1, \ldots, Z_b) = \frac{d}{dV_{cnt}(Z_b)} \cdots \frac{d}{dV_{cnt}(Z_2)} W_h^{cnt}(\underline{T}, Z_1). \qquad (4.253)$$

Recall that the generating function for the one-loop functions is

$$w(\underline{g}, z) = \sum_{h=0}^{\infty} \frac{1}{N^{2h}} w_h(\underline{g}, z), \qquad (4.254)$$

where each $w_h(\underline{g}, z)$ has a formal Laurent expansion in z, and $w_h(\underline{g}, z)$ satisfies the loop equation

$$\oint_C \frac{d\omega}{2\pi i} \frac{V'(\omega) - 2w_0(\omega)}{z - \omega} w_h(\omega) = \sum_{h'=1}^{h-1} w_{h'}(z) w_{h-h'}(z) + \frac{d}{dV(z)} w_{h-1}(z).$$

$$(4.255)$$

Here we have dropped the coupling constants $\underline{g}$ from our notation. Consider now the approach to a given multi-critical point $\underline{g}_0$ as above. In the double scaling limit the "continuum part" of $w(z)$ is given by

$$W^{cnt}(\underline{T}, Z) = \sum_{h=0}^{\infty} G^{2h} W_h^{cnt}(\underline{T}, Z), \qquad G^2 = \frac{1}{N^2 a^{2m+1}}, \qquad (4.256)$$

where each $W_h^{cnt}(\underline{T}, Z)$ has a formal expansion in inverse powers $Z^{-3/2-n}$ and G is the renormalized string coupling constant introduced earlier (see Eq. (4.235)). The one-loop functions $W_h^{cnt}(\underline{T}, Z)$ obey the following continuum loop equation:

$$\oint_{C_1} \frac{d\Omega}{2\pi i} \frac{V_{cnt}'(\Omega) - 2W_0^{cnt}(\Omega)}{Z - \Omega} W_h^{cnt}(\Omega) \qquad (4.257)$$

$$= \sum_{h'=1}^{h-1} W_{h'}^{cnt}(Z) W_{h-h'}^{cnt}(Z) + \left(\frac{d}{dV^{cnt}(Z)} W_{h-1}^{cnt}(Z) + \frac{\delta_{h,1}}{16Z^2} \right),$$

for each $h > 0$. This equation is obtained by substituting Eqs. (4.250) and (4.252) in Eq. (4.255) and setting

$$\omega = c_+^0 + a\Omega \qquad (4.258)$$

such that $z - \omega = a(Z - \Omega)$. The contour C in Eq. (4.255) encloses the cut along $[c_-(\underline{g}), c_+(\underline{g})]$ of $w_h(z)$ but does not encircle z. The change of variable given by Eq. (4.258) transforms this cut to $]-\infty, -\sqrt{\Lambda}]$ on the real axis in the limit $a \to 0$. The transformed curve C_1 encircles the transformed cut as well as the cut of $V_{cnt}'(\Omega)$ coming from $\sqrt{\Omega}$, but has winding number 0 with respect to Z.

We have thus managed to turn the original combinatorial identities into continuum formulas. As may be expected from the above definitions the $h = 0$ case requires special consideration. The equation for $w_0(z)$ is

$$\oint_C \frac{d\omega}{2\pi i} \frac{V'(\omega) w_0(\omega)}{z - \omega} = w_0(z)^2. \qquad (4.259)$$

If we use Eq. (4.250) to rewrite the numerator $V'(\omega)w_0(\omega)$ in the integrand above as

$$V'(\omega)w_0(\omega) = \left(\frac{1}{4}(V'(\omega))^2 + w_0(\omega)^2 \right) \qquad (4.260)$$

$$+ a^{2m-1} \left(-\frac{1}{4}(V_{cnt}'(\Omega))^2 - (W_0^{cnt}(\Omega))^2 + V_{cnt}'(\Omega) W_0^{cnt}(\Omega) \right),$$

use the fact that $(V'(\omega))^2$, as well as $(V_{cnt}'(\Omega))^2$, have no singularities except for the term $T_0^2/4\Omega$, and the fact that $w_0(\omega)$ and $W_0(\Omega)$ fall off at infinity, we obtain

$$\oint_C \frac{d\omega}{2\pi i} \frac{V'(\omega)w_0(\omega)}{z - \omega} = w_0(z)^2 \tag{4.261}$$

$$+ a^{2m-1} \left(-\frac{T_0^2}{16Z} - \left(W_0^{cnt}(\Omega) \right)^2 + \oint_{C_1} \frac{d\Omega}{2\pi i} \frac{V'_{cnt}(\Omega) W_0^{cnt}(\Omega)}{Z - \Omega} \right).$$

By Eq. (4.259), the left-hand side of Eq. (4.261) equals $w_0(z)^2$ and we obtain

$$\oint_{C_1} \frac{d\Omega}{2\pi i} \frac{V'_{cnt}(\Omega) W_0^{cnt}(\Omega)}{Z - \Omega} = \left(W_0^{cnt}(Z) \right)^2 + \frac{T_0^2}{16Z}. \tag{4.262}$$

We see that except for the term $T_0^2/16Z$ the continuum loop equation is similar to Eq. (4.259).

Recall that Eq. (4.259) for $h = 0$ and Eq. (4.255) for $h > 0$ can be combined into one equation for $w(z)$:

$$\oint_C \frac{d\omega}{2\pi i} \frac{V'(\omega)}{z - \omega} w(\omega) = w(z)^2 + \frac{1}{N^2} \frac{dw(z)}{dV(z)}. \tag{4.263}$$

Likewise the continuum loop equations (4.262) and (4.257) can be written as

$$\oint_{C_1} \frac{d\Omega}{2\pi i} \frac{V'_{cnt}(\Omega)}{Z - \Omega} W^{cnt}(\Omega) = W^{cnt}(Z)^2 + \frac{T_0^2}{16Z} + G^2 \frac{dW^{cnt}(Z)}{dV^{cnt}(Z)} + \frac{G^2}{16Z^2}. \tag{4.264}$$

Although Eq. (4.264) was derived by studying the approach to a specific multi-critical point of order m and was derived as an expansion in h, it has no reference to either m or h, and is formulated entirely in terms of continuum variables. Solving this equation under the condition that the solution $W^{cnt}(\underline{T}, Z)$ has an asymptotic expansion of the form (4.256) is tantamount to solving quantum gravity coupled to the particular matter fields discussed in the previous section.

Since $W^{cnt}(Z) = O(Z^{-3/2})$ for Z large the equality of the $1/Z$ terms on the two sides of the loop equation (4.262) implies

$$\oint_{C_1} \frac{d\Omega}{2\pi i} V'_{cnt}(\Omega) W^{cnt}(\Omega) = \frac{T_0^2}{16}. \tag{4.265}$$

Equivalently, setting

$$W^{cnt}(Z) = \sum_{n=0}^{\infty} \frac{W_n(\underline{T})}{Z^{n+\frac{3}{2}}}, \tag{4.266}$$

where $W_n(\underline{T}) = \sum_h G^{2h} W_h^{cnt}(\underline{T}, n)$ in the notation from Eqs. (4.249) and

(4.256), we have

$$-\sum_{n=1}^{\infty}\left(n+\frac{1}{2}\right) T_n W_{n-1}(\underline{T}) = \frac{T_0^2}{16}.$$
(4.267)

This equation is known as the *pre-string equation*. Let us introduce the notation

$$D = \frac{\partial}{\partial T_0}$$
(4.268)

and

$$R(\underline{T}, Z) = \frac{1}{\sqrt{Z}} - 4D W^{cnt}(\underline{T}, Z).$$
(4.269)

We have the expansion

$$R(\underline{T}, Z) = \sum_{n=0}^{\infty} \frac{R_n(\underline{T})}{Z^{n+\frac{1}{2}}},$$
(4.270)

where

$$R_0(\underline{T}) = 1, \qquad R_n(\underline{T}) = -4D W_{n-1}(\underline{T}), \quad n > 0.$$
(4.271)

Finally, it is convenient to introduce a special symbol for $R_1(\underline{T})$:

$$u(\underline{T}) = R_1(\underline{T}).$$
(4.272)

Like $W^{cnt}(\underline{T}, Z)$ the function $R(\underline{T}, Z)$ admits a genus expansion

$$R(\underline{T}, Z) = \sum_{h=0}^{\infty} G^{2h} R^{(h)}(\underline{T}, Z).$$
(4.273)

We use the notation $R_n^{(h)}(\underline{T})$ for the genus h component of $R_n(\underline{T})$ and likewise the notation $u_h(\underline{T})$ for $R_1^{(h)}(\underline{T})$.

Differentiating Eq. (4.267) with respect to T_0 leads to the so-called *string equation*

$$\sum_{n=0}^{\infty}\left(n+\frac{1}{2}\right) T_n R_n(\underline{T}) = 0.$$
(4.274)

For genus $h > 0$ we have constructed the functions $\mathscr{L}_h(\underline{T})$ such that

$$W_h^{cnt}(Z) = \frac{d}{dV_{cnt}(Z)}\mathscr{L}_h,$$
(4.275)

i.e.

$$R_{n+1}^{(h)} = 4\frac{\partial^2 \mathscr{L}_h}{\partial T_0 \partial T_n}, \qquad u_h = 4\frac{\partial^2 \mathscr{L}_h}{\partial T_0^2},$$
(4.276)

and acting with D we obtain

$$DR_{n+1}^{(h)} = \frac{\partial u_h}{\partial T_n}. \tag{4.277}$$

We have not yet constructed $\mathscr{L}_0(\underline{T})$ explicitly, so the arguments given above are not yet valid for $h = 0$. Below we will show that Eq. (4.277) is also satisfied for $h = 0$, and construct $\mathscr{L}_0(\underline{T})$, and it follows that we can write (4.277) in the compact form

$$DR(Z) = -\frac{du}{dV_{cnt}(Z)}, \tag{4.278}$$

where $R(Z)$ is defined by Eq. (4.273) and $u = R_1$ in this formal power series in G^2.

We now turn to the proof of Eq. (4.277) for $h = 0$. The loop equation (4.259) has a solution given by Eq. (4.68), which, together with the boundary conditions (4.69), expresses $w_0(z)$ in terms of the coupling constants $\underline{g}$. The arguments which lead from Eq. (4.259) to Eq. (4.68) (see Eqs. (4.64)–(4.68)) can be repeated for the continuum loop equation (4.262) and we obtain

$$W_0^{cnt}(Z) = \frac{1}{2} \oint_{C_1} \frac{d\Omega}{2\pi i} \frac{V_{cnt}'(\Omega)}{Z - \Omega} \frac{\sqrt{Z + \sqrt{\Lambda}}}{\sqrt{\Omega + \sqrt{\Lambda}}}. \tag{4.279}$$

In the same way as the values of $c_+(\underline{g})$ and $c_-(\underline{g})$ are determined as functions of the coupling constants $\underline{g}$ by the boundary conditions (4.69) arising from the requirement that $w_0(z) = 1/z + \cdots$, the cosmological constant Λ is uniquely determined as a function of the coupling constants $\underline{T}$ by demanding that $W^{cnt}(Z) = O(Z^{-3/2})$ for large Z. This implies

$$\oint_{C_1} \frac{d\Omega}{2\pi i} \frac{V_{cnt}'(\Omega)}{\sqrt{\Omega + \sqrt{\Lambda}}} = 0; \tag{4.280}$$

see Eq. (4.242). If we apply D to Eqs. (4.279) and (4.280) we obtain

$$
\begin{aligned}
DW_0^{cnt}(Z) = {} & \frac{1}{4}\sqrt{Z + \sqrt{\Lambda}} \int_{C_1} \frac{d\Omega}{2\pi i} \frac{\Omega^{-1/2}}{Z - \Omega} \frac{1}{\sqrt{\Omega + \sqrt{\Lambda}}} \\
& - \frac{1}{4} \frac{D\sqrt{\Lambda}}{\sqrt{Z + \sqrt{\Lambda}}} \int_{C_1} \frac{d\Omega}{2\pi i} \frac{V_{cnt}'(\Omega)}{(\Omega + \sqrt{\Lambda})^{3/2}},
\end{aligned}
$$

from which we conclude that

$$DW_0^{cnt}(Z) = \frac{1}{4}\left(\frac{1}{\sqrt{Z}} - \frac{1}{\sqrt{Z + \sqrt{\Lambda}}} \right). \tag{4.281}$$

It follows from the definition of $R(\underline{T}, Z)$ and $u(\underline{T})$ that

$$u_0(\underline{T}) = -\frac{1}{2}\sqrt{\Lambda} \qquad (4.282)$$

and

$$R^{(0)}(\underline{T}, Z) = \frac{1}{\sqrt{Z - 2u_0(\underline{T})}}, \qquad (4.283)$$

from which we obtain

$$R_n^{(0)}(\underline{T}) = c_n(2u_0(\underline{T}))^n. \qquad (4.284)$$

The string equation for $h = 0$ then reduces to Eq. (4.242), as it should since both equations express the asymptotic behaviour of the one-loop function. The explicit form of the string equation for $h = 0$ becomes

$$\sum_{n=0}^{\infty} \left(n + \frac{1}{2}\right) T_n c_n \left(2u_0(\underline{T})\right)^n = 0, \qquad (4.285)$$

from which $u_0(\underline{T})$ is implicitly determined. It admits a unique solution such that $u_0(T_0 = 0, T_1, \ldots) = 0$. This is most conveniently seen by making the change in variable $T_1 = \tilde{T}_1 - 2/3$, in which case we can write

$$u_0 = \frac{1}{2} T_0 + \frac{3}{2}\frac{1}{2} \tilde{T}_1(2u_0) + \frac{5}{2}\frac{3}{8} T_2\left(2u_0\right)^2 + \cdots, \qquad (4.286)$$

which can be solved iteratively in powers $T_0^{n_0} \tilde{T}_1^{n_1} T_2^{n_2} \ldots$, such that

$$n_0 + n_1 + \cdots \le n \quad \text{and} \quad n_1 + 2n_2 + \cdots \le n - 1$$

after the nth iteration. It follows that

$$
\begin{aligned}
u_0(\underline{T}) &= \frac{1}{2}T_0 + \frac{3}{4}T_0\tilde{T}_1 + \left(\frac{9}{8}T_0\tilde{T}_1^2 + \frac{15}{16}T_0^2 T_2\right) \\
&\quad + \left(\frac{27}{16}T_0\tilde{T}_1^3 + \frac{135}{32}T_0^2\tilde{T}_1 T_2 + \frac{35}{32}T_0^3 T_3\right) + \cdots.
\end{aligned}
\qquad (4.287)
$$

It can now easily be checked that Eq. (4.277) is satisfied for $h = 0$ and we can use Eq. (4.277) and the fact that $u_0(T_0 = 0, T_1, \ldots) = 0$ to prove that

$$\mathscr{L}_0(\underline{T}) = \frac{1}{4} \int_0^{T_0} dT_0' (T_0 - T_0') \, u_0(T_0', T_1, \ldots) \qquad (4.288)$$

satisfies

$$W_0^{cnt}(\underline{T}, Z) = \frac{d\mathscr{L}_0}{dV_{cnt}(Z)}. \qquad (4.289)$$

We omit the details of this straightforward calculation.

Note that $R^{(0)}(\underline{T}, Z)$ is a function of the coupling constants $\underline{T}$ solely through u_0, which is determined by the string equation. We now use Eqs. (4.274) and (4.277) to prove that this is true for arbitrary genus h

in the sense that $R(\underline{T}, Z)$ is a function of u and the derivatives $D^n u$ only, and that the loop equation for arbitrary coupling constants is formally equivalent to an infinite set of non-linear differential equations. We apply the operator $\Delta_Z D$, where

$$\Delta_Z \equiv G^2 D^3 + (2u - Z)D + (Du),$$

to both sides of the loop equation (4.264). In order to facilitate the use of $\Delta_Z D$ it is convenient to rewrite the loop equation (4.264) as

$$\oint \frac{d\Omega}{2\pi i} \frac{\tilde{V}'(\Omega)\tilde{W}(\Omega)}{Z - \Omega} = \tilde{W}^2(Z) + G^2 \frac{d\tilde{W}(Z)}{dV_{cnt}(Z)} + \frac{5G^2}{16Z^2}, \qquad (4.290)$$

where we have introduced the notation

$$\tilde{W}(Z) = W^{cnt}(Z) - \frac{T_0}{4\sqrt{Z}}, \qquad \tilde{V}'(Z) = V'_{cnt}(Z) - \frac{T_0}{2\sqrt{Z}}. \qquad (4.291)$$

By this change of variables

$$[\Delta_Z, \tilde{V}'(Z)] = 0, \qquad [\Delta_Z, D] = 0, \qquad R(Z) = -4D\tilde{W}(Z),$$

and applying $\Delta_Z D$ on the left-hand side of Eq. (4.290) we obtain

$$-\frac{1}{4} \oint \frac{d\Omega}{2\pi i} \frac{\tilde{V}'(\Omega)\Delta_Z R(\Omega)}{Z - \Omega} = -\frac{1}{4} \oint \frac{d\Omega}{2\pi i} \frac{\tilde{V}'(\Omega)\Delta_\Omega R(\Omega)}{Z - \Omega} - \frac{1}{8}. \qquad (4.292)$$

The last equation follows from $\Delta_Z = \Delta_\Omega - (Z - \Omega)D$ and

$$\frac{1}{4} \oint \frac{d\Omega}{2\pi i} \tilde{V}'(\Omega)DR(\Omega) = \frac{1}{4} \sum_{n=1}^{\infty} \left(n + \frac{1}{2}\right) T_n DR_n = -\frac{1}{8},$$

where the last equality is a consequence of the string equation (4.274). In order to apply $\Delta_Z D$ on the right-hand side of Eq. (4.290) we use

$$[\Delta_Z D, \frac{d}{dV_{cnt}(Z)}] = 2DRD^2 + D^2RD, \qquad (4.293)$$

which follows from Eq. (4.278). Using Eq. (4.293) we obtain

$$\Delta_Z D \left(\tilde{W}^2(Z) + G^2 \frac{d\tilde{W}(Z)}{dV_{cnt}(z)} + \frac{5G^2}{16Z^2}\right) \qquad (4.294)$$

$$= -\frac{\tilde{W}\Delta_Z R}{2} - \frac{G^2}{4} \frac{d\Delta_Z R}{dV_{cnt}(z)} + \frac{2G^2RD^2R - G^2(DR)^2 + (2u - Z)R^2}{8}.$$

Combining Eqs. (4.292) and (4.294) we obtain

$$-2 \oint \frac{d\Omega}{2\pi i} \frac{\tilde{V}'(\Omega)\Delta_\Omega R(\Omega)}{Z - \Omega} + 4\tilde{W}(Z)\Delta_Z R(Z) + 2G^2 \frac{d\Delta_Z R}{dV_{cnt}(Z)} \qquad (4.295)$$

$$= 1 + 2G^2RD^2R - G^2(DR)^2 + (2u - Z)R^2.$$

This equation is satisfied if R satisfies the non-linear differential equation

$$G^2(DR)^2 - 2G^2RD^2R + (Z - 2u)R^2 = 1, \qquad (4.296)$$

since differentiation of Eq. (4.296) with respect to T_0 yields the following third-order linear equation:

$$\left(G^2D^3 + (2u - Z)D + (Du)\right)R = 0, \quad \text{i.e.} \quad \Delta_Z R(Z) = 0. \qquad (4.297)$$

If we insert the expansion (4.270) in this equation we obtain

$$DR_{n+1} = \left(G^2D^3 + 2uD + (Du)\right)R_n, \qquad (4.298)$$

which allows a recursive solution for $R_n(u, Du, \ldots)$:

$$R_0 = 1, \quad R_1 = u, \quad R_2 = \frac{3}{2}u^2 + G^2D^2u, \quad \ldots \qquad (4.299)$$

This completes the proof of our claim that $R(\underline{T}, Z)$ depends solely on the coupling constants through u. The polynomials R_n are the so-called Gelfand–Dikii polynomials and the infinite set of non-linear differential equations Eq. (4.277), together with Eq. (4.298), is called the KdV-hierarchy of differential equations. In the langauge of integrable systems, imposing, in addition, the string equation turns the hierarchy into a two-reduced hierarchy. Eq. (4.296) is closely related to a Schrödinger equation in the variable T_0. Recall that the Schrödinger equation

$$\left[-\hbar^2 D^2 + v(T_0)\right]\Psi(T_0) = E\Psi(T_0) \qquad (4.300)$$

can be written as

$$\hbar^2 \left[(DR_E)^2 - 2R_E D^2 R_E\right] + 4(v(T_0) - E)R_E^2 = 1, \qquad (4.301)$$

if the two independent solutions $\Psi_{\pm}(T_0)$ to (4.300) are parametrized in the following way:

$$\Psi_{\pm}(T_0) = \sqrt{R_E(T_0)}\, e^{\pm S_E(T_0)/\hbar}, \quad DS_E(T_0) = \frac{1}{2R_E(T_0)}. \qquad (4.302)$$

Eqs. (4.301) and (4.302) are (one version) of the *WKB-equations*. Eq. (4.301) becomes identical to Eq. (4.296) if we make the identifications

$$Z = -4E, \quad u(T_0) = -2v(T_0), \quad R(Z) = R_E \quad G = \hbar. \qquad (4.303)$$

This interpretation of Eq. (4.296) highlights that $R(\underline{T}, Z)$ is known if the "potential" $-u(\underline{T})/2$ is known. In addition, it gives a simple interpretation of the genus expansion of $R(u, Z)$. *It is the WKB-expansion in G^2 of the quantity R which enters in the WKB-decomposition of the wave functions $\Psi_{\pm}(T_0)$ in Eq. (4.302).* In particular, we recognize the genus zero contribution $\sqrt{R^{(0)}(u, Z)}$, Eq. (4.283), as the the first term in the WKB expansion of R_E in the classically forbidden region ($E < v(T_0)$):

$$\sqrt{R_E^{(0)}} \sim \frac{1}{(v(T_0) - E)^{\frac{1}{4}}}.$$

Let us finally note that $R_E(T_0)/\hbar$ is the diagonal element of the resolvent of $\hat{H} = -\hbar^2 D^2 + v(T_0)$, as is seen immediately by constructing the Green function for $\hat{H} - E$ in terms of $\Psi_{\pm}(T_0)$, i.e.

$$\begin{aligned} R(u, Z) &= G \left\langle T_0 \left| \frac{1}{-G^2 D^2 - \frac{1}{2} u(\underline{T}) + \frac{1}{4} Z} \right| T_0 \right\rangle \\ &= G \int_0^\infty ds \left\langle T_0 \left| e^{s(G^2 D^2 + \frac{1}{2} u(\underline{T}) - \frac{1}{4} Z)} \right| T_0 \right\rangle. \end{aligned} \tag{4.304}$$

Gelfand and Dikii used standard properties of the resolvent to give an alternative proof of the relation between the Schrödinger operator $\hat{H} - E$ and Eq. (4.296).

For a given choice of coupling constants $\underline{T}$ the problem of solving two-dimensional quantum gravity has now been reduced to a determination of $u(\underline{T})$. The string equation is a non-linear differential equation which, for fixed couplings $\underline{T}$, gives us the possibility of determining $u(\underline{T})$. One simple choice of coupling constants is given by Eq. (4.198) and it corresponds to[*]

$$T_0, T_m \neq 0, \qquad T_k = 0, \quad k \neq 0, m. \tag{4.305}$$

In this case the string equation becomes

$$\left(m + \frac{1}{2}\right) T_m R_m(u) + \frac{1}{2} T_0 = 0. \tag{4.306}$$

Let us consider only pure gravity, i.e. $m = 2$:

$$\frac{5}{2} T_2 \left(G^2 D^2 u + \frac{3}{2} u^2\right) + \frac{1}{2} T_0 = 0. \tag{4.307}$$

By a rescaling,

$$u = 2G^2 \tilde{u}/\lambda^2, \quad T_0 = \lambda t, \quad \lambda^5 = -30 T_2 G^4,$$

we arrive at the so-called *Painlevé equation*

$$\frac{1}{3} \frac{d^2}{dt^2} \tilde{u} + \tilde{u}^2 - t = 0. \tag{4.308}$$

---

[*] The choice of coupling constants given by Eq. (4.198) is one where $c_-(g)$ changes, i.e. we have to use $c_-(g)$ instead of c_-^0 in Eq. (4.236). This implies that $M_m^0(g) = O(a^m)$ for $k < m$ and, consequently, $T_0 \neq 0$, while $T_k = 0$ for $1 \leq k < m$ when $a \to 0$.

Note that the variable t in Eq. (4.308) is proportional to the variable t introduced in Eq. (4.192) since Eqs. (4.282) and (4.285) allow us, for fixed T_2, to identify T_0 and Λ.

The solutions to the Painlevé equation have an asymptotic expansion of the form

$$\tilde{u} \sim \sum_h \frac{c_h}{t^{\frac{5}{2}(h-1)+2}}, \tag{4.309}$$

where $c_0 = -1$ and the coefficients c_h obey the following recursion relation:

$$c_{h+1} = \frac{25h^2 - 1}{24} c_h + \frac{1}{2} \sum_{k=1}^{h} c_k c_{h+1-k}. \tag{4.310}$$

From this relation it can be seen that $c_h \sim (2h)!$ for large h. Since

$$u \sim \frac{\partial^2 \mathcal{Z}}{\partial t^2},$$

this clearly exhibits the asymptotic nature of the expansion (4.190) for the partition function of pure gravity.

Note that the expansion (4.309) starts with a negative term for genus zero. This is the correct sign of $\chi(t) = \partial^2 \mathcal{Z}_0 / \partial t^2$, i.e. of the part of the full partition function for closed surfaces which scales and is therefore non-analytic when expressed in the original coupling constants g_j. Each differentiation $-\partial/\partial t$ can be viewed as an insertion of a puncture on the closed spherical surfaces. Since we sum over surfaces with a positive weight we should obtain a positive result. If the sum diverges at the critical point it must diverge to $+\infty$. The sums over spherical surfaces with zero, one and two punctures do not diverge, since $\gamma = -1/2$, but the sum over surfaces with three punctures does diverge. This implies that the coefficient of the leading non-analytic term of $\chi(t)$ must be negative since $-\partial\chi/\partial t$ is the three-point function.

The family of solutions to the Painlevé equation is well known (see, for instance, [234]. The question arises as to whether one of them qualifies as a *non-perturbative* definition of the sum over genus. Unfortunately, this seems not to be the case. The real valued solutions with the correct asymptotic behaviour all have singularities which are not physically acceptable. The solutions which are free of singularities suffer from imaginary non-perturbative terms. We conclude that the solutions to the Painlevé equations do not yield a physically acceptable solution to two-dimensional quantum gravity where topological fluctuations are allowed. We shall return to this issue from a slightly different angle in Subsections 4.8.3 and 4.8.4 of this chapter.

4.7   The two-point function

In our discussion of two-dimensional quantum gravity we have so far mostly considered the "global" aspects of the theory. In the following we shall substantiate the formal discussion given in Section 4.1 of "local" two-point functions (volume–volume correlations), using the framework of dynamical triangulations. The main goal of this section is to explain a suggestive calculation which leads to a closed form for the two-point function in the scaling limit. This function obeys the usual scaling relations but with non-standard exponents.

4.7.1   *General considerations*

The geodesic distances associated with a given equivalence class of metrics on a fixed manifold M is a central concept in the definition of the two-point function. Each triangulation represents such an equivalence class. Using Regge's prescription to assign a metric to a piecewise linear manifold defined by an abstract triangulation, the geodesic distance between any pair of points in M is defined. This notion of geodesic distance is rather inconvenient for our purposes and it is not entirely in the spirit of viewing the length of the lattice links a as a cutoff. From this point of view it is more natural to use a graph-theoretical definition of geodesic distance which allows us to talk about distances between the basic building blocks of triangulations, i.e. vertices, links, triangles and more general polygons. In view of the universality properties of two-dimensional gravity that we have seen in the present chapter it is desirable to have a notion of distance that applies to general graphs. The definition we give in the following is by no means unique. It is chosen from a number of candidates because it fits naturally to the problem we are about to address. We define the *geodesic distance* between two *links* in a given triangulation as the number of links in the shortest path between the corresponding links on the *dual lattice*. Recall that there is a one-to-one relation between links on the original lattice and the dual lattice and we can view a path on the dual lattice as one whose links connect the centres of neighbouring elementary polygons in the surface.

For simplicity, we restrict our attention in the following to unrestricted triangulations (all polygons are triangles) with planar topology. Let $\mathscr{T}(2;r)$ denote the collection of all triangulations of S^2 with two marked links where the geodesic distance between the marked links is r. If μ is the bare cosmological constant, we define the two-point function by

$$G_\mu(r) = \sum_{T \in \mathscr{T}(2;r)} e^{-\mu N_t(T)}. \qquad (4.311)$$

This can also be written as

$$G_\mu(r) = \sum_N e^{-\mu N} G(r, N),$$ (4.312)

where

$$G(r; N) = \#\{T \in \mathcal{T}(2; r) : N_t(T) = N\}.$$ (4.313)

We regard $G_\mu(r)$ as the discretized version of the continuum two-point function $G(R; \Lambda)$, while $G(r; N)$ can be viewed as the discretization of $G(R; V)$, see Section 4.1.

It is not difficult to verify that the two-point function $G_\mu(r)$ falls off exponentially with r if $\mu > \mu_0$. More precisely, the limit

$$m(\mu) = \lim_{r \to \infty} \frac{-\log G_\mu(r)}{r}$$ (4.314)

exists for all $\mu > \mu_0$ and defines a positive concave increasing function.

This can be proved by essentially the same method as was used to prove the existence of a mass in the hyper-cubic random surface models. For completeness we outline the argument. A pair of triangulations T_1, T_2 from $\mathcal{T}(2; r_1)$ and $\mathcal{T}(2; r_2)$ can be used to produce a triangulation $T_1 \oplus T_2$ in $\mathcal{T}(2; r_1 + r_2)$ by cutting the triangulations open along one of their marked links and then gluing them together. We have $N_t(T_1 \oplus T_2) = N_t(T_1) + N_t(T_2)$ and by introducing suitable boundary conditions[*] on the triangulations at the marked links we can ensure that different pairs T_1, T_2 give rise to different $T_1 \oplus T_2$. This is illustrated in Fig. 4.10. It follows that

$$G_\mu(r_1 + r_2) \ge G_\mu(r_1) G_\mu(r_2),$$ (4.315)

up to corrections due to the boundary conditions. The statements about the two-point function and the mass now follow by standard subadditivity arguments.

We expect the following behaviour of the two-point function:

$$G_\mu(r) \sim e^{-m(\mu)r}, \qquad r \gg 1/m(\mu)$$ (4.316)

$$G_\mu(r) \sim r^{1-\eta}, \qquad 1 \ll r \ll 1/m(\mu),$$ (4.317)

and for the susceptibility we expect

$$\chi(\mu) = \sum_r G_\mu(r) \sim (\mu - \mu_0)^{-\gamma}$$ (4.318)

for μ close to μ_0. We note that $\chi(\mu) \sim \mathcal{Z}''(\mu)$, in accordance with Eq. (4.174), since both quantities can be viewed as being obtained by

---

[*] This boundary condition is analogous to the "no pockets" condition used in the discussion of the hyper-cubic random surface models.

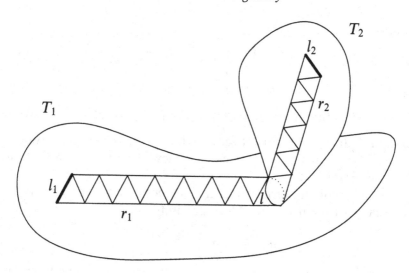

Fig. 4.10. Two triangulations with marked links separated by distances r_1 and r_2 can be glued together to a triangulation with a geodesic distance $r_1 + r_2$ between the marked links. This is done by cutting along a marked link in each of the triangulations to obtain a boundary of length two and then gluing the two boundary components together.

summing the Boltzmann factor $e^{-\mu N_t(T)}$ over all triangulations with two marked triangles or links. From the results given in the previous section we know that $\gamma = -1/2$.

In order to obtain a continuum two-point function it is necessary that the mass m scales to zero at μ_0. We shall argue below that this is actually the case. For the moment we assume the scaling of the mass, i.e. we assume that there is $\nu > 0$, such that

$$m(\mu) \sim (\mu - \mu_0)^\nu, \tag{4.319}$$

and explore the consequences. Since the relation between the lattice spacing a and μ is given by $\mu - \mu_0 = \Lambda a^2$ (see Eq. (4.159)) we can introduce a continuum mass M by

$$m(\mu(a)) = Ma^{2\nu}, \tag{4.320}$$

and a continuum distance R by

$$R = ra^{2\nu}. \tag{4.321}$$

In this way the continuum propagator $G(R;\Lambda)$ is unambiguously fixed by Eqs. (4.316)–(4.317):

$$G(R;\Lambda) = \lim_{a \to 0} a^{2\nu(1-\eta)} G_{\mu(a)}(Ra^{-2\nu}). \tag{4.322}$$

Since $m(\mu)r = MR$ we deduce from Eqs. (4.316) and (4.317) that the asymptotic behaviour of the continuum propagator $G(R;\Lambda)$ is of the form

$$G(R;\Lambda) \sim R^{1-\eta}, \qquad\qquad R \ll M, \qquad (4.323)$$
$$G(R;\Lambda) \sim e^{-MR}, \qquad\qquad R \gg M^{-1}. \qquad (4.324)$$

The remarks made in Section 4.1 about the relations $d_h = 1/v$ and $G(R,V) \sim V^{\gamma-2}R^{d_h-1}$ carry directly over to $G(r,N)$. For later reference we list the relations for $G(r;N)$ equivalent to Eqs. (4.316)–(4.318):

$$G(r;N) \sim e^{-c(r/N^v)^{\frac{1}{1-v}}} e^{\mu_0 N}, \qquad r \gg N^v, \qquad (4.325)$$
$$G(r;N) \sim N^{\gamma-2} r^{d_h-1} e^{\mu_0 N}, \qquad r \ll N^v, \qquad (4.326)$$
$$\chi(N) = \sum_r G(r;N) \sim N^{\gamma-1} e^{\mu_0 N}, \qquad N \gg 1. \qquad (4.327)$$

We remark that Eq. (4.325) describes the tail of the function and r must of course be smaller than the maximal distance between two links in a surface made up of N triangles. This maximal distance is clearly proportional to N. The term $e^{\mu_0 N}$ in $G(r;N)$ is an entropy factor describing the leading behaviour of $G(r;N)$ as $N \to \infty$ with r fixed. The exponent (anomalous dimension) η is determined by the short-distance behaviour of $G(r;N)$ as given by Eq. (4.326). For $r \ll m(\mu)^{-1}$ we obtain

$$G_\mu(r) = \sum_N e^{-\mu N} G(r;N) \sim r^{d_h-1} \sum_{N=r^{d_h}}^{\infty} N^{\gamma-2} \sim r^{\gamma d_h-1}. \qquad (4.328)$$

This shows that $\eta = 2 - \gamma d_h$, i.e.

$$\gamma = v(2-\eta), \qquad (4.329)$$

which we recognize as Fisher's scaling relation. Of course this relation could be derived directly from Eqs. (4.316)–(4.318), but the argument given here shows that the canonical and grand canonical definitions of the Hausdorff dimension agree.

4.7.2 A differential equation for the geodesic two-loop function

As in the case of the wave functionals of the previous sections we shall treat the two-point function from a combinatorial point of view. In order to formulate the problem it is convenient to introduce some definitions.

We have already defined the geodesic distance $d(\ell,\ell')$ between two links ℓ and ℓ' in an arbitrary triangulation. We extend this definition and say that the geodesic distance between a link ℓ and a set of links $\mathscr{L}$ is $\min_{\ell'\in\mathscr{L}} d(\ell,\ell')$. Furthermore, we say that a loop $\mathscr{L}_1$ has a geodesic distance r to another loop $\mathscr{L}_2$ if *all* links $\ell \in \mathscr{L}_1$ have a geodesic distance r to the loop $\mathscr{L}_2$. Note that the definition is *not symmetric* in $\mathscr{L}_1$ and $\mathscr{L}_2$. However, it is the most natural definition for the application we have

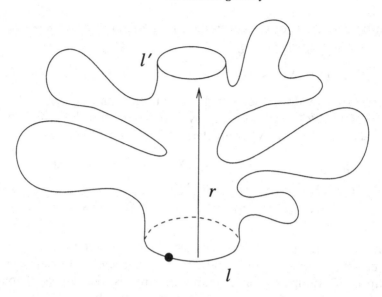

Fig. 4.11. A typical surface contributing to $G_\mu(l, l'; r)$. The "dot" on the entrance loop represents the marked link.

in mind. Let $\mathcal{T}(l_1, l_2; r)$ denote the subclass of triangulations in $\mathcal{T}(l_1, l_2)$ which satisfy the following condition: The loop $\mathcal{L}_1$ of length l_1, which we shall call the *entrance loop*, has one marked link. The other loop, $\mathcal{L}_2$, which we shall call the the *exit loop*, has geodesic distance r from the entrance loop. The exit loop has no marked link. The triangulations in $\mathcal{T}(l_1, l_2; r)$ may be viewed as cylinders of height r; see Fig. 4.11. Triangulations in $\mathcal{T}(l_1, l_2; r)$ are related to triangulations in $\mathcal{T}(l_1)$, the class of triangulations with a boundary of length l_1 and one marked link, as shown in Fig. 4.12. As can be seen from Fig. 4.12 the elements in $\mathcal{T}(l_1, l_2; r)$ are obtained by chopping off a disk T' from the disk $T \in \mathcal{T}(l_1)$ such that a loop of length l_2 and geodesic distance r to $\mathcal{L}_1$ is created. As illustrated in the diagram the loop $\mathcal{L}_2$ might only be one of many loops in T which have a geodesic distance r to $\mathcal{L}_1$.

We now define the *geodesic two-loop function* by

$$G_\mu(l_1, l_2; r) = \sum_{T \in \mathcal{T}(l_1, l_2; r)} e^{-\mu N_t(T)}, \qquad (4.330)$$

We expect the geodesic two-loop function to fall off exponentially as $r \to \infty$ with the same mass as the two-point function, and this is verified by the calculations given below. It is convenient to introduce the generating function

$$G_\mu(z_1, z_2; r) = \sum_{l_1, l_2 = 1}^{\infty} z_1^{-(l_1 + 1)} z_2^{-(l_2 + 1)} G_\mu(l_1, l_2; r). \qquad (4.331)$$

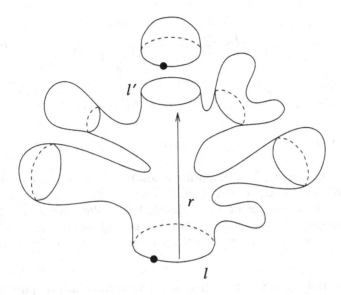

Fig. 4.12. Illustration of the relation between a triangulation in $\mathcal{T}(l, l'; r)$ and a triangulation in $\mathcal{T}(l)$. One obtains the elements in $\mathcal{T}(l, l'; r)$ by chopping off a disc with boundary length l', i.e. an element in $\mathcal{T}(l')$. The diagram shows that many loops in a specific $T' \in \mathcal{T}(l)$ can have the geodesic distance r to the entrance loop $\mathcal{L}_1$ of length l.

The initial value condition, corresponding to $r = 0$, satisfied by the geodesic two-loop function is

$$G_\mu(l_1, l_2; 0) = \delta_{l_1, l_2}, \tag{4.332}$$

which in terms of the generating function reads

$$G_\mu(z_1, z_2; 0) = \frac{1}{z_1 z_2} \frac{1}{z_1 z_2 - 1}. \tag{4.333}$$

We note that

$$G_\mu(l_1, l_2; r) = \oint_{C_1} \frac{dz_1}{2\pi i} z_1^{l_1} \oint_{C_2} \frac{dz_2}{2\pi i} z_2^{l_2} G_\mu(z_1, z_2; r), \tag{4.334}$$

where the contours C_1 and C_2 enclose the singularities of $G_\mu(z_1, z_2; r)$ [263, 372].

Our goal now is to derive a differential equation for the generating function $G_\mu(z_1, z_2; r)$. Let us consider a set of elementary deformations of the surfaces that contribute to the geodesic two-loop function. We start at the marked link on the entrance loop in a triangulation T and remove the triangle containing this link. Remembering the degenerate structure of the unrestricted triangulations, the marked link might not belong to any triangle. In this case it is a double-link in the parlance of

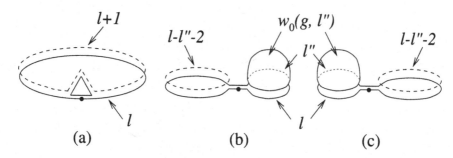

Fig. 4.13. The "peeling" decompositions are the same as already used in the derivation of the loop equations. A marked link on the entrance boundary can either belong to a triangle or to a "double-link". We have drawn the three possible situations in a way which makes the "peeling aspect" more transparent. The dashed curve indicates the new entrance loop after one peeling step.

Section 4.2 and we remove both links. The triangulation splits into two parts under this operation, T_1 containing the exit loop and T_2 with the topology of a sphere with one hole. We use some convention to define a new marked link in T_1. Note that the distance from the exit loop to the entrance loop in T_1 is still r but the boundary of T_1 is shorter than that of T. If the marked link in T is not a double-link the surface is connected after the removal of the triangle but the distance from the exit loop to the entrance loop is not necessarily well defined any more. The elementary deformations are illustrated in Fig. 4.13. After sufficiently many applications of the deformations we have removed all triangles and links between the entrance loop and the exit loop in any given triangulation and one can think of the construction as the peeling of a fruit with outgrowths that are chopped off in the peeling process. One can constrast this process with the one where, instead of peeling the surface, we "slice" it like an onion in rings of thickness 1, as shown in Fig. 4.14. By definition the entrance and the exit loop are separated by r "slicing steps". Heuristically, one can view an elementary deformation performed on a loop of length l as $1/l$th of one slicing.

The equation satisfied by $G_\mu(l, l'; r)$ under elementary deformations is

$$G_\mu(l, l'; r) = g G_\mu'(l+1, l'; r) + 2 \sum_{l''=0}^{l-2} w_0(g, l'') G_\mu(l - l'' - 2, l'; r). \quad (4.335)$$

The first term on the right-hand side of the above equation corresponds to the case when the surface is connected after the deformation, while the second term arises from the cases when it splits in two. The prime on the G' indicates that the surfaces contributing to this quantity do not really satisfy the constraint imposed on surfaces in $\mathscr{T}(l+1, l'; r)$. The

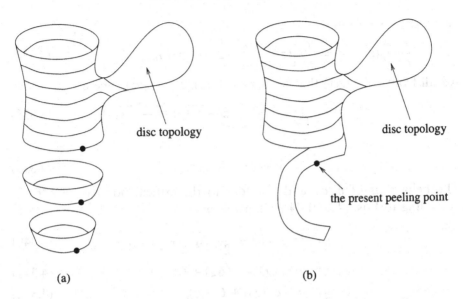

Fig. 4.14. Decomposition of a surface by (a) slicing and (b) peeling.

application of the one-step peeling l times should on average correspond to cutting a slice of thickness 1 from the surface. It is therefore tempting to take a leap of faith and identify the change caused by one elementary deformation with

$$\left[\frac{\partial}{\partial r}G_\mu(l,l',r)\right]\frac{1}{l},$$

forgetting that r is really an integer variable. It follows that we can write

$$\frac{\partial}{\partial r}G_\mu(l,l';r) = -lG_\mu(l,l',r) + glG_\mu(l+1,l';r)$$

$$+2\sum_{l''=0}^{l-2} lw_0(g,l'')G_\mu(l-l''-2,l';r), \qquad (4.336)$$

where we have dropped the prime on the G. This equation is manifestly incorrect even if we replace the differentiation with respect to r by a finite difference operator. However, *we conjecture that this equation captures the asymptotic behaviour of the loop functions* for large r and the foregoing discussion should be regarded as a motivation for this conjecture. At the end of this section we mention another method for studying the two-point function in the scaling limit which gives results identical to those obtained from Eq. (4.336).

Written in terms of the generating functional $G_\mu(z,z';r)$, Eq. (4.336) becomes

$$\frac{\partial}{\partial r} G_\mu(z, z'; r) = \frac{\partial}{\partial z}\Big[(z - gz^2 - 2w_0(g, z))G_\mu(z, z'; r)\Big]. \tag{4.337}$$

Recall from Eq. (4.61) that $w_0(g, z) = V'(z)/2 + f(g, z)$, where

$$f(g, z) = \frac{g}{2}(z - c(g))\sqrt{(z - c_+(g))(z - c_-(g))}, \tag{4.338}$$

and

$$V'(z) = z - gz^2. \tag{4.339}$$

The behaviour of c, c_+ and c_- close to the critical point $g_0 = e^{-\mu_0}$ is, according to Eqs. (4.107)–(4.109), given by

$$c(g) = z_0\left(1 + \frac{1}{2}\sqrt{\Delta g}\right) + O(\Delta g), \tag{4.340}$$

$$c_+(g) = z_0(1 - \sqrt{\Delta g}) + O(\Delta g), \tag{4.341}$$

$$c_-(g) = c_-(g_0) + O(\Delta g), \tag{4.342}$$

where $\Delta g = g_0 - g$. It should be noted that the non-universal term $V'(z)/2$ cancels in Eq. (4.337) such that the differential equation becomes

$$\frac{\partial}{\partial r} G_\mu(z, z'; r) = -2\frac{\partial}{\partial z}\Big[f_\mu(z)G_\mu(z, z'; r)\Big], \tag{4.343}$$

where we have introduced the notation $f_\mu(z) = f(g, z)$, $g = e^{-\mu}$. The combinatorial arguments for the peeling are similar to those used for the derivation of the loop equations so they can be generalized to an arbitrary potential $V(z)$. By writing $w_0(g, z) = V'(z)/2 + f(g, z)$ the cancellation of the non-universal potential term $V'(z)/2$ will still occur. The function $f(g, z)$ depends on the potential, but this does not affect the scaling limit.

4.7.3 Solution of the differential equation

We now discuss the solution of the first-order partial differential equation (4.343). The characteristic equation is

$$\frac{d\hat{z}}{dr} = -2f(\hat{z}), \qquad \hat{z}(z, r = 0) = z, \tag{4.344}$$

with the following integral:

$$r = \int_{\hat{z}(z,r)}^{z} \frac{dz'}{2f_\mu(z')} = \left[-\frac{1}{\delta_0} \sinh^{-1}\sqrt{\frac{\delta_1}{1 - c/z'} - \delta_2}\right]_{z'=\hat{z}(z,r)}^{z'=z}. \tag{4.345}$$

This expression can be inverted and we find

$$\frac{1}{\hat{z}(z,r)} = \frac{1}{c} - \frac{\delta_1}{c} \frac{1}{\sinh^2\left(\delta_0 r + \sinh^{-1}\sqrt{\frac{\delta_1}{1-c/z} - \delta_2}\right) + \delta_2}, \tag{4.346}$$

where δ_0, δ_1 and δ_2 are all positive and defined by

$$\delta_0 = \frac{g}{2}\sqrt{(c-c_+)(c-c_-)} = O\left((\Delta g)^{\frac{1}{4}}\right), \tag{4.347}$$

$$\delta_1 = \frac{(c-c_+)(c-c_-)}{c(c_+-c_-)} = O\left((\Delta g)^{\frac{1}{2}}\right), \tag{4.348}$$

$$\delta_2 = -\frac{c_-(c-c_+)}{c(c_+-c_-)} = O\left((\Delta g)^{\frac{1}{2}}\right). \tag{4.349}$$

It can readily be checked that $\hat{z} \to c$ as $r \to \infty$.

The solution to Eq. (4.343) with the boundary condition (4.333) is

$$G_\mu(z,z';r) = \frac{f_\mu(\hat{z})}{f_\mu(z)}\frac{1}{\hat{z}z'-1}\frac{1}{\hat{z}z'}, \tag{4.350}$$

where $\hat{z} = \hat{z}(z,r)$ is given by Eq. (4.346).

In principle, $G_\mu(l_1,l_2;r)$ can be calculated from Eqs. (4.332), (4.350) and (4.346). We shall be content with observing that the exponential decay of $G_\mu(l_1,l_2;r)$ is independent of l_1 and l_2. For large r one can check that

$$G_\mu(l_1,l_2;r) = C\,\delta_0\delta_1\,e^{-2\delta_0 r} + O(e^{-4\delta_0 r}). \tag{4.351}$$

where the constant C depends on c_-, c_+, c, l_1 and l_2.

The two-point function $G_\mu(r)$ can be expressed in terms of the geodesic two-loop function and the one-loop function as we now describe. Consider a surface with an entrance loop of length 1 and an exit loop of arbitrary length l_2. We can close this surface up by gluing a disc with a boundary of length l_2 to the exit loop and closing the entrance loop by a single triangle. Up to boundary-dependent corrections, which we ignore, the two-point function is obtained by multiplying the two-loop function $G_\mu(1,l_2;r)$ by the one-loop function $w_0(g,l_2)$ and summing over l_2. This is illustrated in Fig. 4.15 and the corresponding equation is

$$G_\mu(r) = \sum_{l_2=1}^{\infty} G_\mu(1,l_2;r)\,l_2 w_0(g,l_2) \tag{4.352}$$

$$= \oint_{C_{z'}} \frac{dz'}{2\pi i z'}\left[z^2 G_\mu\left(z,\frac{1}{z'};r\right)\right]\left[\frac{\partial}{\partial z'}z'w_0(z')\right]\Big|_{z=\infty}.$$

The additional factor l_2 in Eq. (4.352) is due to the fact that the exit loop is unmarked and we can glue the marked one-loop cap to it in l_2 ways.

The contour integral can be evaluated by using the expression (4.350) for $G_\mu(z,z';r)$. The integration contour can be deformed to infinity and

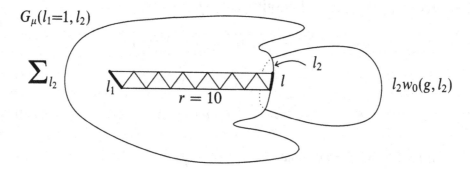

$G_\mu(l_1{=}1, l_2)$

$\displaystyle\sum_{l_2}$ l_1 $r = 10$ l l_2 $l_2 w_0(g, l_2)$

Fig. 4.15. The two-point function represented as a sum over geodesic two-loop functions multiplied by one-loop functions.

the only contribution comes from the pole at $\hat{z}(z, r)$, i.e.

$$
\begin{aligned}
G_\mu(r) &= \left. \frac{z^2 f(\hat{z})}{f(z)} \frac{\partial (\hat{z} w_0(\hat{z}))}{\partial \hat{z}} \right|_{z=\infty} \\
&= \left. -\frac{1}{g} \frac{\partial \hat{z}}{\partial r} \frac{\partial (\hat{z} w_0(\hat{z}))}{\partial \hat{z}} \right|_{z=\infty},
\end{aligned}
\tag{4.353}
$$

where the last equality in Eq. (4.353) follows from the characteristic equation (4.344). As long as $c - c_+ \sim \sqrt{\Delta g}$ is small, $z = \infty$ and r is larger than a few lattice spacings, it follows from Eq. (4.346) that $\hat{z}$ is close to c, and the large r behaviour of $G_\mu(r)$ is determined by the term

$$
\left. \frac{\partial \hat{z}(z, r)}{\partial r} \right|_{z=\infty} \approx 2c \delta_0 \delta_1 \frac{\cosh \delta_0 r}{\sinh^3 \delta_0 r}.
\tag{4.354}
$$

It follows that

$$
G_\mu(r) = \text{const.}\, \delta_0 \delta_1 \frac{\cosh(\delta_0 r)}{\sinh^3(\delta_0 r)} (1 + O(\delta_0)).
\tag{4.355}
$$

Using the behaviour of δ_0 and δ_1 in the scaling limit, Eq. (4.355) can be written as

$$
G_\mu(r) \sim a^{3/2} \frac{\cosh\left[\beta \Lambda^{\frac{1}{4}} a^{\frac{1}{2}} r \right]}{\sinh^3 \left[\beta \Lambda^{\frac{1}{4}} a^{\frac{1}{2}} r \right]}.
\tag{4.356}
$$

where $\beta = \sqrt{6} g_0$, and we have used the notation $\Delta g = g_0 \Lambda a^2$.

We may summarize our results as follows:

1. $G_\mu(r)$ falls off as $e^{-2(\mu-\mu_0)^{\frac{1}{4}} \beta r}$ as $r \to \infty$, i.e. the critical exponent $\nu = \frac{1}{4}$ and the Hausdorff dimension $d_h = 4$.

2. $G_\mu(r)$ behaves as r^{-3} for $1 \ll r \ll (\mu - \mu_0)^{-\frac{1}{4}}$, i.e. the anomalous dimension $\eta = 4$.

3. The wave-function renormalization is $a^{2\nu(\eta-1)}$ and Fisher's scaling relation is fulfilled.

4. Any two-loop function $G_\mu(l_1, l_2; r)$ has the same asymptotic behaviour as $G_\mu(r)$, as long as l_1, l_2 stay finite, in the limit $a \to 0$.

We remark that the anomalous dimension η is out of the range $[0, 2]$ normally encountered in quantum field theory [204]. We could have taken the continuum limit at various stages of the above calculations and obtained

$$G(R; \Lambda) = C\Lambda^{3/4} \frac{\cosh[\Lambda^{\frac{1}{4}}R]}{\sinh^3[\Lambda^{\frac{1}{4}}R]}, \qquad (4.357)$$

where C is a constant and we have rescaled the continuum distance R by a factor β; see Eq. (4.321).

4.7.4 A transfer matrix approach

As claimed previously, the result given by Eq. (4.357) can be obtained by an independent derivation using the so-called transfer matrix formalism. Let us briefly describe the basic ideas in this construction since they might be useful in other contexts as well.

Following [263] we may, at least from a heuristic point of view, uniquely decompose any triangulation in $\mathscr{T}(l_1, l_2, r)$ into r rings of thickness 1, i.e. into elements belonging to various $\mathscr{T}(l, l', 1)$, by applying the slicing procedure described above (see Fig. 4.14). This yields, in particular, the identity

$$G_\mu(l_1, l_2; r_1 + r_2) = \sum_{l=1}^{\infty} G_\mu(l_1, l; r_1) \, G_\mu(l, l_2; r_2), \qquad (4.358)$$

which in terms of the generating function reads

$$G_\mu(z_1, z_2; r_1 + r_2) = \oint_C \frac{dz}{2\pi i z} \, G_\mu\left(z_1, \frac{1}{z}; r_1\right) G_\mu(z, z_2; r_2). \qquad (4.359)$$

The sum in Eq. (4.358) can be viewed as a sum over intermediate states depending on diffeomorphism classes of one-dimensional (induced) metrics, i.e. on the length of the intermediate loop. The decomposition is illustrated in Fig. 4.16. It follows that we can in principle determine $G_\mu(l_1, l_2, r)$ in terms of $G_\mu(l, l', 1)$ and view the latter as a transfer matrix. We now argue that *it is possible to determine $G_\mu(l, l', 1)$ by purely combinatorial arguments*.

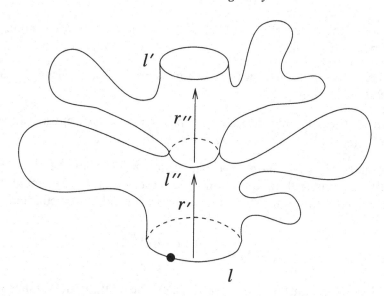

Fig. 4.16. A typical surface which explains the composition law for $G_\mu(l, l'; r)$.

If $G_\mu^{(0)}(z_1, z_2)$ denotes the generating function corresponding to a cylinder where the exit loop is at a geodesic distance 1 from the entrance loop and both boundaries are unmarked, we have

$$G_\mu(z_1, z_2; r = 1) = -\frac{\partial}{\partial z_1} z_1 G_\mu^{(0)}(z_1, z_2), \qquad (4.360)$$

since the differentiation will contribute a factor l_1 for each entrance loop of length l_1 which makes up for the marking. In Fig. 4.17 we show a "typical" triangulation contributing to $G_\mu^{(0)}(z_1, z_2)$. If we introduce the usual relation $g = e^{-\mu}$ it is possible to write

$$
\begin{aligned}
G_\mu^{(0)}(z_1, z_2) &= \sum_{n=1}^{\infty} \frac{1}{n} \left(\frac{g}{z_1 z_2^2} + \frac{g w_0(z_1)}{z_1^2 z_2} + \frac{g z_1 (w_0(z_1) - 1)}{z_1^2} + \frac{w_0(z_1)}{z_1^2} \right)^n \\
&= -\log \left(1 - \frac{g}{z_1 z_2^2} - \frac{g w_0(z_1)}{z_1^2 z_2} - \frac{g z_1 (w_0(z_1) - 1)}{z_1^2} - \frac{w_0(z_1)}{z_1^2} \right).
\end{aligned}
$$

The summation index n is the length of the entrance loop and the factor $1/n$ is the symmetry factor present since there is no marked link at the boundary. The four terms correspond to the four different situations that can be encountered in the one-step deformation from the entrance to the exit loop. They have the following graphical representations (where blobs denote any triangulation of the disk, the dotted lines indicate entrance

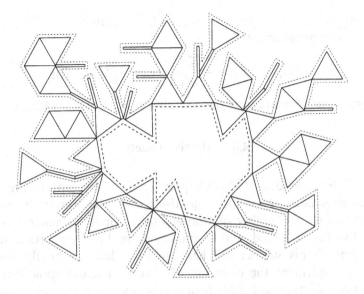

Fig. 4.17. A typical graph contributing to $G_\mu^{(0)}(z_1, z_2)$. A thin broken line represents an entrance loop and a thick broken line represents an exit loop.

links and the dashed lines exit links):

$$\triangle \;=\; \frac{g}{z_1 z_2^2}, \tag{4.361}$$

$$\text{(img)} \;=\; \frac{g w_0(z_1)}{z_1^2 z_2}, \tag{4.362}$$

$$\text{(img)} \;=\; \frac{g(z_1 w_0(z_1) - 1)}{z_1^2}, \tag{4.363}$$

$$\text{(img)} \;=\; \frac{w_0(z_1)}{z_1^2}. \tag{4.364}$$

An elementary, but rather tedious, calculation leads to the result

$$G_\mu(z_1, z_2; r = 1) = \frac{1}{a}\frac{1}{Z_1 + Z_2} - \frac{1}{\sqrt{a}}\frac{\partial}{\partial Z_1}\left[\frac{W(Z_1)}{Z_1 + Z_2}\right] + O(1) \tag{4.365}$$

in the scaling limit, where

$$z_1 = z_0(1 + a Z_1), \quad z_2 = z_0(1 + a Z_2) \quad \text{and} \quad g = g_0(1 - \Lambda a^2).$$

As usual, $W(Z) = (Z - \frac{1}{2}\sqrt{\Lambda})\sqrt{Z + \sqrt{\Lambda}}$ is the universal part of the one-loop function. If we introduce the *continuum* generating function for the two-loop functions $G(L_1, L_2; R)$, where the two loops of (continuum) length L_1 and L_2 are separated a geodesic distance R by

$$G(Z_1, Z_2; R) = a\, G_\mu(z_1, z_2; r), \qquad R = r\sqrt{a}, \tag{4.366}$$

the decomposition law expressed by Eq. (4.359) with $r_1 = 1$ reduces to the following differential equation:

$$\frac{\partial}{\partial R} G(Z_1, Z_2; R) = -\frac{\partial}{\partial Z_1} [W(Z_1) G(Z_1, Z_2; R)]. \qquad (4.367)$$

We recognize this equation as the continuum limit of Eq. (4.343).

4.8   Matrix models

In the previous sections of this chapter we have, by purely combinatorial arguments, derived the loop equations which determine the generating function for the number of triangulations (or more general graphs). Historically, the so-called matrix models have been important for the development of this subject and here we shall describe briefly how it is possible to implement the counting of unrestricted triangulations using the formalism of quantum field theory. This approach may also suggest a way giving a *non-perturbative* definition of the sum over all genera.

4.8.1   *Counting triangulations using matrix models*

Let ϕ be a Hermitian $N \times N$ matrix with matrix elements $\phi_{\alpha\beta}$ and consider for $k = 0, 1, 2, \ldots$ the integral

$$\int d\phi \, e^{-\frac{1}{2}\operatorname{Tr} \phi^2} \frac{1}{k!} \left(\frac{1}{3}\operatorname{Tr} \phi^3\right)^k, \qquad (4.368)$$

where

$$d\phi = \prod_{\alpha \leq \beta} d\operatorname{Re} \phi_{\alpha\beta} \prod_{\alpha < \beta} d\operatorname{Im} \phi_{\alpha\beta}. \qquad (4.369)$$

We can regard ϕ as a zero-dimensional matrix-valued field so the integral can be evaluated in the standard way by doing all possible Wick contractions of $(\operatorname{Tr} \phi^3)^k$ and using

$$\langle \phi_{\alpha\beta} \phi_{\alpha'\beta'} \rangle = C \int d\phi \, e^{-\frac{1}{2}\sum_{\alpha\beta} |\phi_{\alpha\beta}|^2} \phi_{\alpha\beta} \phi_{\alpha'\beta'} = \delta_{\alpha\beta'} \delta_{\beta\alpha'}, \qquad (4.370)$$

where C is a normalization factor. The evaluation of the expression (4.368) can be interpreted graphically by associating to each factor $\operatorname{Tr} \phi^3$ an oriented triangle and to each term $\phi_{\alpha\beta} \phi_{\beta\gamma} \phi_{\gamma\alpha}$ contributing to the trace a labelling of its vertices by α, β, γ in cyclic order, such that the matrix element $\phi_{\alpha\beta}$ is associated with the oriented link whose endpoints are labelled by α and β in accordance with the orientation. Eq. (4.370) can then be interpreted as a gluing of the link labelled by $(\alpha\beta)$ to an oppositely oriented copy of the same link, see Fig. 4.18.

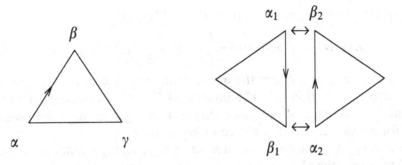

Fig. 4.18. The matrix representation of triangles which converts the gluing along links to a Wick contraction.

In this way the integral (4.368) can be represented as a sum over closed, possibly disconnected, tringulations with k triangles. Triangulations with an arbitrary genus arise in this representation. By summing over k in (4.368) it follows by standard arguments that the (formal) logarithm of the corresponding integral is represented as a sum over all closed and *connected* triangulations. The contribution of a given triangulation can be determined by observing that in the process of gluing we pick up a factor of N whenever a vertex becomes an internal vertex in the triangulation. Thus, the weight of a triangulation T is simply

$$N^{N_v(T)}.$$

If we make the substitution

$$\text{Tr } \phi^3 \to \frac{g}{\sqrt{N}} \text{Tr } \phi^3, \tag{4.371}$$

the weight of T is replaced by

$$g^k N^{N_v(T)-k/2} = g^k N^{\chi(T)},$$

where $\chi(T)$ is the Euler characteristic of T. Note also that the factor $(k!)^{-1}$ in (4.368) is cancelled in the sum over different triangulations because of the $k!$ possible permutations of the triangles, except for triangulations with non-trivial automorphisms, in which case the symmetry factor C_T^{-1} survives. With the identifications

$$\frac{1}{\kappa} = \log N, \qquad \mu = -\log g$$

we conclude that

$$\mathscr{Z}(\mu, \kappa) = \log \frac{Z(g, N)}{Z(0, N)} \tag{4.372}$$

where $\mathscr{L}(\mu, \kappa)$ is defined by Eq. (4.42)–(4.43) and

$$Z(g, N) = \int d\phi \, \exp\left(-\frac{1}{2}\text{Tr } \phi^2 + \frac{g}{3\sqrt{N}}\text{Tr } \phi^3\right). \qquad (4.373)$$

The variables g and N are the indeterminates in the generating functions introduced in Section 4.2. The integral (4.373) is of course divergent and should at this stage of our discussion just be regarded as a shorthand for the formal power series in the coupling constant g.

It is straightforward to generalize the preceding arguments to the case of general unrestricted triangulations where arbitrary polygons are allowed. Instead of one coupling constant g we have a set $\underline{g} = \{g_1, g_2, g_3, \ldots\}$, as in Section 4.2. In this case Eq. (4.373) is replaced by

$$Z(\underline{g}, N)) = \int d\phi \, e^{-N\text{Tr } V(\phi)}, \qquad (4.374)$$

where the potential V, which depends on all the coupling constants $\underline{g}$, is given by

$$V(\phi) = \frac{1}{2}\phi^2 - \sum_{j=1}^{\infty} \frac{g_j}{j} \phi^j. \qquad (4.375)$$

In Eq. (4.375) we have scaled $\phi \to \sqrt{N}\phi$ for later convenience. It is not difficult to check that the obvious generalization of Eq. (4.372) holds and the weight of a triangulation T is given by

$$C_T^{-1} N^{\chi(T)} \prod_{j\geq 1} g_j^{N_j(T)},$$

where $N_j(T)$ is the number of j-gons in T. Eq. (4.374) is of course a representation of a formal power series which is obtained by expanding the exponential of the non-quadratic terms as a power series in the coupling constants and then performing the Gaussian integrations term by term.

Differentiating $\log Z(\underline{g}, N)$ with respect to the coupling constants g_j, one obtains the expectation values of products of traces of powers of ϕ. These expectations have a straightforward interpretation in terms of triangulations. Denoting the expectation with respect to the measure

$$Z(\underline{g}, N)^{-1} e^{-N\text{Tr } V(\phi)} \, d\phi$$

by $\langle \cdot \rangle$ we see, for example, that $\langle N^{-1}\text{Tr } \phi^n \rangle$ is given by the sum over all connected triangulations of arbitrary genus whose boundary is an n-gon with one marked link. Similarly,

$$\frac{1}{N^2} \langle \text{Tr } \phi^n \text{Tr } \phi^m \rangle - \frac{1}{N^2} \langle \text{Tr } \phi^n \rangle \langle \text{Tr } \phi^m \rangle \qquad (4.376)$$

is given by the sum over all connected triangulations whose boundary consists of two components with n and m links. More generally, the relation to the combinatorial problem discussed in Section 4.3. is given by

$$w(\underline{g}, z_1, \ldots, z_b) = N^{b-2} \sum_{k_1, \ldots, k_b} \frac{\left\langle \operatorname{Tr} \phi^{k_1} \cdots \operatorname{Tr} \phi^{k_b} \right\rangle_{conn}}{z_1^{k_1+1} \cdots z_b^{k_b+1}}, \tag{4.377}$$

where the subscript *conn* indicates the connected part of the expectation $\langle \cdot \rangle$. One can rewrite Eq. (4.377) as

$$w(g, z_1, \ldots, z_b) = N^{b-2} \left\langle \operatorname{Tr} \frac{1}{z_1 - \phi} \cdots \operatorname{Tr} \frac{1}{z_b - \phi} \right\rangle_{conn}. \tag{4.378}$$

The one-loop function is related to the density $\rho(\lambda)$ of eigenvalues of ϕ defined by

$$\rho(\lambda) = \left\langle \sum_{i=1}^{N} \delta(\lambda - \lambda_i) \right\rangle, \tag{4.379}$$

where λ_i, $i = 1, \ldots, N$ denote the N eigenvalues of the matrix ϕ. With this definition we have

$$\frac{1}{N} \left\langle \operatorname{Tr} \phi^n \right\rangle = \int_{-\infty}^{\infty} d\lambda \, \rho(\lambda) \lambda^n, \quad n \geq 0. \tag{4.380}$$

Hence,

$$w(z) = \int_{-\infty}^{\infty} d\lambda \, \frac{\rho(\lambda)}{z - \lambda}. \tag{4.381}$$

In the limit $N \to \infty$ the support of ρ is confined to a finite interval $[c_-, c_+]$ on the real axis. In this case $w(z)$ will be an analytic function in the complex plane, except for a cut along the interval $[c_-, c_+]$. Note that $\rho(\lambda)$ is determined from $w(z)$ by

$$2\pi i \rho(\lambda) = \lim_{\varepsilon \to 0} (w(\lambda - i\varepsilon) - w(\lambda + i\varepsilon)). \tag{4.382}$$

We would like to mention that the combinatorics of the restricted set of graphs with even-sided polygons which we considered in Subsection 4.2.4 also has a matrix model representation. Let us replace the Hermitian matrix ϕ by a general complex matrix Φ and let the action be

$$V(\Phi^\dagger \Phi) = \operatorname{Tr} \Phi^\dagger \Phi - \sum_{j=1}^{\infty} \frac{g_j}{j} \operatorname{Tr} (\Phi^\dagger \Phi)^j. \tag{4.383}$$

Defining the expectation $\langle \cdot \rangle$ as before with $d\phi$ replaced by

$$d\Phi = \prod_{\alpha, \beta} d \operatorname{Re} \Phi_{\alpha\beta} \, d \operatorname{Im} \Phi_{\alpha\beta},$$

the loop functions are defined as in Eq. (4.378), with ϕ replaced by $\Phi^\dagger \Phi$. In the same way as the Hermitian matrix model defined by Eq. (4.374) described the gluing of arbitrary polygons along their links, the complex matrix model describes the gluing of "checker board" polygons, i.e. polygons where the links have alternating black and white colours, under the restriction that we can only glue black links to white links. Black links are represented by Φ, white links by $\Phi^\dagger$ and the restriction on gluing is reflected in the rules for Wick contractions analogous to Eq. (4.370):

$$\langle \Phi^\dagger_{\alpha\beta} \Phi^\dagger_{\alpha'\beta'} \rangle_0 = \langle \Phi_{\alpha\beta} \Phi_{\alpha'\beta'} \rangle_0 = 0, \qquad \langle \Phi^\dagger_{\alpha\beta} \Phi_{\alpha'\beta'} \rangle_0 = \delta_{\alpha\beta'} \delta_{\beta\alpha'},$$

where $\langle \cdot \rangle_0$ denotes Gaussian exceptation values, i.e. expectation values calculated with all $g_j = 0$ in Eq. (4.383). The *planar* checker board graphs are identical to the planar graphs constructed from even-sided polygons with the additional restriction that all boundary components should have an even number of links*.

4.8.2 *The loop equations*

It is a standard method in quantum field theory to derive identities by a change of variable in functional integrals. Here we apply this method to the matrix models and explore the invariance of the matrix integral (4.374) under infinitesimal field redefinitions of the form

$$\phi \to \phi + \varepsilon \phi^n, \tag{4.384}$$

where ε is an infinitesimal parameter. One can show that to first order in ε the measure $d\phi$ defined by Eq. (4.369) transforms as

$$d\phi \to d\phi \left(1 + \varepsilon \sum_{k=0}^{n} \mathrm{Tr}\, \phi^k \, \mathrm{Tr}\, \phi^{n-k} \right). \tag{4.385}$$

The action transforms according to

$$\mathrm{Tr}\, V(\phi) \to \mathrm{Tr}\, V(\phi) + \varepsilon \mathrm{Tr}\, \phi^n V'(\phi) \tag{4.386}$$

to first order in ε. We can use these formulas to study the transformation of the measure under more general field redefinitions of the form

$$\phi \to \phi + \varepsilon \sum_{k=0}^{\infty} \frac{\phi^k}{z^{k+1}} = \phi + \varepsilon \frac{1}{z - \phi}. \tag{4.387}$$

This field redefinition only makes sense if z is on the real axis outside the support ρ. In the limit $N \to \infty$ this is possible for z outside the interval $[c_-, c_+]$. Under the field redefinitions (4.387) the transformations of the

---

* The identity is not true for higher-genus graphs, but we do not expect such differences in gluing at the cutoff scale to be important in the scaling limit.

measure and the action are given by

$$d\phi \rightarrow d\phi \left(1 + \varepsilon \, \mathrm{Tr} \, \frac{1}{z - \phi} \, \mathrm{Tr} \, \frac{1}{z - \phi} \right), \qquad (4.388)$$

$$\mathrm{Tr} \, V(\phi) \rightarrow \mathrm{Tr} \, V(\phi) + \varepsilon \, \mathrm{Tr} \, \left(\frac{1}{z - \phi} V'(\phi) \right). \qquad (4.389)$$

The integral (4.374) is of course invariant under this change of the integration variables. By use of Eqs. (4.388) and (4.389) we obtain the identity

$$\int d\phi \left\{ \left(\mathrm{Tr} \, \frac{1}{z - \phi} \right)^2 - N \mathrm{Tr} \, \left(\frac{1}{z - \phi} V'(\phi) \right) \right\} e^{-N \mathrm{Tr} \, V(\phi)} = 0. \qquad (4.390)$$

The contribution to the integral coming from the first term in $\{\cdot\}$ in Eq. (4.390) is, by definition,

$$N^2 w^2(z) + w(z, z). \qquad (4.391)$$

The contribution from the second term inside $\{\cdot\}$ in Eq. (4.390) can be written as an integral over the one-loop function as follows:

$$\frac{1}{N} \left\langle \mathrm{Tr} \, \frac{V'(\phi)}{z - \phi} \right\rangle = \int d\lambda \, \rho(\lambda) \frac{V'(\lambda)}{z - \lambda} = \oint_C \frac{d\omega}{2\pi i} \frac{V'(\omega)}{z - \omega} w(\omega), \qquad (4.392)$$

where the second equality follows from Eq. (4.382). The curve C encloses the support of ρ but not z. It is essential for the existence of C that ρ have compact support. We can then write Eq. (4.390) in the form

$$\oint_C \frac{d\omega}{2\pi i} \frac{V'(\omega)}{z - \omega} w(\omega) = w^2(z) + \frac{1}{N^2} w(z, z), \qquad (4.393)$$

where z is outside the interval $[c_-, c_+]$ on the real axis. Since both sides of Eq. (4.393) can be analytically continued to $\mathbb{C} \setminus [c_-, c_+]$ the equation holds in this domain.

We recognize Eq. (4.393) as the loop equation derived in Section 4.3 and we have already discussed the solution of interest to us. To leading order in $1/N^2$ we have

$$w(z) = w_0(z) = \frac{1}{2} \left(V'(z) - M(z) \sqrt{(z - c_+)(z - c_-)} \right) \qquad (4.394)$$

and from Eq. (4.382) the corresponding eigenvalue density is

$$\rho(\lambda) = \frac{1}{2\pi} M(\lambda) \sqrt{(c_+ - \lambda)(\lambda - c_-)}. \qquad (4.395)$$

This shows that the compactness of the support of the eigenvalue density is related to the fact that the number of planar graphs only grows exponentially with the number of polygons. In the case of a purely Gaussian potential

$$V(\phi) = \frac{1}{2}\phi^2. \tag{4.396}$$

we found in Section 4.2 that

$$w_0(z) = \frac{1}{2}\left(z - \sqrt{(z-2)(z+2)}\right), \tag{4.397}$$

which yields

$$\rho(\lambda) = \frac{1}{2\pi}\sqrt{(2-\lambda)(2+\lambda)}. \tag{4.398}$$

This is Wigner's *semicircle law* for the eigenvalue distribution of random Hermitian matrices [384, 385].

There is an obvious formal similarity between the transformations defined by (4.384) and the conformal transformations

$$z \to z + \varepsilon z^{n+1}, \qquad n \geq -1, \tag{4.399}$$

in two dimensions. The transformation of a complex function $f(z)$ under the conformal transformation (4.399) is given in terms of the differential operators

$$\mathscr{L}_n = z^{n+1}\frac{d}{dz}, \qquad [\mathscr{L}_n, \mathscr{L}_m] = (m-n)\mathscr{L}_{m+n}, \tag{4.400}$$

by

$$f(z) \to f(z) + \varepsilon\mathscr{L}_n f(z). \tag{4.401}$$

This analogy can be carried further since we can bring down the term $\text{Tr}\,\phi^n$ in the matrix integral by diffentiating the action with respect to the coupling constant g_n. This implies that Eq. (4.390) is formally equivalent to

$$L(z)Z(\underline{g}, N) = 0, \tag{4.402}$$

where the operator $L(z)$ is defined by

$$L(z) = \sum_{n=-1}^{\infty} \frac{L_n}{z^{n+1}}, \tag{4.403}$$

and

$$\begin{aligned}
L_n &= \frac{1}{N^2}\sum_{k=1}^{\infty}k(n-k)\frac{\partial^2}{\partial g_k \partial g_{n-k}} - 2n\frac{\partial}{\partial g_n} \\
&\quad + \sum_{k=1}^{\infty}(k+n)g_k\frac{\partial}{\partial g_{n+k}} - \delta_{-1,n}Ng_1 + \delta_{0,n}N^2. \tag{4.404}
\end{aligned}$$

In Eq. (4.404) derivatives with respect to the g_js where $j \leq 0$ are to be omitted. The L_ns defined by Eq. (4.404) satisfy the same commutation relations (4.400) as do the $\mathscr{L}_n$s. The double scaling limit of Eq. (4.402) has played an important role in revealing the relation between the matrix models and integrable systems; for a review see [182].

4.8.3 *Non-perturbative quantum gravity?*

So far we have regarded the matrix integral as a slick technical device to take care of the combinatorics of triangulations. Here we address the question of whether the matrix model allows us to give a meaning to the sum over all genera. Let us restrict the following discussion to pure gravity, i.e. we assume that $g_j \geq 0$ for $j \geq 3$. In this case the matrix integral (4.374) is never convergent and only has a meaning as an asymptotic expansion in the coupling constants g_j, $j \geq 3$.

It is a very interesting question whether it is possible to make sense of the matrix integral and in this way define two-dimensional quantum gravity in a non-perturbative way even when summing over topologies. As explained earlier, this is a necessity in string theory. In the case of gravity it is not entirely clear that such a summation is required by any physical principles if we disregard a possible connection between gravity and string theory. It is nevertheless tempting to investigate the possibility. In the general case of random surfaces we have already mentioned that our regularized (discretized) approach has little to say about the sum over topologies. It is a non-perturbative regularization of the string path integral for a fixed topology, but perturbative in topology. In the special case of strings in $d = 0$, i.e. for pure two-dimensional gravity, the formula (4.373) seems to offer a possibility of defining the sum over all topologies. An obvious first suggestion is to define the matrix integral by analytic continuation. Applying the rotation $\phi \to e^{i\pi/6}\phi$ we can define a convergent matrix integral which has the *same* perturbation expansion as (4.373) by

$$\tilde{Z}(g, N) = e^{-i\pi N^2/6} \int d\phi \, \exp\left(-e^{i\pi/3}\mathrm{Tr}\,\phi^2 + i\frac{g}{3\sqrt{N}}\mathrm{Tr}\,\phi^3\right). \quad (4.405)$$

Contrary to the quantity (4.373) this integral is well defined. In the *double scaling limit the second derivative of $\mathscr{X} = \log \tilde{Z}(g, N)$ with respect to g is a solution to the Painlévé equation (4.308)*. Not surprisingly, it suffers from the disease already mentioned for the regular solutions to the Painlévé equation, that is, of not being real. It can be shown [132, 133] that it has an imaginary part which in the double scaling limit has the form

$$\mathrm{Im}\log\tilde{Z}(g, N) \sim \exp\left(-\mathrm{const.}\frac{\Lambda^{5/4}}{G}\right), \quad (4.406)$$

for small G. We have used the notation for Λ and G appropriate for the double scaling limit (see Eqs. (4.159), (4.160), (4.164) and (4.190)), i.e.

$$N\left(\frac{g_0}{g} - 1\right)^{5/4} \sim \frac{\Lambda^{5/4}}{G}.$$

Since the imaginary part of $\log \tilde{Z}(g, N)$ is exponentially suppressed for small G it will be absent in the perturbative genus expansion. It thus seems that Eq. (4.405) does not offer a satisfactory definition of a convergent sum over topologies, but it shows the potential power of the discretized approach that one is able to address these questions at all. In the next subsection we discuss another aspect of this problem.

4.8.4　*The Kontsevich model*

The Kontsevich model is defined by the partition function

$$Z_N^K[M] = \frac{\int d\phi \, \exp\left\{-N\mathrm{Tr}\left(\frac{M\phi^2}{2} + \frac{i\phi^3}{6}\right)\right\}}{\int d\phi \, \exp\left\{-N\mathrm{Tr}\left(\frac{M\phi^2}{2}\right)\right\}}, \tag{4.407}$$

where the integration is over $N \times N$ Hermitian matrices ϕ and M is a symmetric, positive definite $N \times N$ matrix with eigenvalues m_i, $i = 1, \ldots, N$. As for the ordinary Hermitian matrix model there is a graphical interpretation of the series obtained by expanding the exponential of the cubic term. The difference between this model and the ordinary Hermitian matrix model with a cubic potential is that the weight of triangulations is slightly different. First, we have a factor $i/2$ for each triangle due to the choice of "coupling constant" in front of the cubic term. Next, in the process of gluing we pick up a factor $1/(m_i + m_j)$ from each link whose endpoints are labelled by i and j since the propagator is

$$\langle \phi_{\alpha\beta} \phi_{\alpha'\beta'} \rangle = \delta_{\alpha\beta'} \delta_{\alpha'\beta} \frac{2}{m_\alpha + m_{\alpha'}}. \tag{4.408}$$

As for the ordinary Hermitian matrix model the integral (4.407) has a large N expansion which is at the same time an expansion in the topology of the triangulations. Defining

$$N^2 F_N^K[M] = \log Z_N^K[M] \tag{4.409}$$

we can write

$$F_N^K[M] = \sum_{h=0}^{\infty} \frac{1}{N^{2h}} F_{h,N}^K[M], \tag{4.410}$$

where

$$F_{h,N}^K[M] = \sum_{T \in \mathscr{T}^{(h)}} \frac{1}{C_T} \left(\frac{i}{2}\right)^{N_t(T)} \prod_{(ij) \in L(T)} \frac{2}{m_i + m_j}. \tag{4.411}$$

It is a remarkable fact [278, 242] that the partition function only depends on the parameters t_k defined as

$$t_k = \frac{1}{k + \frac{1}{2}} \frac{1}{N} \text{Tr} \, M^{-(2k+1)}, \qquad k \geq 0. \tag{4.412}$$

Let us introduce a "potential" $V_K(Z)$ by

$$V_K(Z) = \sum_{k=0}^{\infty} Z^{k+1/2} t_k \tag{4.413}$$

and a "loop insertion operator" by

$$\frac{d}{dV_K(Z)} = -\sum_{k=0}^{\infty} \frac{1}{Z^{k+3/2}} \frac{d}{dt_k}, \tag{4.414}$$

and define the multi-loop functions as (see Eq. (4.253))

$$W^K(Z_1, \ldots, Z_b) = \frac{d}{dV_K(Z_b)} \cdots \frac{d}{dV_K(Z_1)} F_N^K. \tag{4.415}$$

If the loop insertion operator acts on functions $f(t_0, t_1, \ldots)$ it follows from Eq. (4.412) by use of the chain rule that

$$\frac{d}{dV_K(Z)} = N \sum_i \frac{1}{2m_i} \frac{\partial}{\partial m_i}\Big|_{m_i^2 = Z}. \tag{4.416}$$

So far these multi-loop functions have no physical significance. It is now possible to mimic the derivation of the loop equation for the Hermitian matrix model in Subsection 4.8.2 by using the invariance of $Z_N^K[M]$ under the change of variable $\phi \to \phi + \varepsilon \phi^{n+1}$, $n \geq -1$. In this way one obtains a set of differential operators L_n in the variables m_i such that

$$L_n Z_N^K = 0, \qquad [L_n, L_m] = (n - m) L_{n+m}, \quad n, m \geq -1, \tag{4.417}$$

which is analogous to Eq. (4.402). In the same way as Eq. (4.402) is equivalent to the loop equation for $w(z)$, it can be shown that Eqs. (4.416) and (4.417) lead to the following loop equation for $W^K(Z)$:

$$\oint_{C_1} \frac{d\Omega}{2\pi i} \frac{V_K'(\Omega)}{Z - \Omega} W^K(\Omega) = (W^K(Z))^2 + \frac{1}{N^2} \left(\frac{dW^K(Z)}{dV^K(Z)} + \frac{1}{16Z^2} \right) + \frac{t_0^2}{16Z}, \tag{4.418}$$

where

$$V_K'(Z) = \sum_{n=0}^{\infty} \left(n + \frac{1}{2} \right) t_n Z^{n-1/2},$$

and the curve C_1 in the complex plane encircles the cut of $W^K(\Omega)$ and the cut corresponding to the term $1/\sqrt{\Omega}$ in $V_K'(\Omega)$, but not Z. We shall not present the rather complicated but elementary calculation which leads to

Eq. (4.418) here. We emphasize that this loop equation is identical to the continuum loop equation (4.264) derived by combinatorial arguments in Section 4.6 if we identify G in (4.264) with $1/N$ in (4.418). This implies the following result.

Theorem 4.4 *The multi-loop functions $W^K(Z_1, ..., Z_b)$ of the Kontsevich model agree with the continuum multi-loop functions $W^{cnt}(Z_1, ..., Z_b)$ of two-dimensional quantum gravity.*

One of the magic features of the Kontsevich model is that it provides a representation of the so-called *intersection indices* on the moduli spaces $\mathcal{M}_{h,b}$ of Riemann surfaces of genus h with b punctures, i.e. b marked points x_i, $i = 1, ..., b$. The details of this subject are outside the scope of this book but let us just outline the main concepts.

The cotangent spaces at the punctures x_i define line bundles $\mathcal{L}_i$ over $\mathcal{M}_{h,b}$ whose first Chern classes $c_1(\mathcal{L}_i)$ are 2-forms on $\mathcal{M}_{h,b}$. Unfortunately, it is not straightforward to give an intuitive picture of these line bundles. For a set $\alpha_1, ..., \alpha_b$ of non-negative integers for which

$$\sum_{i=1}^{b} \alpha_i = \dim \mathcal{M}_{h,b} = 3h - 3 + b \quad (\equiv d_{h,b}), \qquad (4.419)$$

it is possible to form the integral

$$\langle \alpha_1 \cdots \alpha_b \rangle_h = \int_{\bar{\mathcal{M}}_{h,b}} c_1(\mathcal{L}_1)^{\alpha_1} \cdots c_1(\mathcal{L}_b)^{\alpha_b}, \qquad (4.420)$$

where the products under the integral are exterior products and $\bar{\mathcal{M}}_{h,b}$ denotes a certain compactification (the so-called Deligne–Knudsen–Mumford compactification) of $\mathcal{M}_{h,b}$. The numbers $\langle \alpha_1 \cdots \alpha_b \rangle$ are topological invariants, called intersection indices, and one of the achievements of Kontsevich was to prove the following formula for the generating function for the intersection indices:

$$\sum_{\alpha_1 + \cdots + \alpha_b = d_{h,b}} \langle \alpha_1 \cdots \alpha_b \rangle_h \prod_{i=1}^{b} \frac{(2\alpha_i - 1)!!}{m_i^{2\alpha_i + 1}} = \sum_{\Gamma} \frac{2^{-N_v(\Gamma)}}{C_\Gamma} \prod_{e \in L(\Gamma)} \frac{2}{m_f + m_{f'}}, \qquad (4.421)$$

where the sum on the right-hand side runs over all connected ϕ^3 graphs Γ of genus h and with b faces, $N_v(\Gamma)$ denotes the number of vertices in Γ, the product on e runs over all links in Γ and f and f' denote the faces of Γ to which e belongs. Finally, C_Γ is the order of the automorphism group of the graph Γ. We have thus seen that two-dimensional gravity is closely related, via the Kontsevich model, to intersection theory on the moduli space of Riemann surfaces. In physics the latter theory has been called *topological gravity* [281, 392].

It follows from Theorem 4.4 that *the intersection indices of the Kontsevich model are precisely the rational numbers encountered in the double scaling limit of the partition function* (see Eqs. (4.229)–(4.230)). It is an interesting mathematical problem whether, more generally, the rational numbers $\langle \alpha_i; \beta_i | \alpha\beta\gamma \rangle_h$ which appear when $\mathscr{Z}_h$ is expressed in terms of the moments M_k, J_k and d *before* the double scaling limit is taken (see Eq. (4.150)), can be given a topological interpretation. There are indications that this is so (see [123, 122]).

One may ask whether the Kontsevich model provides a definition of the sum over all topologies. The answer is, unfortunately, in the negative. The problem is, as before, that the matrix integral (4.407), although well defined, is not real for matrices M which correspond to pure gravity. It leads to non-perturbative complex contributions to $\log Z_N^K[M]$.

4.9 More on matter and gravity

4.9.1 *Coupling matter fields to gravity*

In this final section on two-dimensional gravity we elaborate on the idea of coupling matter fields to two-dimensional gravity by extending the definition of certain statistical mechanical models from regular lattices to arbitrary triangulations and then averaging over triangulations. Suppose that a statistical model on a fixed regular triangulation has a second-order phase transition at a critical point where a mass parameter goes to zero, assume a scaling limit exists, and assume further that this transition is carried over to the model where we average over triangulations. Then we wish to regard the associated continuum limit as a realization of a coupling between the original continuum theory and two-dimensional quantum gravity. This is in accordance with our general philosophy that averaging over triangulations is the discretized counterpart of integrating over equivalence classes of metrics.

Let us spell this out in more detail. We assume that we are given an ensemble $\{\mathscr{S}_k\}$ of statistical mechanical systems which are to be thought of as the same model $\mathscr{S}$ but defined on different triangulations T_k which we assume to have a fixed topology. The physical characteristics of each $\mathscr{S}_k$ in general depends both on the system $\mathscr{S}$ and the underlying triangulation T_k. Let us denote by $\{\sigma_i^k\}$ the dynamical variables of $\mathscr{S}_k$ and let $S_k(\{\sigma_i^k\})$ denote the corresponding action. We regard the triangulations as a dynamical variable that is summed over on the same footing as the field variables $\{\sigma_i^k\}$ so the expectation value of an observable $\mathcal{O}$ in $\{\mathscr{S}_k\}$

is given by

$$\langle \mathcal{O} \rangle = \frac{\sum_k \rho_k \sum_{\{\sigma_i^k\}} \mathcal{O}[\{\sigma_i^k\}] \, P_k[\{\sigma_i^k\}]}{\sum_k \rho_k \sum_{\{\sigma_i^k\}} P_k[\{\sigma_i^k\}]}, \qquad (4.422)$$

where ρ_k denotes the weight of the statistical system $\mathscr{S}_k$ in the ensemble $\{S_k\}$ and $P_k[\{\sigma_i\}]$ is the Boltzmann factor of the matter configuration $\{\sigma_i^k\}$ in the system $\mathscr{S}_k$ given by

$$P_k[\{\sigma_i^k\}] = e^{-S_k[\{\sigma_i^k\}]}.$$

This average over the triangulations is called the *annealed average* and is appropriate when one averages over dynamical variables. There is another notion of averaging, particularly important in the theory of disordered systems where the background is random but inert, called *quenched average* defined by

$$\langle \mathcal{O} \rangle_{quenched} = \sum_k \rho_k \frac{\sum_{\{\sigma_i^k\}} \mathcal{O}[\{\sigma_i^k\}] P_k[\{\sigma_i^k\}]}{\sum_{\{\sigma_i^k\}} P_k[\{\sigma_i^k\}]}. \qquad (4.423)$$

The quenched average will not be discussed further in this section.

We shall refer to the statistical system defined by the expectation (4.422) as *the system $\mathscr{S}$ coupled to two-dimensional quantum gravity* provided the collection of triangulations $\{T_k\}$, together with the weights ρ_k, yield pure gravity in the scaling limit when no matter fields are present. This means that ρ_k will depend at least on a cosmological, as well as on a gravitational, coupling constant. Here we shall only consider the simplest case where

$$\rho_k = e^{-\mu |T_k|}$$

and μ is the bare cosmological constant. We can ignore the gravitational coupling when we consider triangulations of a fixed topology. If the model $\mathscr{S}$ has a second-order phase transition with a corresponding scaling limit on a regular lattice and the transition survives in the model defined by Eq. (4.422) and likewise yields a continuum limit, then this limiting theory will be called the continuum limit of $\mathscr{S}$ coupled to quantum gravity.

4.9.2   The Ising model

We have already encountered special cases of the scenario described above. The bosonic string in $\mathbb{R}^d$ considered in Chapter 3 can from this point of view be regarded as two-dimensional gravity coupled to d scalar fields. The statistical system $\mathscr{S}$ is in this case critical, i.e. the mass is zero, and there are no coupling constants in its action.

In Section 4.5.2 we discussed the Ising model on a random triangulation. In this case the only coupling constants in the system $\mathscr{S}$ are the inverse temperature β and the external magnetic field H. We considered the limiting case $H \to i\pi/2$ corresponding to a system of hard dimers and the second-order transition associated with the Lee–Yang edge singularity. We found that this model interacting with gravity, in accordance with Eq. (4.422), exhibited a phase transition and that the corresponding scaling limit could be interpreted as a certain non-unitary conformal field theory coupled to gravity.

If $H = 0$ it is of course well known that the Ising model can be solved exactly on a regular lattice. It has a second-order phase transition and all the critical exponents can be calculated. The corresponding continuum theory is a free Majorana field of spin $\frac{1}{2}$, which in the massless case is a conformal field theory with central charge $\frac{1}{2}$.

In order to define the Ising model on an arbitrary triangulation it is convenient to place the spins at the centres of the triangles and let each spin interact with its three nearest neighbours. Restricting attention to triangulations of the sphere we can write the partition function of the Ising model interacting with gravity as

$$Z(\mu, \beta) = \sum_{T \in \mathcal{T}} C_T^{-1} e^{-\mu|T|} \sum_{\{\sigma_i\}} e^{\frac{1}{2}\beta \sum_{(ij)} (\sigma_i \sigma_j - 1)}, \qquad (4.424)$$

where the sum in the exponent is over all nearest neighbour pairs of triangles in T. From Eq. (4.424) we see that for each $\beta \geq 0$ there is a critical value $\mu_0(\beta)$ of the cosmological constant such that $Z(\mu, \beta)$ is real analytic in μ for $\mu > \mu_0(\beta)$ but singular at $\mu = \mu_0(\beta)$. The critical exponent $\gamma(\beta)$ of the susceptibility (often called the string susceptibility in order not to be confused with the magnetic susceptibility) is, in accordance with the definition used for pure gravity, defined by

$$Z(\mu, \beta) \sim (\mu - \mu_0)^{-\gamma(\beta)+2} \qquad (4.425)$$

as $\mu \to \mu_0$. It turns out [269, 92] that even after coupling the Ising model to gravity as above, one can solve it exactly in the sense that the critical temperature β_0, as well as the critical exponents, can be determined. It would take us too far afield here to describe the details of the calculation leading to this exact solution, which is obtained by a generalization of the matrix model discussed in Section 4.8. Let us describe the main features of the solution.

At low temperatures (large β) the system is magnetized, i.e. the spins are on average aligned. In the limit $\beta \to \infty$ all triangulations have the same magnetization density, which means that the spins do not interact with the metric degrees of freedom and the system behaves as pure gravity. Similarly, in the high-temperature limit, the spins are randomly oriented, the system is unmagnetized, and the spins do not interact with the triangulations. Thus, the interaction between the matter and metric

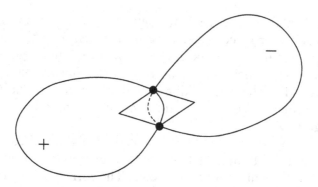

Fig. 4.19. Two spin clusters separated by only two links which form a "bottleneck" on the surface. The two triangles which contain one of the (two) spin-boundary links are also shown.

degrees of freedom is strongest at intermediate temperatures. One can argue plausibly as follows that this interaction should lead to a change in the critical behaviour of the magnetic system. The magnetic energy is proportional to the length of the boundaries of clusters of spins that are aligned. On a regular lattice a spin cluster of area A has a boundary of length at least $\sqrt{A}$. This is not so on a random triangulation, where an arbitrarily large spin cluster can have a boundary consisting of a few links; see Fig. 4.19. One therefore expects the interaction to favour triangulations with an abundancy of "bottlenecks" since the spins fluctuate more easily on such triangulations. The back-reaction of the spins on the geometry of triangulations grows stronger when β increases from zero and the exact solution shows that there is a critical value β_0 where it effects a shift of the (string) susceptibility exponent $\gamma(\beta)$ from $-\frac{1}{2}$ for $\beta < \beta_0$ to $-\frac{1}{3}$ at $\beta = \beta_0$. For $\beta > \beta_0$ the susceptibility exponent agains assumes the pure gravity value $-\frac{1}{2}$. Not only do the spins affect the critical behaviour of the metric degrees of freedom, the fluctuating geometry also influences the critical properties of the spin system. Intuitively one would expect it to soften the transition and that turns out to be the case. The second-order transition of the Ising model on a regular lattice becomes a third-order transition and the specific heat exponent α is shifted from 0 to -1. The critical exponents of the magnetization and magnetic susceptibility are likewise subject to change. The exponents obtained at β_0 all agree with the results of the continuum calculations referred to in Section 4.5.3 for a $c = \frac{1}{2}$ minimal conformal field theory coupled to two-dimensional quantum gravity.

We would like to mention that the method used to solve the Ising model on a random triangulation is to map it onto a matrix model with two interacting Hermitian matrices. This technique can be extended to more general multi-matrix models and used to describe the coupling of arbitrary unitary minimal conformal field theories with $c < 1$ to quantum gravity. The values one finds for the critical exponents are in agreement with those obtained in the continuum approach and they agree in particular with the formula (4.223) relating γ and c. Although interesting we shall not discuss these models further here, but see, e.g., [182] for a review. Instead, we turn to a discussion of the more problematic situation when $c > 1$ and the known continuum methods break down as can be seen from the fact that γ given by Eq. (4.223) becomes complex for $c > 1$. Moreover this is the relevant case for strings embedded in physical space.

4.9.3 Multiple-spin systems

In general there are many different physical systems of a given central charge $c > 1$ and it is not known whether their scaling behaviour after coupling to gravity is determined by c alone as is the case for the minimal conformal models with $c < 1$. It is therefore of interest to investigate and compare various systems of a given central charge coupled to gravity in order to gain information about the nature of the transition from $c < 1$ to $c > 1$.

Since d independent scalar fields have central charge $c = d$ we have already considered one such model, i.e. the bosonic string embedded in $\mathbb{R}^d$. In that case we found strong evidence that the model is dominated by branched polymer-like surfaces, at least for large d. As we have seen above for the Ising model, dominance of branched polymers can be seen as a sign of a strong interaction between the matter fields and the metric degrees of freedom represented by the triangulations. This brings us to another realization of a model of central charge d coupled to quantum gravity which is given by $n = 2d$ independent Ising models. In this case the partition function can be written

$$Z_n(\mu, \beta) = \sum_{T \in \mathcal{T}} C_T^{-1} e^{-\mu |T|} Z_T^n(\beta), \qquad (4.426)$$

where $Z_T(\beta)$ is the partition function of a single Ising model on the triangulation T. On a fixed regular lattice the multiple Ising model does not lead to any new physics but on dynamical triangulations the Ising models interact via their coupling to the geometry. These models provide an interesting supplement to the random surface models studied earlier and they have advantages from a technical point of view. The dominating triangulations in the limit $n \to \infty$ have been identified [221] and they are the same as those found in the large d limit of the bosonic string in Chapter 3.

Below we consider a simple *mean field theory* associated with the limit $n \to \infty$ which is exactly solvable and gives a rather clear picture of the interaction between matter and gravity. We assume, and this is confirmed by numerical simulations, that there is a critical value β_0 of the inverse temperature below which the magnetization is zero and above which the system is magnetized and $\gamma(\beta_0) > 0$ for $n > 2$, or at least for n large. Moreover, for β either sufficiently small or sufficiently large the critical behaviour is that of pure gravity. Furthermore, numerical simulations indicate that below β_0 there is a region where the important triangulations have a pronounced branched polymer structure. It then seems reasonable to attempt a description of the model for large n in a region around β_0 as well as for large $\beta > \beta_0$ in terms of an effective single Ising model on a random triangulation with the restriction that the only allowed spin configurations are those where all spin clusters have boundary components of minimal length. The leading term in the standard low-temperature (i.e. large β) expansion of spin models comes precisely from a summation over spin clusters with boundary components of minimal length. The crucial difference between an ordinary spin system and the ones considered here is that a cluster of flipped spins can be large even if the boundary is minimal, as already emphasized and illustrated in Figs. 4.19 and 4.20. In the following we consider the class of triangulations with spherical topology where the shortest closed loops are of length two; see again Figs. 4.19 and 4.20. A part of a triangulation T which is connected to the rest of T by a loop of length two is sometimes called a *baby universe*. We say that a baby universe is *minimal* if it does not properly contain any other baby universe. We can characterize the allowed spin configurations as follows. On all minimal baby universes the spins are aligned and the only allowed phase boundaries are unions of loops of length two. If we remove minimal baby universes and close up the resulting loops of length two by identifying the two boundary links, then the spins on the new minimal baby universes obtained in this way are also aligned, and so on. This gives us a method to relate the one-loop function of our mean field model to the one-loop function of pure gravity, since by continuing to remove minimal baby universes we reach finally a base surface with all spin aligned.

Let us denote the one-loop function, for a loop of length two, in the reduced model by

$$G(\mu, \beta) = \sum_{T \in \mathcal{T}(2)} e^{-\mu N_t(T)} \sideset{}{'}\sum_{\{\sigma_i\}} \exp\left(\frac{\beta}{2}\sum_{(ij)}(\sigma_i\sigma_j - 1)\right), \qquad (4.427)$$

where $\mathcal{T}(2)$ denotes the class of triangulations where the boundary is a loop consisting of two links and $\sum_{\{\sigma_i\}}'$ denotes the sum over the restricted class of spin configurations. For a given β the model defined by Eq. (4.427) has a critical cosmological constant $\mu_0(\beta)$ such that the sum is

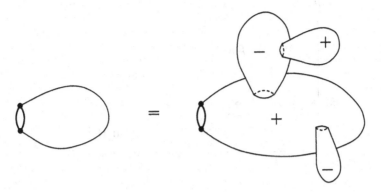

Fig. 4.20. A graphical representation of Eq. (4.430). The complete one-loop function allows a recursive decomposition into baby universes with definite spin assignment.

convergent for $\mu > \mu_0(\beta)$ and divergent for $\mu < \mu_0(\beta)$. The susceptibility $\chi(\mu, \beta)$ is defined, as before, as

$$\chi(\mu, \beta) = -\frac{\partial G}{\partial \mu} \qquad (4.428)$$

and the string susceptibility exponent $\gamma(\beta)$ is determined by

$$\chi(\mu, \beta) \sim \frac{1}{(\mu - \mu_0(\beta))^{\gamma(\beta)}}. \qquad (4.429)$$

The corresponding quantities in pure gravity will be denoted by $G_0(\mu)$, $\chi_0(\mu)$, $\gamma^{(0)}$ and the critical cosmological constant by μ_0.

Decomposing an arbitrary surfaces into baby universes and a base surface on which the baby universes sit, and summing first over spin configurations on the baby universes, we obtain (see Fig. 4.20)

$$G(\mu, \beta) = \sum_{T \in \mathcal{T}(2)} e^{-\mu N_t(T)} \left(1 + e^{-2\beta} G(\mu, \beta)\right)^{N_l(T)-2}. \qquad (4.430)$$

In Eq. (4.430) the factor $e^{-2\beta}$ represents the coupling of a baby universe across the phase boundary to the rest of the surface and the factor 1 in the parentheses corresponds to the empty baby universe. We can thus write[*]

$$G(\mu, \beta) = \sum_{T \in \mathcal{T}(2)} e^{-\bar{\mu} N_t(T)} = G_0(\bar{\mu}), \qquad (4.431)$$

---

[*] Here we have dropped a boundary-dependent correction factor which does not affect the subsequent argument.

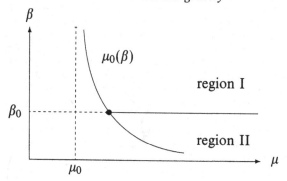

Fig. 4.21. The phase diagram of the reduced Ising model. The partition function is defined (convergent) above the critical line $\mu_0(\beta)$. The infinite volume limit is obtained by keeping β fixed and approaching the critical line. As $\beta \to \infty$ the critical coupling $\mu_0(\beta)$ approaches μ_0.

where

$$\bar{\mu} = \mu - \frac{3}{2} \log \left(1 + e^{-2\beta} G(\mu, \beta) \right).$$ (4.432)

Note that the last equation can be written as

$$\mu = \bar{\mu} + \frac{3}{2} \log \left(1 + e^{-2\beta} G_0(\bar{\mu}) \right),$$ (4.433)

which expresses μ in terms of known functions of $\bar{\mu}$ and β since pure gravity can be solved.

From Eqs. (4.431)–(4.433) we obtain

$$\chi(\mu, \beta) = \chi_0(\bar{\mu}) \frac{\partial \bar{\mu}}{\partial \mu}$$ (4.434)

and

$$\frac{\partial \bar{\mu}}{\partial \mu} = \frac{e^{2\beta} + G_0(\bar{\mu})}{e^{2\beta} - (\frac{3}{2}\chi_0(\bar{\mu}) - G_0(\bar{\mu}))}.$$ (4.435)

It is clear that $G_0(\mu_0)$ and $\chi_0(\mu_0)$ are both finite since the susceptibility exponent $\gamma^{(0)} = -1/2 < 0$ for pure gravity. This implies that there exists a β_0 such that the denominator in Eq. (4.435) is different from zero for all $\mu \geq \mu_0(\beta)$ provided $\beta > \beta_0$. Fig. 4.21 shows the phase diagram in the (μ, β)-plane. Let us first consider region I in Fig. 4.21 where $\beta > \beta_0$. The only possibility for a singularity of $\chi(\beta, \mu)$ at the boundary arises if $\bar{\mu}(\mu_0(\beta)) = \mu_0$ and in this case the leading singularity is due to the factor $\chi_0(\bar{\mu})$ in Eq. (4.434). We conclude that

$$\bar{\mu}(\mu_0(\beta)) = \mu_0 \quad \text{and} \quad \gamma(\beta) = \gamma^{(0)} \quad \text{for} \quad \beta > \beta_0.$$ (4.436)

This is the magnetized phase where the spin fluctuations are small and the geometry of the triangulations is not affected by the spins.

Next consider region II in Fig. 4.21, where $\beta < \beta_0$. The number β_0 can be characterized as the largest β for which $\partial\mu/\partial\bar{\mu}$ has a zero, i.e. (by Eq. (4.435)) the largest β for which the equation

$$e^{2\beta} = \frac{3}{2}\chi_0(\bar{\mu}) - G_0(\bar{\mu}) \tag{4.437}$$

has a solution. If we define $\bar{\mu}_0(\beta)$ by

$$\left.\frac{\partial\mu}{\partial\bar{\mu}}\right|_{\bar{\mu}_0(\beta)} = 0 \quad \text{for} \quad \beta \le \beta_0, \tag{4.438}$$

then $\bar{\mu}_0(\beta)$ obviously solves Eq. (4.437) and we have

$$\bar{\mu}_0(\beta_0) = \mu_0, \qquad \bar{\mu}_0(\beta) > \mu_0 \quad \text{for} \quad \beta < \beta_0, \tag{4.439}$$

so

$$e^{2\beta_0} = \frac{3}{2}\chi_0(\mu_0) - G_0(\mu_0).$$

If we use Eqs. (4.434) and (4.435) this implies that $\chi(\mu, \beta)$ diverges as $\mu \to \mu_0(\beta)$. In fact, since $\bar{\mu}_0(\beta) > \mu_0$, both $\chi_0(\bar{\mu})$ and $G_0(\bar{\mu})$ will be regular around $\bar{\mu}_0(\beta)$ and we can Taylor expand the right-hand side of Eq. (4.434) and obtain

$$\chi(\mu, \beta) \sim \frac{1}{\bar{\mu} - \bar{\mu}_0(\beta)} \sim \frac{1}{\sqrt{\mu - \mu_0(\beta)}}, \tag{4.440}$$

where we have used Eq. (4.438), which implies that

$$\mu = \mu_0(\beta) + \text{const.} \, (\bar{\mu} - \bar{\mu}_0(\beta))^2 + O((\bar{\mu} - \bar{\mu}_0(\beta))^3). \tag{4.441}$$

We conclude that $\gamma(\beta) = 1/2$ for $\beta < \beta_0$. In this phase baby universes are rampant, we have effectively a theory of branched polymers and the total magnetization of the system is zero.

Let us finally consider the system at the critical point β_0. This point is characterized by the fact that $\bar{\mu}_0(\beta_0)$ coincides with μ_0. Although the singularity of $\chi(\mu, \beta_0)$ for $\mu \to \mu_0(\beta_0)$ is still dominated by the zero of $\partial\mu/\partial\bar{\mu}$ we can no longer Taylor expand $\mu(\bar{\mu}, \beta_0)$ around $\bar{\mu}_0(\beta_0) (= \mu_0)$ since the functions in Eq. (4.435) are singular at μ_0. On the other hand, we can use the known singular behaviour of the functions G_0 and χ_0 at μ_0 to deduce from Eq. (4.435) that

$$\frac{\partial\mu}{\partial\bar{\mu}} \sim (\bar{\mu} - \bar{\mu}_0(\beta_0))^{-\gamma^{(0)}}. \tag{4.442}$$

Integrating with respect to $\bar{\mu}$ we obtain

$$\mu - \mu_0(\beta_0) \sim (\bar{\mu} - \bar{\mu}_0(\beta_0))^{-\gamma^{(0)}+1}. \tag{4.443}$$

From Eq. (4.434) we obtain, using Eq. (4.443),

$$\chi(\mu, \beta_0) \sim \frac{1}{(\bar{\mu} - \bar{\mu}_0(\beta_0))^{-\gamma^{(0)}}}$$

$$\sim \frac{1}{(\mu - \mu_0(\beta_0))^{-\gamma^{(0)}/(-\gamma^{(0)}+1)}}, \tag{4.444}$$

which shows that

$$\gamma(\beta_0) = \frac{\gamma^{(0)}}{\gamma^{(0)} - 1} = \frac{1}{3}. \tag{4.445}$$

The model can easily be extended to include the coupling to an external magnetic field and one can explicitly verify that the system is magnetized for $\beta > \beta_0$ and has zero total magnetization for $\beta < \beta_0$. Furthermore, the transition is of third order as for the full Ising model on dynamical triangulations. The model under consideration captures the essential features observed numerically for a large number of Ising models on random triangulations. It also strongly suggests the existence of a region below β_0 where the surfaces consist of numerous magnetized baby universes with different spin orientations such that the total magnetization is zero. The spin transition observed at β_0 is, according to this picture, the alignment of the spins of the baby universes.

For a finite number of Ising models the mean field approximation above clearly fails for $\beta \to 0$, since in the infinite temperature limit γ equals $-1/2$, whereas the mean field approximation yields $\gamma = 1/2$. If we assume that for sufficiently large n there exists an interval below β_0 where $\gamma = 1/2$, it is an interesting and unsolved question as to how the transition to $\gamma = -1/2$ takes place. Is there a single transition, a continuous change or a cascade of transitions, and how are these transitions characterized?

An important ingredient in the preceding derivation was to relate the mean field model to pure gravity. This aspect of the argument can be generalized to the full multiple Ising system as well as to more general matter systems coupled to two-dimensional gravity. It is in fact straightforward to effect the renormalization group step by summing over outgrowths (baby universes). In this way quantities relating to the full model can be expressed in terms of corresponding quantities in an effective model. In particular, the susceptibilities of the two models obey a relation similar to Eq. (4.434), from which one can conclude that the susceptibility of the effective theory is not divergent at its critical point, i.e. the corresponding critical exponent $\bar{\gamma}$ is negative. Assuming that the susceptibility exponent γ of the original model is positive in addition to some weak technical conditions, one may derive the relation

$$\gamma = \frac{\bar{\gamma}}{\bar{\gamma} - 1}, \tag{4.446}$$

which generalizes Eq. (4.445). We refer to [151] for the details of this argument. Let us just note here that if we take the fact that $\bar{\gamma} < 0$ as an indication that the effective model is equivalent, at a second-order phase transition point, to a unitary conformal field theory with central charge $\bar{c} < 1$, then only a discrete set of values of $\bar{\gamma}$ is available. In the unitary case we have in fact $\bar{\gamma} = -m^{-1}$ for some integer $m \geq 2$ corresponding to the central charge

$$\bar{c} = 1 - \frac{6}{m(m+1)},$$

which yields

$$\gamma = \frac{1}{m+1}. \tag{4.447}$$

We thus see that the result obtained above for what we called the mean field model is simply the first in the series of possibilities given by Eq. (4.447). The second case, corresponding to a single Ising model coupled to the mean field model has been dealt with in [249]. The transmutation of critical exponents described by Eq. (4.446) can also be obtained in the context of matrix models [127, 280].

In view of the generality of these arguments it is interesting that they seem to predict only a discrete set of possible values for γ in the range from 0 to $\frac{1}{2}$. In all cases these models are governed by a $\bar{c} < 1$ model resulting from summing over baby universes. This mechanism may provide a clue to understanding the transition from $c < 1$ to $c > 1$. One of the missing ingredients in our present understanding is the dependence of $\bar{c}$ on c. In the mean field model we saw that $\bar{c} = 0$ in the limit $c \to \infty$.

4.10   Notes

For a discussion of the theoretical background and some of the technical and conceptual problems encountered in path integral quantization of Euclidean gravity we refer to the papers by Hawking [225, 226]. The Hartle–Hawking wave functionals are introduced in [223]. See [222] for an early discussion of simplicial manifolds in the context of quantum gravity.

In this chapter we have taken a combinatorial approach to two-dimensional quantum gravity. In spirit it is quite close to Tutte's original idea in his famous paper *A Census of Planar Triangulations* from 1962 [363], but it uses the loop equations and the so-called moments to extract the asymptotic behaviour of the number of triangulations with a given number of triangles. The loop equations were first studied in the context of the large N expansion of two-dimensional matrix models. These models attracted attention as toy models for quantum chromodynamics in the late 1970s and early 1980s. This line of research was triggered by the seminal paper [356] of t'Hooft on two-dimensional quantum chromodynamics. Loop or Dyson–Schwinger equations for the matrix models are derived in [370] and [302] and loop equations are considered in a more general context in [288]. It is well known that a major part of the contribution to high-order perturbation theory in bosonic field theory comes from the number of Feynman diagrams. The counting of these became important and matrix models were an important tool in the counting of planar graphs as well as graphs of higher genus. Important papers from this line of development are [96], [241] and [82]. Recent work on estimating the asymptotic number of maps on surfaces (which are analogous to general triangulations) can be found in a series of papers by Bender and collaborators [76, 77, 78]. The explicit solution of the loop equations for the planar graphs follows [45, 50].

The idea of representing fluctuating geometry in quantum gravity in terms of planar graphs goes back to Weingarten, who considered planar graphs embedded in hyper-cubic lattices [374, 375]. A matrix model formulation can be found in these papers; see also [165]. A general discussion of the relation between graphs and quantum gravity can be found in [188]. The relation between planar graphs and two-dimensional quantum gravity is emphasized by David [129] and Kazakov [268], while two-dimensional gravity coupled to Gaussian fields is studied in [23, 271, 128]; see also [27, 24, 93]. It is furthermore possible to obtain exact results in the special case of $d = -2$ as mentioned in Chapter 3 [271, 130], and for the Ising model coupled to two-dimensional gravity via the two matrix model [269, 92]. For some time the exact results obtained were viewed as a curiosity, in particular because they did not agree with perturbative continuum calculations extrapolated from $d = -\infty$ [398]. Some non-perturbative continuum results were available [319, 198], but only after the derivation of the continuum formulas for γ by Knizhnik, Polyakov and Zamolodichov (KPZ) [277] and the related results by David and by Distler and Kawai (DDK) [131, 143] was it generally acknowledged that the dynamically triangulated models constituted a regularization of two-dimensional quantum gravity coupled to matter. Additional interest arose due to the work of Douglas and Shenker, Kazakov and Brezin, and Gross and Migdal, [146, 97, 209], where the possibility of a summation over genera was first suggested and the concept of the double scaling limit was introduced. Multi-critical behaviour was first discussed by Kazakov [270] and although his tentative identification of the models with unitary matter coupled to gravity was not correct, the work was nevertheless instrumental in providing a link between the double scaling limit and the KdV hierarchies. Numerous papers have been written which discuss aspects of the KdV hierarchies and the "integrability" of two-dimensional quantum gravity coupled to matter. Here we have made no attempt to cover this vast topic, and our treatment of multi-critical models and the continuum loop equations should be viewed as an introduction which emphasizes the statistical mechanics aspect of the theory. We have followed [350] in the study of the $m = 3$ multi-critical model and [140, 286] in the study of the continuum loop equations, while the general technique of representing the loop functions in terms of moments can be found in [20]. The discussion of the complex matrix model uses [45], while the graphical interpretation of the complex model is taken from [304]. The special features of the complex matrix model in terms of moments can be found in [49]. The introduction of the variables T_k in terms of the moments M_k can be found in [47]. An extensive discussion of various choices of coupling constants T_k is given in [287]. The revival of the loop equations in the context of two-dimensional gravity and matrix models is due to David [132], while the relation between the loop equations, the Virasoro generators and the KdV hierarchy was discovered in [193]; see also [194].

Matrix models have a prominent role in the literature and one sometimes gets the impression that the matrix models themselves contain the key to two-dimensional gravity. One of the purposes of the presentation here is to show that matrix models are nothing but a convenient device to formulate certain combinatorial identities as Dyson–Schwinger equations. The Kontsevich model [278] has a different status since it provides a representation of the intersection indices on the moduli space of punctured Riemann surfaces. An account of this theory, readable for physicists, can be found in [242]. The model put on a firm foundation the statement that so-called topological gravity, as defined by Witten [281, 392] in terms of intersection indices on punctured Riemannian surfaces, is identical to the part of two-dimensional gravity defined by the loop functions $W^{cnt}(Z_1,\ldots,Z_n)$ in Section 4.6. The connection between ordinary matrix models and the Kontsevich model has been studied in a number of papers [197, 294, 295, 210, 274].

The idea of using a transfer matrix in two-dimensional gravity and the derivation of a partial differential equation for the loop–loop correlation function was first presented in [263]. In [372] the alternative presentation in terms of peeling can be found. Related work is also presented in [214]. The introduction of a reparametrization invariant two-point function in quantum gravity can be found in [34] and the scaling of the two-point function is discussed in [55]. The transfer matrix formalism suggests the existence of a Hamiltonian for two-dimensional quantum gravity and it has triggered a lot of interesting work [236, 237, 195]. While it seems without doubt that the fractal dimension of the surfaces which dominate pure two-dimensional quantum gravity is four and that the proper time introduced in the Hamiltonian formalism can be identified with the geodesic distance, it is not well understood at present how to view the relation between geodesic distance and the proper time related to the Hamiltonian in cases where conformal field theories are coupled to quantum gravity. Numerical simulations [118, 46] seem to favour a Hausdorff dimension $d_h = 4$ for the central charge c satisfying $0 \le c \le 1$.

The continuum formalism breaks down when the central charge $c > 1$. The borderline case $c = 1$ is quite interesting and belongs to the solvable cases, although it requires special treatment. The model was first solved in [96], although the interpretation as quantum gravity coupled to a scalar field, or one-dimensional non-critical string theory or two-dimensional critical string theory was only given afterwards [272]. Due to lack of space we have chosen not to treat the $c = 1$ case at all. The reason is that a proper treatment requires somewhat different methods than the ones used for $c < 1$ as well as for $c > 1$. Let us only mention that the double scaling limit was first discussed in [98, 211, 201]. An interesting hybrid of $c = 0$ and $c = 1$ should be mentioned, the so-called Marinari–Parisi model [289], which is based on an older idea by Greensite and Halpern [208]. The idea is to use stochastic quantization. In ordinary field theory the method of stochastic quantization produces results identical to those obtained by the standard path integral methods, but in the case of a potential unbounded from below, a matrix integral will be ill-defined, or only defined perturbatively around the Gaussian part, while the method of stochastic quantization yields a well-defined theory, which in addition has a perturbative expansion which agrees with the expansion of the ordinary matrix integral around the Gaussian part. The reason that the matrix model of pure gravity reduces in this case to a one-dimensional matrix integral, corresponding to $c = 1$, is that the stochastic time enters as an additional (fictitious) dimension, but contrary to the situation for genuine $c = 1$ matter coupled to gravity, the matrix potential is bounded from below. In this way one obtains a non-perturbative definition of two-dimensional gravity in the sense that it reproduces the perturbative expansion in genus, the summation over genera is well defined and (contrary to the analytic continuation of the ordinary zero-dimensional matrix integral discussed in the text) produces real amplitudes. Unfortunately, it is not clear to what extent one can use the stabilization of unbounded potentials by stochastic quantization as a general principle in field theory. It should be mentioned that one can prove that the partition function obtained in this way differs from the real part of the usual partition function obtained by analytic continuation [37, 347]. See [48] for a general discussion of the use of stochastic quantization in two-dimensional gravity.

The treatment of large c presented here follows [35, 151]. A systematic low temperature expansion for $c \to \infty$ can be found in [377, 376], where it is also proved that the $c \to \infty$ limit of multiple Ising models agrees with the $q \to \infty$ limit of q-state Potts models in the low-temperature limit. Closely related results are found in [221] by considering the dominant ϕ^3 graphs in the $c \to \infty$ limit, as well as by the use of modified matrix models [127, 280, 16]. In addition, numerical studies exist of the large c and the large q limits which support the mean field treatment presented in Section 4.9 [61, 116, 36, 51, 52]. For interesting continuum formulations of models which allow $c > 1$ we refer to [296].

5

Monte Carlo simulations
of random geometry

In previous chapters we have advocated the point of view that the techniques and concepts of statistical mechanics can be applied in a natural way to quantum gravity coupled to matter. Computer simulations are important tools in the investigations of physical systems where perturbative methods break down. In the study of critical phenomena numerical approaches are frequently called for and it is natural to ask whether they can be applied to random manifolds. In the present chapter we answer this question in the affirmative and explain the basic ideas in Monte Carlo simulations of random geometry.

Pure two-dimensional gravity and the Ising model coupled to two-dimensional gravity are good tests for the numerical approach. In fact they can serve as a test case in the same way as the ordinary Ising model is used as a laboratory for numerical experiments in statistical mechanics. These models can be solved exactly as described in the previous chapter; their intrinsic geometry as well as the behaviour of the matter sector are non-trivial and they allow us to study analytically the back-reaction of matter on gravity.

5.1 Basic principles

Let us briefly recall the framework of Monte Carlo simulations in statistical mechanics. For a more complete discussion we refer to [89]. For illustration consider an Ising spin system with Hamiltonian[*]

$$H[\{\sigma_i\}] = -\sum_{(ij)} \sigma_i \sigma_j. \tag{5.1}$$

---

[*] We do not distinguish between the Hamiltonian of a statistical system and the action of the corresponding Euclidean field theory.

Here σ_i denotes the spin variables located on lattice sites i, (ij) signifies neighbouring sites and $\{\sigma_i\}$ is a *spin configuration*, i.e. an assignment of spins to each lattice site. The partition function is

$$Z(\beta) = \sum_{\{\sigma_i\}} e^{-\beta H[\{\sigma_i\}]}, \tag{5.2}$$

where the sum runs over all spin configurations. The expectation value of an observable $\mathcal{O}[\{\sigma_i\}]$ is defined as

$$\langle \mathcal{O} \rangle = \frac{1}{Z(\beta)} \sum_{\{\sigma_i\}} \mathcal{O}[\{\sigma_i\}] \, e^{-\beta H[\{\sigma_i\}]}. \tag{5.3}$$

Monte Carlo simulations of spin systems attempt to calculate expectation values such as (5.3) by creating a sequence of statistically independent spin configurations $\{\sigma_i^a\}$, $a = 1, \dots, N$ with a probability distribution

$$P[\{\sigma_i^a\}] = Z(\beta)^{-1} \, e^{-\beta H[\{\sigma_i^a\}]} \tag{5.4}$$

and evaluating the quantity (5.3) as

$$\langle \mathcal{O} \rangle = \frac{1}{N} \sum_{a=1}^{N} \mathcal{O}[\{\sigma_i^a\}] \tag{5.5}$$

with N sufficiently large. Of course, we only obtain a strict equality in the limit $N \to \infty$.

The spin configurations are usually generated by a stochastic process t (Markov chain) which can be regarded as a random walk in configuration space with a transition function

$$t[\{\sigma_i\}, \{\sigma_i'\}]. \tag{5.6}$$

This means that if we have a given spin configuration $\{\sigma_i\}$ after a certain number of steps, the function (5.6) is the probability of stepping from $\{\sigma_i\}$ to $\{\sigma_i'\}$ in the next step. The transition function should be chosen such that

(i) any configuration can be reached in a finite number of steps

(ii) the asymptotic probability distribution for the spin configurations is independent of the starting configuration $\{\sigma_i^0\}$ and converges, as the number of steps goes to infinity, to the Boltzmann distribution (5.4).

Condition (i) is usually called *ergodicity*. Ergodicity is usually not a problem for spin systems and the convergence of the Markov chain can be ensured by a general rule called *detailed balance* which requires the transition function t to satisfy

$$P[\{\sigma_i^a\}] \, t[\{\sigma_i^a\}, \{\sigma_i^b\}] = P[\{\sigma_i^b\}] \, t[\{\sigma_i^b\}, \{\sigma_i^a\}]. \tag{5.7}$$

The detailed balance condition is sufficient, but not necessary, for convergence to the correct probability distribution P.

Let us describe this in a little more detail. Denote the initial distribution of the spin configurations by $P_0[\{\sigma_i\}]$. After n steps in the stochastic process we have generated a new distribution $P_n[\{\sigma_i\}]$ and the following relation holds between $P_n[\{\sigma_i\}]$ and $P_{n+1}[\{\sigma_i\}]$:

$$P_{n+1}[\{\sigma_i^a\}] = \sum_{\{\sigma_i^b\}} P_n[\{\sigma_i^b\}] t[\{\sigma_i^b\}, \{\sigma_i^a\}] \tag{5.8}$$

$$= P_n[\{\sigma_i^a\}] + \sum_{\{\sigma_i^b\}} \left(P_n[\{\sigma_i^b\}] t[\{\sigma_i^b\}, \{\sigma_i^a\}] - P_n[\{\sigma_i^a\}] t[\{\sigma_i^a\}, \{\sigma_i^b\}] \right),$$

where we have used the fact that

$$\sum_{\{\sigma_i^b\}} t[\{\sigma_i^a\}, \{\sigma_i^b\}] = 1. \tag{5.9}$$

It follows that P_{n+1} is normalized if P_n is. According to Eq. (5.8) detailed balance implies that the normalized Boltzmann distribution $P[\{\sigma_i\}]$ is a *stationary* probability distribution under t. We refer to [173] for a proof of the convergence of P_n to P as $n \to \infty$.

A general procedure exists for choosing the transition function so that Eq. (5.7) is satisfied:

$$t[\{\sigma_i^a\}, \{\sigma_i^b\}] = C \min \left\{ 1, \frac{P[\{\sigma_i^b\}]}{P[\{\sigma_i^a\}]} \right\}, \tag{5.10}$$

where C is a normalization constant. This choice is called a *Metropolis updating* of the system. It might not give the fastest convergence to the equilibrium distribution but it has the advantage of being general. In general, one sets the transition functions $t[\{\sigma_i^a\}, \{\sigma_i^b\}]$ equal to zero except in special cases where the relation between the two spin configurations is simple. For example, we would obtain a convergent and ergodic algorithm if we set $t[\{\sigma_i^a\}, \{\sigma_i^b\}] = 0$ unless $\{\sigma_i^a\}$ and $\{\sigma_i^b\}$ are identical or differ by a single spin flip.

The configuration space is usually large and $P[\{\sigma_i\}]$ often has a sharp peak. Once we end with a configuration $\{\sigma_i^{a_0}\}$ in the neighbourhood of this peak, attempts to move further will, according to Eq. (5.10), be rejected with a probability close to 1 for most configurations. It is more efficient to choose a transition function t which only implements small changes in a configuration, such that the new configuration $\{\sigma_i^a\}$ has a probability $P[\{\sigma_i^a\}]$ which is not much smaller than $P[\{\sigma_i^{a_0}\}]$. As a general rule the most efficient updating is obtained if one can choose a t such that the probability of stepping to a new configuration is roughly 1/2 on average. This ensures a compromise between the desire to implement as large a change as possible on a given configuration $\{\sigma_i^a\}$, and the possibility of this change being accepted by the algorithm.

Let us consider a spin system which has a second- (or higher-) order phase transition. Close to the transition point there will be large-scale fluctuations and the simplest Metropolis algorithms, which use only local changes of the spin configurations, are inefficient in creating new, statistically independent configurations. This phenomenon is called *critical slowing down* and a great deal of ingenuity has gone into the creation of algorithms which beat the critical slowing down. For spin systems like the Ising model, so-called *cluster algorithms* [354, 395] have been very successful. In essence these algorithms trace the spin clusters, i.e. regions where the spins point in the same direction. By flipping the spin in all clusters simultaneously, one can induce a global change in the spin configuration with no cost in energy. The art of choosing algorithms which induce changes in the configurations which beat the critical slowing down has two facets: one should understand the nature of the critical fluctuations in order to design efficient updatings of the configurations and one should choose the updating in such a way that convergence to the desired distribution is guaranteed. The second aspect might be non-trivial if the changes in configurations in single steps are complicated.

5.2 Updating geometry

If we wish to apply Monte Carlo simulations to quantum gravity coupled to matter we have to update the matter system for a given triangulation and we have to update the triangulation for a given matter field configuration. The updating of matter fields on a given random triangulation presents no problem and can be done just as on a regular lattice. The Metropolis algorithm can of course be applied and cluster algorithms work well. The updating of geometry is a new feature encountered in the Monte Carlo study of random surfaces and random higher-dimensional manifolds. In order to implement the Monte-Carlo method in this context we need to introduce elementary *moves* which change the triangulations in small steps that correspond to the flipping of a single spin discussed above. These moves have to be ergodic in the chosen class of triangulations, i.e. successive application of the moves should allow us to transform a given triangulation into any other triangulation belonging to the class under consideration. For two-dimensional triangulations of a fixed topology such a set of moves is quite simple and they are illustrated in Fig. 5.1.

The moves described above are not the most efficient local updatings of triangulations. A set of ergodic moves which are more convenient from a numerical point of view are illustrated in Fig. 5.2. Let us discuss in some

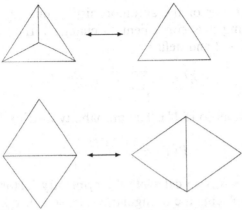

Fig. 5.1. The three moves which are ergodic in the class of two-dimensional triangulations with fixed topology. The first diagram illustrates two moves where one either inserts or deletes a vertex of order 3. The second move alone, called the *flip*, is ergodic in the class of triangulations with a fixed number of triangles and fixed topology.

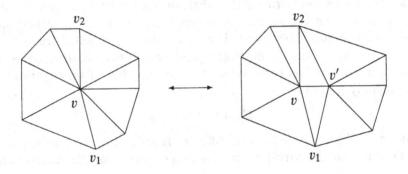

Fig. 5.2. A set of ergodic moves in the class of type III triangulations.

detail the algorithm for updating triangulations using the moves shown in Fig. 5.2. It involves new aspects compared to the standard Metropolis algorithm, which will be important in the next chapter where we study higher-dimensional gravity. If for simplicity we consider pure gravity, no matter fields are involved and the space of configurations is the space of triangulations of a fixed topology. In the following we consider the class $\mathcal{T}$ of type III triangulations of S^2, i.e. the class of regular triangulations of S^2. As before, the partition function is given by

$$Z(\mu) = \sum_{T \in \mathcal{T}} \frac{1}{C_T} e^{-\mu N_t(T)}, \qquad (5.11)$$

where C_T is the order of the automorphism group of the triangulation T. In the following it is convenient to change variables from $N_t(T)$ to $N_v(T) = N_t(T)/2 + 2$ and define

$$\tilde{Z}(\tilde{\mu}) = \sum_{T \in \mathcal{T}} \frac{1}{C_T} e^{-\tilde{\mu} N_v(T)}, \tag{5.12}$$

which is proportional to (5.11). The probability distribution is given by

$$P(T) = \frac{1}{C_T} \frac{e^{-\tilde{\mu} N_v(T)}}{\tilde{Z}(\tilde{\mu})}. \tag{5.13}$$

There is no simple way to calculate the symmetry factor C_T, but we can dispose of it by replacing the triangulations by so-called *labelled triangulations*. Oriented type III triangulations are uniquely defined by a labelling of their vertices, and by the incidence relation of links and triangles, expressed in terms of the vertex labels. Two triangulations, defined by the the same set of vertex labels $\{a_i\}$, are isomorphic if there exists a bijective map $\phi : \{a_i\} \mapsto \{a_i\}$, which maps neighbours onto neighbours. The factor C_T is the number of such maps. Given a triangulation and a labelling of the vertices we can permute the labels in $N_v(T)!$ ways. In general we will have to redefine the neighbour assignments by such a permutation and we can distinguish the graphs if we view them as labelled triangulations. Some of the permutations might preserve the neighbour assignments if the triangulation has some symmetry. The number of these permutations is precisely C_T, so

$$C_T \mathcal{N}(T) = N_v(T)!, \tag{5.14}$$

where $\mathcal{N}(T)$ is the number of different labellings of T. This implies that we can rewrite the partition function as a sum over labelled triangulations,

$$\sum_{T \in \mathcal{T}} \frac{1}{C_T} e^{-S_T} = \sum_{\tilde{T} \in \tilde{\mathcal{T}}} \frac{1}{N_v(\tilde{T})!} e^{-S_{\tilde{T}}}, \tag{5.15}$$

where we have used the notation $\tilde{\ }$ for labelled triangulations. It is worth emphasizing the following interpretation of labelled triangulations: different labellings of the same triangulation is the discrete analogy of different choices of parametrization of the same manifold and in this way the factor $N_v(\tilde{T})!$ is the discretized analogue of the volume of the group of *volume preserving diffeomorphisms*. The division by $N_v(\tilde{T})!$ ensures that we effectively sum over inequivalent parametrizations. The concept of triangulations was introduced in such a way that non-isomorphic triangulations represent inequivalent metric assignments, but for computer simulations it is convenient to work with labelled triangulations since we avoid dealing with the factor $1/C_T$ for closed manifolds and we must anyway label the triangulations in order to represent them in the computer.

The Eq. (5.7) for detailed balance corresponding to the move shown in Fig. 5.2 is

$$\frac{e^{-\bar{\mu}N}}{N!} \, t[\tilde{T}_a(N), \tilde{T}_b(N+1)] = \frac{e^{-\bar{\mu}(N+1)}}{(N+1)!} \, t[\tilde{T}_b(N+1), \tilde{T}_a(N)], \qquad (5.16)$$

where $\tilde{T}_a(N)$ and $\tilde{T}_b(N+1)$ are labelled triangulations with N and $N+1$ vertices, respectively. The transition probability t can be written as a product of two factors: one determined by the way we implement the move in the computer and another one chosen such that Eq. (5.16) is satisfied. We implement the move from $\tilde{T}_a(N)$ to $\tilde{T}_b(N+1)$ in the following way. Select a vertex v randomly. Denote the order of the vertex by $\sigma(v)$, i.e. it has $\sigma(v)$ neighbouring vertices cyclically ordered by the orientation of the triangulation. Select one of the neighbours, v_1, randomly, and a number n_0 such that $1 \leq n_0 \leq \sigma(v) - 1$. Denote by v_2 the n_0th neighbour to v counted from v_1 in the positive direction. These three vertices are the ones shown in Fig. 5.2. The probability of this choice is

$$P_1(N \to N+1) = \frac{1}{N} \frac{1}{\sigma(v)(\sigma(v)-1)}. \qquad (5.17)$$

We can now insert a new vertex v' with label $N+1$ in accordance with Fig. 5.2. After this insertion we have that $\sigma(v)$ changes to $\sigma(v) - n_0 + 2$ and $\sigma(v') = n_0 + 2$. The inverse move is implemented by choosing a random vertex v' among the $N+1$ possible vertices in $\tilde{T}_b(N+1)$. Next choose a vertex v randomly among its $\sigma(v')$ neighbours and delete v' and the link between v and v', as indicated in Fig. 5.2. We assume for the moment that this move is allowed, i.e. that we stay within the type III triangulations when we remove v'. By deleting v' we change the order of v from $\sigma(v)$ to $\sigma(v) + \sigma(v') - 4$. If the label of v' is less than $N+1$ we attach this label to the vertex that was previously labelled by $N+1$, and perform a corresponding reassignment of links and neighbours. The probability of choosing v, given v', is

$$P_2(N+1 \to N) = \frac{1}{\sigma(v')}. \qquad (5.18)$$

Note the asymmetry between Eq. (5.17) and Eq. (5.18) with respect to the factor $1/N$. We now obtain the second factor in $t[\tilde{T}_a(N), \tilde{T}_b(N+1)]$ by setting

$$t[\tilde{T}_a(N), \tilde{T}_b(N+1)] = P_1(N \to N+1)P_{up}, \qquad (5.19)$$

where

$$P_{up} = \frac{e^{-\bar{\mu}N}}{N+1} \frac{\sigma(v)(\sigma(v)-1)}{\sigma(v) - n_0 + 2}. \qquad (5.20)$$

The factor P_{up} should be thought of in the following way: after having chosen v we select a random number r with uniform probability between 0 and 1. If $r < P_{up}$ we accept the move, otherwise not. In a similar way we determine $t[\tilde{T}_a(N), \tilde{T}_b(N+1)]$ by writing

$$t[\tilde{T}_b(N+1), \tilde{T}_a(N)] = P_2(N+1 \to N)P_{down}, \qquad (5.21)$$

where

$$P_{down} = \frac{e^{\tilde{\mu}}(N+1)}{N} \frac{\sigma(v)}{(\sigma(v) + \sigma(v') - 4)(\sigma(v) + \sigma(v') - 5)}, \qquad (5.22)$$

and the probability interpretation of P_{down} is similar to the one for P_{up}.

As mentioned above, it might happen that the removal of v' creates a triangulation which is not a type III triangulation. This happens if $\sigma(v_1) = 3$ or $\sigma(v_2) = 3$. In this case one counts it as a genuine attempt to perform a move, but rejects it, i.e. we put $P_{down} = 0$. By such a choice, detailed balance is still satisfied since no inverse move exists originating from a type III triangulation.

In summary, the Monte Carlo procedure can be described as follows. Choose randomly either to delete or to insert a vertex. Select vertices v and v' as described above and perform the move with probability P_{up} or P_{down}, respectively, depending on whether we have chosen to insert or delete a vertex. From a practical point of view it is not convenient to allow N to be unrestricted and there is no violation of detailed balance even though we require $N_{min} \le N \le N_{max}$ by the choice $P_{up}(N) = 0$ for $N \ge N_{max}$ and $P_{down}(N) = 0$ for $N \le N_{min}$.

In the next chapter we discuss Monte Carlo simulations of higher-dimensional gravity and a set of ergodic moves in the class of triangulations of d-dimensional manifolds where moves do not in general conserve the number of simplexes (or subsimplexes). The two-dimensional case is special in the sense that the "flip" shown in Fig. 5.1 is ergodic in the class of two-dimesional triangulations with a fixed topology and a fixed number of triangles or vertices. In many applications it is convenient to perform the computer simulations at fixed volume and it allows one to use the powerful finite-size scaling relations, as will be discussed in the next section.

Let us consider the situation where matter fields are coupled to gravity. The system is characterized by its matter configuration $\{\sigma_i^a\}$ as well as the triangulation T_a, and the equation of detailed balance (5.7) reads

$$P(\{\sigma_i^a\}, T_a)\, t([\{\sigma_i^a\}, T_a], [\{\sigma_i^b\}, T_b]) = P(\{\sigma_i^b\}, T_b)\, t([\{\sigma_i^b\}, T_b], [\{\sigma_i^a\}, T_a]).$$
$$(5.23)$$

On triangulations with a fixed number of triangles the transition described by t is a combination of a flip-transition for the triangulation and an updating of the spin configuration. The transition amplitude for a flip

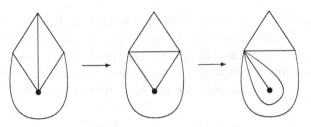

Fig. 5.3. A sequence of flips changing the order of a vertex from 3 to 1.

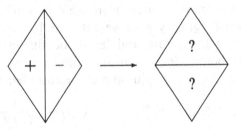

Fig. 5.4. The ambiguity in spin assignment after a flip if the triangles involved in the flip have opposite spin orientation.

is always taken to be one when a link is selected randomly and a flip is attempted. In most cases it is allowed. Occasionally it might lead to configurations which do not belong to the class of triangulations under consideration. An example is shown in Fig. 5.3 where a flip can generate vertices of order two. We might not want these. Even if we allow vertices of order two, further flips might lead to vertices of order one (see Fig. 5.3). Again we might not want these, but even if they are allowed a flip might result in an isolated vertex. The way to deal correctly with such situations is simply to ignore the attempt to make the flip when a configuration in a forbidden class is generated and instead choose (randomly) a new link and attempt a new flip. While this takes care of the geometric part of the updating it remains to update the matter fields when we change the geometry. The reason is that a change in connectivity entails a change in the assignments of neighbours and therefore a change in the matter action even if no spins are actually flipped. This is illustrated in Fig. 5.4: two triangles disappear and are replaced by two new triangles by the flip. The two spins of the original triangles are randomly assigned to the two new triangles and the transition function is calculated according to the Metropolis scheme given by Eq. (5.10).

Using the ideas outlined above it is possible to perform extensive Monte Carlo simulations of spin systems on random surfaces. By the same methods one can perform simulations of Gaussian fields coupled to gravity which we discussed in Chapter 3, and one can add more complicated matter actions. The numerical results described in Chapter 3 for string models with extrinsic curvature action were generated by such methods. An important tool in analysing the results is finite-size scaling.

5.3   Finite-size scaling

Finite-size scaling is concerned with the behaviour of a finite system in the presence of a divergent correlation length at a phase transition point in the corresponding infinite system. Here we only outline the ideas needed for two-dimensional gravity and we refer to [88, 67] for reviews. Let β be the inverse temperature and let β_c be the inverse of the critical temperature. As before, let us consider the simplest spin system, i.e. an Ising spin system. At β_c the spin–spin correlation length diverges:

$$\langle \sigma_i \sigma_{i+n} \rangle \sim e^{-n/\xi(\beta)}, \qquad \xi(\beta) \sim \frac{1}{|\beta - \beta_c|^\nu}. \qquad (5.24)$$

For a system with a finite number of degrees of freedom there will of course never be a genuine phase transition. Instead there will be a so-called pseudo-critical point where the correlation length coincides with the linear size of the system. Finite-size scaling is based on the assumption that close to this point the free energy will be a function of "reduced" variables:

$$F(\beta, H, L) = F\left(\frac{L}{\xi(\beta)}, H\xi^y(\beta)\right), \qquad (5.25)$$

where H is the external magnetic field and $L \approx \xi(\beta)$ is the linear size of the system. The constant y is related to the critical exponents of the system, as we describe below. The free energy per volume can be written

$$f(\beta, h) = L^{-d} F(|\beta - \beta_c|^\nu L, H L^y) \qquad (5.26)$$

and from this equation we can derive the known scaling relations and the finite-size behaviour of f. The energy density, the specific heat, the magnetization per unit volume and the magnetic susceptibility can all be calculated from f by differentiation with respect to β and H:

$$\varepsilon = \frac{\partial f}{\partial \beta}, \qquad c_v = \frac{\partial^2 f}{\partial \beta^2}, \qquad (5.27)$$

$$m = \frac{\partial f}{\partial H}, \qquad \chi = \frac{\partial^2 f}{\partial H^2}. \qquad (5.28)$$

The standard critical exponents of the magnetic system are defined by

$$c_v \sim |\beta - \beta_c|^{-\alpha}, \qquad \chi \sim |\beta - \beta_c|^{-\gamma_m}, \qquad m \sim |\beta - \beta_c|^{\beta_m}, \qquad (5.29)$$

where β_m and γ_m are not to be confused with the inverse temperature β and the string susceptibility exponent γ, respectively.

From the assumed form of f, given by Eq. (5.26), and the further assumption that

$$|\beta - \beta_c| L^{\frac{1}{\nu}} = O(1), \qquad |\beta_c^* - \beta_c| L^{\frac{1}{\nu}} = O(1), \qquad (5.30)$$

where β_c^* is a pseudo-critical point (usually defined as the location of the peak in the specific heat or its derivatives), we obtain the relations

$$\varepsilon(\beta_c^*) \sim L^{\frac{1}{\nu}-d}, \qquad c_v(\beta_c^*) \sim L^{\frac{2}{\nu}-d}, \qquad (5.31)$$

$$m(\beta_c^*) \sim L^{y-d}, \qquad \chi(\beta_c^*) \sim L^{2y-d}. \qquad (5.32)$$

From the above equations we obtain the identifications

$$\alpha = 2 - \nu d, \qquad \beta_m = \nu(d - y), \qquad \gamma_m = \nu(2y - d), \qquad (5.33)$$

and the scaling relations

$$\alpha + 2\beta_m + \gamma_m = 2, \qquad 2\beta_m + \gamma_m = \nu d \qquad (5.34)$$

follow.

Eqs. (5.31) and (5.32) are the most important results for our purposes. Formulated in terms of the volume $V = L^d$ they read:

$$\varepsilon(\beta_c^*) \sim V^{\frac{\alpha-1}{\nu d}}, \qquad c_v(\beta_c^*) \sim V^{\frac{\alpha}{\nu d}}, \qquad (5.35)$$

$$m(\beta_c^*) \sim V^{-\frac{\beta_m}{\nu d}}, \qquad \chi(\beta_c^*) \sim V^{\frac{\gamma_m}{\nu d}}. \qquad (5.36)$$

The derivations presented above can be made more solid by appealing to renormalization group arguments and they are valid in any dimensions d.

In two dimensions second-order phase transitions are closely related to conformal invariance and finite-size scaling for a statistical system can be inferred from the scaling of operators in the corresponding conformal field theory. In the case of critical spin systems the interesting operators of the corresponding continuum theory are the spin operator Φ_σ and the energy operator Φ_ε. Assume they have scaling dimensions $\Delta_\sigma^{(0)}$ and $\Delta_\varepsilon^{(0)}$, respectively. For dimensional reasons we obtain

$$\int_V d^d x_1 \cdots d^d x_n \langle \Phi_{i_1}(x_1) \cdots \Phi_{i_n}(x_n) \rangle \sim V^{n - \Delta_{i_1}^{(0)} - \cdots - \Delta_{i_n}^{(0)}}, \qquad (5.37)$$

where V is the volume of the system and each Φ_i is either Φ_σ or Φ_ε. Eq. (5.37) allows us to express the critical exponents in terms of $\Delta_\sigma^{(0)}$ and $\Delta_\varepsilon^{(0)}$. In the case of the energy density we obtain

$$\varepsilon(V) \sim V^{-\Delta_\varepsilon^{(0)}},$$

while for the magnetization we find

$$m(V) \sim V^{-\Delta_\sigma^{(0)}}.$$

For the specific heat and the spin susceptibility we obtain:

$$c(V) \;=\; \frac{1}{V} \int d^d x \, d^d y \langle \Phi_\varepsilon(x) \Phi_\varepsilon(y) \rangle \;\sim\; V^{1-2\Delta_\varepsilon^{(0)}}, \qquad (5.38)$$

$$\chi(V) \;=\; \frac{1}{V} \int d^d x \, d^d y \langle \Phi_\sigma(x) \Phi_\sigma(y) \rangle \;\sim\; V^{1-2\Delta_\sigma^{(0)}}. \qquad (5.39)$$

In the above formulas we have used the notation

$$\varepsilon(V) \equiv \varepsilon(\beta^*), \quad m(V) \equiv m(\beta_c^*), \quad c_v(V) \equiv c_v(\beta^*), \quad \chi(V) \equiv \chi(\beta^*);$$

see Eqs. (5.35) and (5.36). If we compare the above equations with Eqs. (5.35) and (5.36) we can write:

$$\alpha = vd(1 - 2\Delta_\varepsilon^{(0)}), \qquad \gamma_m = vd(1 - 2\Delta_\sigma^{(0)}). \qquad (5.40)$$

Finally, the expressions for the exponents $\Delta_\varepsilon^{(0)}$ and $\Delta_\sigma^{(0)}$ in terms of standard critical exponents related to the renormalization group is established by using the fact that the Hamiltonian of the matter system close to the critical point can be expanded in terms of operators with definite scaling properties. The most relevant operators are those with the largest eigenvalues under a scaling $x \to \lambda x$. The two most relevant operators of spin systems under the above scaling have relevant eigenvalues $\lambda^{1/v}$ and λ^y where

$$\frac{1}{v} = 2(1 - \Delta_\varepsilon^{(0)}), \qquad y = (2 - \Delta_\sigma^{(0)}). \qquad (5.41)$$

A priori it is not known whether we can expect relations such as Eqs. (5.35) and (5.36) to be valid after coupling to quantum gravity. As already mentioned, the most general derivation is based on renormalization group arguments which depend on the presence of a divergent correlation length. The concept of correlation length is not readily available in quantum gravity. The scaling *ansatz* by David, Distler and Kawai (DDK) [131, 143] avoids this problem by generalizing Eqs. (5.37)–(5.39) to two-dimensional quantum gravity. One simply makes the replacement

$$\int_V d^2 x \;\to\; \int d^2 x \sqrt{g} \, \delta \left(\int d^2 x \sqrt{g} - V \right),$$

moving the expectation outside the integration in order to include the gravitational average over surfaces of area V. One also replaces the scaling dimension $\Delta^{(0)}$ by the corresponding, postulated, scaling dimension Δ in the theory coupled to quantum gravity. By the use of self-consistent scaling arguments in quantum Liouville theory DDK were able to derive the KPZ-relation between Δ and $\Delta^{(0)}$ which can be written as

$$\Delta = \frac{\sqrt{1 - c + 24\Delta^{(0)}} - \sqrt{1 - c}}{\sqrt{25 - c} - \sqrt{1 - c}}, \qquad (5.42)$$

where c is the central charge of the model. In this way one finds the quantum gravity version of Eq. (5.37)

$$\left\langle \int d^d x_1 \cdots d^d x_n \Phi_{i_1}(x_1) \cdots \Phi_{i_n}(x_n) \right\rangle_V \sim V^{n - \Delta_{i_1} - \cdots - \Delta_{i_n}}, \qquad (5.43)$$

where the average $\langle \cdot \rangle_V$ includes the functional integration over matter fields as well as the functional integration over equivalence classes of Riemannian metrics. Eqs. (5.42) and (5.43) constitute the theoretical basis for finite-size scaling in two-dimensional quantum gravity. We now obtain the gravitationally dressed critical exponents α/vd, β_m/vd and γ_m/vd. If we assume that the so-called *hyper-scaling relation*

$$\alpha = 2 - vd,$$

which follows from Eq. (5.34), is still valid, we also find values for vd, α, β_m and γ_m. In the case of the Ising model these exponents agree with the ones determined directly by the use of dynamical triangulations. Note that only the combination vd is determined by these general scaling arguments. This is linked to the fact that no reparametrization invariant correlation length has been introduced. In the original scaling arguments the dimension of the system, d, entered via $V \sim L^d$. In the context of quantum gravity it is not clear how we should view this relation since we integrate over all manifolds with volume V. Should d be the dimension of the manifold, i.e. two, or should d be a dynamically determined Hausdorff dimension of the manifolds? The predictions of finite-size scaling in quantum gravity, as given by Eqs. (5.42) and (5.43), are nevertheless independent of this ambiguity.

In order to illustrate how well the numerical methods work in practice, we show in Fig. 5.5 the result of Monte Carlo simulations of the Ising model on a random surface. As mentioned above, the exponents β_m and γ_m can be calculated analytically and we see that agreement is excellent.

5.4 Two-dimensional geometry

In the previous section we discussed the measurement of the critical exponents of matter fields coupled to quantum gravity. We now turn to the measurement of the "quantum geometry" of random surfaces. One of the simplest quantities which characterize the fractal structure of two-dimensional gravity is the entropy exponent γ. It was originally introduced as the exponent of the subleading correction to the partition function for surfaces of fixed volume

$$Z(V) \sim e^{cV/a^2} \left(\frac{V}{a^2} \right)^{\gamma - 3}. \qquad (5.44)$$

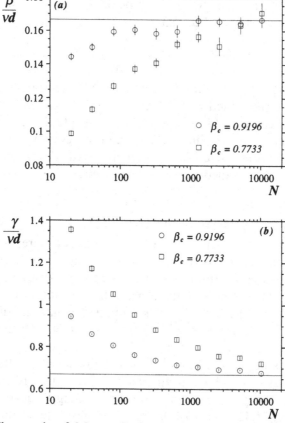

Fig. 5.5. The result of Monte Carlo simulations of Ising spins coupled to dynamical triangulations. Fig. (a) shows the measurement of $\beta_m/\nu d$ at the critical point as a function of the number of triangles N. The squares represent measurements for the class of triangulations without loops of length one or two, the circles represent measurements for triangulations where such configurations are allowed. The horizontal line is the theoretical value $1/6$. Fig. (b) shows similar measurements of $\gamma_m/\nu d$, which has the theoretical value $2/3$. For systems with loops of length one and two there is excellent agreement with the continuum results for quite small systems ($N = 2000$).

In the case of pure gravity $Z(V)$ can be regarded as the number of inequivalent Riemann surfaces with volume V. If matter fields are coupled to gravity the counting is weighted with the effective gravitational action of the matter fields. Of course, the concept of counting only has a precise meaning in a regularized theory, as is clear from Eq. (5.44), where a denotes a cutoff. With dynamical triangulations as regularization, Eq. (5.44) becomes

$$Z(N) \sim e^{\mu_0 N} N^{\gamma-3}, \qquad (5.45)$$

where N is the number of triangles. The exponent γ has a very simple geometrical meaning which is worth emphasizing. Recall that a baby universe is an outgrowth on a triangulation which can be separated from the rest of the triangulation by cutting along a closed link loop of minimal length. Consider triangulations with a fixed number of triangles N. Then the average number of baby universes with volume $n < N/2$, $\langle b(n) \rangle_N$, is given by

$$\langle b(n) \rangle_N \sim \frac{nZ(n)\,(N-n)Z(N-n)}{Z(N)}. \qquad (5.46)$$

In order to make this formula plausible assume for simplicity that we are working with triangulations with minimal loops of length two. The function $Z(N)$ counts the (weighted) number of triangulations consisting of N triangles. How many of these have a structure which allows us to cut the triangulation along a minimal loop into two parts made of n and $N - n$ triangles? In order to find this number consider pairs of closed triangulations constructed from n and $N - n$ triangles, respectively; cut them open along a link, thereby creating boundary loops of length two. Glue the triangulations together along the boundaries. In the process of cutting them open we produce $\frac{3}{2}nZ(n)$ and $\frac{3}{2}(N-n)Z(N-n)$ distinct manifolds with minimal length boundaries, respectively, since we can place the boundary at any link, and the number of links is $3/2$ times the number of triangles. Clearly, these counting arguments are correct only up to combinatorial factors of order one due to non-trivial symmetry factors that may be associated with non-generic triangulations. Now combining Eq. (5.45) with Eq. (5.46) we obtain

$$\langle b(n) \rangle_N \sim N\,[n(1 - n/N)]^{\gamma-2}. \qquad (5.47)$$

This formula is very convenient for measuring γ since it is computationally easy to identify the minimal loops in a given triangulation, and, once identified, it is easy to find the volume of the corresponding baby universes. We can therefore find the distribution $\langle b(n) \rangle_N$ in a Monte Carlo simulation with fixed N. From the measured distribution we can extract γ.

The numerical simulations based on this technique are in perfect agreement with the theoretical results, both for pure gravity, the Ising model coupled to gravity and the 3- and 4-state Potts models coupled to gravity, which have $c = 0$, $1/2$, $4/5$ and 1, respectively, and corresponding susceptibility exponents $\gamma = -1/2$, $-1/3$, $-1/5$ and 0. In Fig. 5.6 we show the values for γ (as a function of β) obtained by fitting to the baby universe distributions for the Ising model and 3-state Potts model. *It is clearly seen how the exponent γ varies near the phase transition point, signalling the back-reaction of matter on quantum gravity.*

In the previous chapter we discussed in detail the geometry of pure

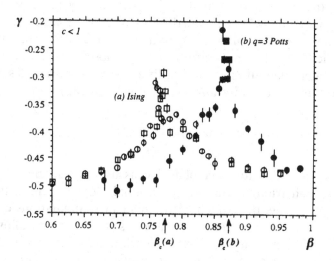

Fig. 5.6. Measured values of γ as a function of β for the Ising model (open symbols) and the 3-state Potts model (filled symbols) for two sizes of the systems, $N = 1000$ (circles) and 2000 (squares).

two-dimensional gravity and we saw that the two-point function, and consequently $\langle n(R) \rangle_V \, dR$, the average volume of spherical shells of radius R and thickness dR, could be calculated analytically. Pure two-dimensional gravity is consequently an excellent laboratory for testing computer simulations of quantum geometry. The question is whether we can reproduce the known continuum results for finite-size systems and whether we can learn something about the finite-size effects. It turns out that computer simulations are very efficient in this respect.

To illustrate this let us discuss measurements of $\langle n(r) \rangle_N$ in discretized two-dimensional gravity, where r is the distance between a triangle and a spherical shell of triangles (as defined in Section 4.7) and N denotes the discretized volume, i.e. the number of triangles. Recall that the continuum function $\langle n(R) \rangle_V$, where R and V refer to continuum geodesic distance and continuum volume, is related to the function $G(R; V)$ by

$$G(R; V) \sim V^{\gamma-2} \langle n(R) \rangle_V, \qquad (5.48)$$

and $G(R; V)$ is the inverse Laplace transform of the two-point function $G(R; \Lambda)$ which we calculated in Section 4.7. We have already discussed the asymptotic behaviour of $G(R; V)$ for small and large R, but it is in fact possible to find an exact expression for $G(R; V)$ in terms of generalized hyper-geometric functions if we expand $G(R, \Lambda)$ as follows:

$$G(R; \Lambda) = \Lambda^{9/4} \sum_{n=1}^{\infty} n^2 \, e^{-2n\Lambda^{\frac{1}{4}}R}. \tag{5.49}$$

In this way we obtain a continuum formula for $\langle n(r) \rangle_V$ expressed as

$$\frac{1}{V} \langle n(R) \rangle_V dR = f(x)dx, \tag{5.50}$$

where $x = R/V^{\frac{1}{4}}$ is a dimensionless variable. The function $f(x)$ can be expressed in terms of generalized hyper-geometric functions and it is normalized such that

$$\int_0^{\infty} f(x)dx = 1. \tag{5.51}$$

We can now compare $f(x)$ directly with the measured function

$$F_N(x) = N^{\frac{1}{4}} \frac{1}{N} \langle n(r) \rangle_N, \qquad x = r/N^{\frac{1}{4}}, \tag{5.52}$$

where N is the number of triangles and r is the geodesic distance along the triangles, as defined in Section 4.7. For N sufficiently large, $F_N(x)$ should be independent of N and equal to $f(x)$. In the Monte Carlo simulations we generate a sequence of statistically independent triangulations with N triangles and spherical topology. For each independent triangulation we can measure the distribution $\langle n(r) \rangle_N$ which for each specific triangulation depends on the triangle chosen as the centre of the spherical shells. From a practical point of view it is therefore an advantage to perform the measurement for many different random choices of "centre triangles" and take the average. The functions $F_N(x)$ obtained in this way are in qualitative agreement with $f(x)$, but the convergence towards the correct distribution $f(x)$ is rather slow as a function of N. One may wonder whether this implies that we have to go to very large N in order to see continuum physics. Fortunately, the discrepancy between $F_N(x)$ and $f(x)$ disappears if we analyse the finite-size effects more closely. It is not to be expected that $F_N(x)$ will agree with $f(x)$ if r is too small, i.e. if $r = 1, 2, ..., r_0$, where r_0 is some fixed number, *independent of* N, since for short distances the effect of the discretization should be detectable for all values of N. This suggests that one should simply discard the first r_0 points by performing a shift in r and N and introduce a new scaling variable defined by

$$x = \frac{r + \alpha}{N^{1/4} + \beta}, \tag{5.53}$$

where α and β are constants. In fact we can obtain Eq. (5.53) from the data in the following way: for a given value of r we determine x such that

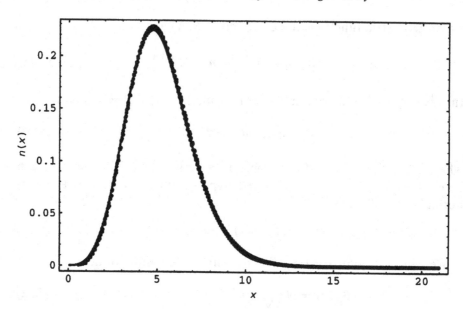

Fig. 5.7. The distributions $F_N(x)$ for $N = 1000$, 2000, 4000, 8000, 16 000 and 32 000, as well as the theoretical distribution $f(x)$.

$F_N(r/N^{1/4}) = f(x)$. We observe a proportionality between x and r except for the smallest values of r ($r < r_0$). In addition, the constant α seems to be approximately independent of N for N in the range 1000-32 000. The slope of the relation between x and r changes with N and it can be fitted well to $N^{1/4} + \beta$. Using Eq. (5.53) we obtain a perfect agreement between the measured and the theoretical distributions all the way down to Ns as small as 1000. The results are shown in Fig. 5.7 for $r_0 = 6$. The diagram represents seven curves, including the theoretical curve $f(x)$.

In the discussion above we could as well have used the "link-distance" r_l between a vertex and a spherical shell of vertices, rather than the "triangle-distance" r between a triangle and a spherical shell of triangles. In fact, finite-size scaling works much better for the link-distance and the constants α and β used in the definition Eq. (5.53) of the scaling variable x are very close to zero if we use the link-distance. We have here used the triangle-distance since in this case the analytic results derived in Chapter 4 allow us to *calculate* the constant of proportionality between the continuum variable $R/V^{1/4}$ and the lattice variable $r/N^{1/4}$. Thus, no "empirical" scale adjustment has been performed in x when we compare the theoretical curve and the Monte Carlo data in Fig. 5.7. The fact that we obtain identical scaling laws using link-distance and the triangle-distance is of course in agreement with universality.

We can now try to apply the same technique for the coupled system of matter and two-dimensional quantum gravity. As an example consider an Ising model coupled to gravity. Away from the critical point we know that the fractal structure of space-time should be identical to that of pure gravity. At the critical point we know that γ changes from $-1/2$ to $-1/3$, i.e. the density of baby universes increases. In order to see whether the Hausdorff dimension also changes, we reverse the produre outlined above for pure gravity. We can measure $\langle n(r) \rangle_N$ and we expect the scaling

$$\frac{1}{N} \langle n(r) \rangle_N = N^\nu F_N(x), \quad x = r/N^\nu. \tag{5.54}$$

The exponent ν equals d_h^{-1} as usual and $F_N(x)$ should be independent of N for N sufficiently large. From this we can extract a value of the exponent ν. Either we use the growth of the peak of $N^{-1}\langle n(r) \rangle_N$ (this is similar to the use of finite-size scaling in the case of the magnetic exponents of the Ising model) or we find the exponent ν which gives the best match of the distributions $F_N(x)$ for different Ns. If we use for r the "triangle-distance" the latter method only works well if we combine it with a shift of the form given by Eq. (5.53). Using instead the "link-distance" the constants α and β in Eq. (5.53) are close to zero, as was the case for pure gravity. The value of ν obtained in this way agrees with the value extracted from the growth of the peak and leads to the conjecture that the Hausdorff dimension of random surfaces coupled to unitary matter fields with central charge $0 \le c \le 1$ is four. Beyond $c = 1$ the numerical simulations indicate that the Hausdorff dimension starts to decrease and that it equals two for large c. This is consistent with our previous arguments that the theory is equivalent to branched polymers for large c. Numerical simulations also suggest that the Hausdorff dimension decreases below four if $c < 0$. As already mentioned, the $c = -2$ system coupled to quantum gravity is special, and using so-called recursive sampling one can simulate very large systems. From the simulations one finds $d_h \approx 3.5$.

There are many interesting questions about the geometry of two-dimensional quantum gravity which remain to be answered. Numerical simulations can never be a substitute for analytical results, but they may, and frequently do, provide us with important hints and guidelines which can inspire further theoretical developments. The success of computer simulations in two-dimensional quantum gravity is also encouraging for the study of quantum gravity in dimensions higher than two. For the higher-dimensional theories, to be described in the next chapter, we still have to rely almost entirely on numerical simulations.

5.5   Notes

For a more extensive introduction to Monte Carlo methods than is given here we refer to [87, 89]. The use of Monte Carlo simulations in Euclidean string theory began with the simulations of the hyper-cubic random surface model [79, 80]. These simulations were performed in a grand canonical ensemble, i.e. the number of plaquettes constituting the random surface was allowed to vary. As described in Chapter 3 this model is trivial in the sense that it reduces to branched polymers in the scaling limit. The simulation of a modified model where extrinsic curvature terms were added to the action produced the non-trivial result $\gamma = 1/4$ for the susceptibility exponent in dimension $d = 4$ [71, 70, 72]. This result is still not well understood, but the same value for γ has been found in Monte Carlo simulations of dynamically triangulated random surfaces with extrinsic curvature terms [19]. It is natural to conjecture that it belongs to the class of theories with $c > 1$ described in Chapter 4, where the possible values of γ were found to be 1/3, 1/4, etc.

The first attempts to perform Monte Carlo simulations of models based on dynamical triangulations can be found in [271, 27, 86, 250, 251]. More extensive computer simulations were reported in [21, 137]. Algorithms were developed both for grand canonical ensembles (where the number of triangles is allowed to vary) and canonical ensembles (where the number of triangles is kept fixed). The emphasis of these early studies was on the bosonic string embedded in d dimensions. With the hindsight of history we now understand that there is not much hope of finding a non-trivial Polyakov string theory for $d \geq 2$ since the string tension is not scaling [22]. However, there might still exist a non-trivial universal theory with well-defined critical indices when the central charge c of matter fields exceeds 1. The only analytical tool available so far has been mean field analysis, but from a numerical point of view nothing prohibits the crossing of the $c = 1$ barrier. A number of simulations of multi-spin and multi-Gaussian fields coupled to two-dimensional quantum gravity have been performed [61, 116, 36, 52].

For a general discussion of finite-size scaling we refer to [88, 67] . The first use of finite-size scaling in the context of dynamical triangulations coupled to spin models can be found in [65, 119], and after this work finite-size scaling was generally accepted as the most reliable way to extract critical exponents for the matter systems. The data shown in the text refer to later, more extensive, simulations [52]. The first measurements of the string susceptibility γ were performed using the grand canonical ensemble [27, 251]. The technique used to extract a reliable value of γ was partly borrowed from related computer simulations of hyper-cubic lattice surfaces [71]. Only later was it realized that the baby universe counting, first defined in [243], allowed one to extract γ from the canonical simulations [40, 51]. This method is superior if γ is not too negative, in which case there will be very few large baby universes. The technique of baby universe surgery has been used successfully in higher-dimensional gravity as well, starting with [39].

The use of the two-point function as a probe of the fractal structure of two-dimensional quantum gravity was advocated in [55]. The verification that the two-point function is a powerful tool from a numerical point of view came in [46] and in [118], where the first study of finite-size scaling of the spin–spin correlation function is found as well. In both articles it was observed that the Hausdorff dimension is $d_h \approx 4$, irrespective of the matter coupled to gravity as long as the central charge $0 \leq c \leq 1$. The measurement of d_h for $c > 1$ is found in [46], while the measurement of d_h for $c = -2$, using recursive sampling, was reported in [267].

6
Gravity in higher dimensions

In this chapter we return to the discussion of quantum gravity which we began in Chapter 4. In the first section we describe some of the technical problems that are encountered in constructing a theory of quantum gravity and some of the ideas that may go into their resolution. We then give a definition of simplicial gravity in arbitrary dimensions and describe a representative sample of the numerical results that have been obtained. It is often convenient to consider the theory in a fixed dimension larger than two. We shall discuss the four-dimensional case since it is physically the most relevant, and will only occasionally consider three-dimensional gravity.

6.1 Basic problems in quantum gravity

Formulating a theory of quantum gravity in dimensions higher than two leads to a number of basic questions, some of which go beyond those encountered in dimension two. Among these are the following:

(i) What are the implications of the unboundedness from below of the Einstein–Hilbert action?

(ii) Is the non-renormalizability of the gravitational coupling a genuine obstacle to making sense of quantum gravity?

(iii) What is the relation between Euclidean and Lorentzian signatures and do there exist analogues of the Osterwalder–Schrader axioms allowing analytic continuation from Euclidean space to Lorentzian space-time?

(iv) What is the role of topology in view, for instance, of the fact that higher-dimensional topologies cannot be classified?

271

We do not have answers to these questions and our inability to deal with them may be an indication that there exists no theory of Euclidean quantum gravity in four dimensions or, possibly, that quantum gravity only makes sense when embedded in a larger theory such as string theory.

Superstring theory has the appealing feature of combining gauge theories and a theory of gravity in a natural way. Distances of the order of the Planck length represent a regime where genuinely string theoretical aspects should be important, if the world were governed by an underlying string theory. However, superstring theory has its own problems at such length scales, where the summation over string loops should be important. This summation has never been defined except in terms of string perturbation theory. It is also a possibility that string theory is simply wrong. We do not as yet have any experimental evidence for supersymmetry which might not be present in our world. If we drop supersymmetry, string theory is probably not a viable theory, as we have discussed in some detail in Chapter 3. In this situation we have an obligation to search for alternative theories of quantum gravity.

Let us discuss in some more detail the problems mentioned above. The Einstein–Hilbert action for a closed four-dimensional manifold M can be written as

$$S(g;\Lambda,G) = -\frac{1}{16\pi G}\int_M d^4\xi\sqrt{g}\,(R - 2\Lambda), \qquad (6.1)$$

where R is the scalar curvature of the metric g on M. The curvature $\tilde{R}$ of the conformally transformed metric

$$\tilde{g}_{ab} = \Omega^2 g_{ab}, \qquad (6.2)$$

where Ω, a positive function on M, is given by

$$\tilde{R} = \Omega^{-2}R - 6\Omega^{-3}\Delta_g\Omega. \qquad (6.3)$$

The action corresponding to $\tilde{g}$ is

$$S(\tilde{g};\Lambda,G) = -\frac{1}{16\pi G}\int_M d^4\xi\,\sqrt{g}\,(\Omega^2 R + 6g^{ab}\partial_a\Omega\partial_b\Omega - 2\Lambda\Omega^4) \qquad (6.4)$$

from which it follows that $S(\tilde{g};\Lambda,G)$ is unbounded from below due to the wrong sign of the quadratic term in the derivatives of Ω. A simple minded way of dealing with this problem is to assume that the Einstein–Hilbert action is only an infrared approximation to the complete theory and add higher-derivative terms of the form R^2 or $R_{ab}R^{ab}$ to the action density. Such terms also tend to alleviate the problem of renormalizability since the propagators will contain higher inverse powers of the momentum. In fact, adding suitable combinations of R^2 terms, one can obtain a theory which is renormalizable by power counting and where the Euclidean action is bounded from below [351, 181]. However, such theories do in general not qualify as unitary quantum field theories after analytic continuation to Minkowski space in the standard fashion.

Several attempts have been made to cure the disease associated with the conformal mode by special analytic continuation of this mode. For different points of view we refer to [200, 297].

A more general way of viewing the Einstein–Hilbert term as the first (infrared leading) term of the full action is due to Weinberg, who introduced the concept of *asymptotic safety* [373]. The idea is to use the renormalization group to obtain a non-trivial ultraviolet stable fixed point from the long-distance behaviour, where the Einstein–Hilbert action yields an effective description of the theory, by fine-tuning a finite number of parameters in a class of bare actions whose associated critical surface has finite co-dimension. The effective Lagrangian description of such a theory in terms of fields appropriate for the infrared limit might then be an infinite series of the form

$$\mathscr{L} = \sqrt{g} \left[\Lambda - \frac{1}{16\pi G} R + f_2 R^2 + f_2' R_{ab} R^{ab} + \cdots \right]. \tag{6.5}$$

This action could be non-polynomial, but the theory might make sense both with Riemannian and Lorentzian signatures.

The weakness of this proposal is the general lack of non-trivial fixed points in four-dimensional field theories. However, some interesting results obtained in $2+\varepsilon$ dimensions lend support to Weinberg's idea [265, 264, 9]. We shall see that four-dimensional quantum gravity formulated in terms of dymamical triangulations might lead to a similar picture.

Purely topological theories might play an important role at the Planck scale. In Chapter 4 we saw the appearance of such theories in two-dimensional gravity. A similar situation arises in three-dimensional gravity, which is closely related to a topological Chern–Simons gauge theory, and enables us to rotate in a natural way between Euclidean and Lorentzian signatures. The three-dimensional case will be discussed in more detail in the following chapter.

In four dimensions an appealing scenario is one where the short-distance quantum gravity is topological, while the effective, long-distance theory is governed by an action which, to lowest order, describes classical Einstein gravity. If this picture is correct, then at sufficiently short distances where topological effects dominate, the equivalence classes of metrics would not be the appropriate dynamical variables.

Almost nothing is known about the problem of fluctuating topologies in quantum gravity. The path integral formalism offers the possibility of summing over diffeomorphism classes of manifolds as well as integrating over the inequivalent Riemannian structures of a given manifold M when formally defining the partition function as

$$Z = \sum_M \int_M \mathcal{D}[g_{ab}] \, e^{-S}. \tag{6.6}$$

In two dimensions the meaning of the sum over M is clear since (smooth or continuous) orientable closed manifolds are uniquely characterized by their Euler characteristic χ and the sum runs over χ or, equivalently, the genus of the surfaces. In spite of this simple prescription surprisingly little progress has been made in defining the sum in Eq. (6.6) using continuum methods. The matrix models allowed us to attempt a non-perturbative definition of the sum, as described in Chapter 4, but this approach is beridden with problems and has not yet led to results which allow a convincing physical interpretation.

We encounter a slight additional problem in three dimensions since there is no parametrization of the diffeomorphism classes of three-dimensional manifolds. But here also the categories of smooth and continuous manifolds are equivalent, contrary to what happens in four and higher dimensions, where there exist continuous manifolds which do not admit smooth structures and some continuous manifolds admit infinitely many inequivalent smooth structures, see e.g. [275].

If we insist on summing over all smooth structures, the sum in (6.6) is, to put it mildly, rather unwieldy. An additional complication might be that four-dimensional manifolds are not algorithmically classifiable, i.e. there exists no finite algorithm which allows us to decide whether two arbitrary four-dimensional manifolds are homeomorphic, see e.g. [293]. This will be discussed in more detail in a subsequent section. On the other hand, there might turn out to be physical requirements which place restrictions on the allowed class of manifolds. Since there seem to be fermions in the world, one could argue that only spin manifolds should be included. If one imposes (rather arbitrarily) the additional restriction that the manifolds should be simply connected as well, it is known, see e.g. [184], that such manifolds are uniquely characterized by their Euler characteristic $\chi(M)$ and signature $\tau(M)$ which in four dimensions are given by

$$\chi(M) = \frac{1}{128\pi^2} \int_M d^4\xi \sqrt{g} \, R_{abcd} R_{a'b'c'd'} \varepsilon^{aba'b'} \varepsilon^{cdc'd'}, \tag{6.7}$$

$$\tau(M) = \frac{1}{96\pi^2} \int_M d^4\xi \sqrt{g} \, R_{abcd} R^{ab}{}_{c'd'} \varepsilon^{cdc'd'}, \tag{6.8}$$

where R_{abcd} is the Riemann curvature tensor. For simply connected spin manifolds the signature τ is a multiple of 16, while χ is an integer ≥ 2. This situation is reminiscent of the one in two dimensions. However, the restriction to simply connected spin manifolds has no physical basis and is probably not natural at this stage in the development of a quantum theory

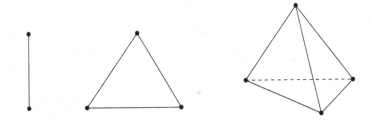

Fig. 6.1. Four simplexes in $\mathbb{R}^3$.

of space-time since there are reasons to believe that violent topological fluctuations play a prominent role in quantum gravity [225].

While the problems we have now described tend to discourage attempts to make sense of the path integral of Euclidean quantum gravity, we shall nevertheless see below that the use of dynamical triangulations, which works so well in two dimensions, allows us to discuss several of the issues above in some detail in any dimension. We also suggest that the simplicial theory might be a candidate for a theory of quantum gravity in dimensions larger than two.

6.2 Simplicial quantum gravity in dimensions $d > 2$

6.2.1 Simplicial complexes and triangulations

We define simplicial quantum gravity in dimensions $d > 2$ as a straightforward generalization of the two-dimensional case. In two dimensions we relied somewhat on intuitive notions but in higher dimensions, where intuition is less reliable, it is useful to define triangulations and related concepts more precisely. We do this in terms of simplicial complexes.

First we define an *n-simplex* σ^n in $\mathbb{R}^D$ as the convex span of $n + 1$ affinely independent points, called *vertices*, in $\mathbb{R}^D$, $D \geq n$. If we denote the vertices by $v_1, v_2, \ldots, v_{n+1}$ then σ^n is the set of points

$$\sigma^n = \left\{ \sum_{i=1}^{n+1} \lambda_i v_i : \lambda_i \geq 0, \ i = 1, \ldots, n+1, \ \sum_{i=1}^{n+1} \lambda_i = 1 \right\}.$$

Note that σ^n is determined by its vertices. A simplex spanned by a subset of the vertices of σ^n is called a *face* or a *subsimplex* of σ^n. The number n is called the *dimension* of σ^n. Four simplexes in $\mathbb{R}^3$ are shown in Fig. 6.1. A (compact) *simplicial complex* in $\mathbb{R}^D$ is a finite collection K of simplexes fulfilling the following two conditions:

(i) If σ' is a face of a simplex $\sigma \in K$, then $\sigma' \in K$.

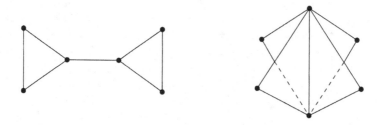

Fig. 6.2. Two examples of simplicial complexes.

(ii) If $\sigma_1, \sigma_2 \in K$ and $\sigma_1 \cap \sigma_2 \neq \emptyset$, then $\sigma_1 \cap \sigma_2$ is a face of σ_1 and σ_2.

The *dimension* of a simplicial complex K is the largest dimension of the simplexes in K. Underlying a simplicial complex K is the topological space

$$|K| = \bigcup_{\sigma \in K} \sigma,$$

which is a polyhedron. We say that K is connected if $|K|$ is connected as a topological space. In Fig. 6.2 are shown two simplicial complexes. We define d-dimensional *pseudo-manifolds* as the class of d-dimensional simplicial complexes which satisfy the following additional requirements:

(a) All simplexes of dimension less than d are contained in at least one d-simplex.

(b) All $(d-1)$-simplexes are contained in at most two d-simplexes (and the ones that only belong to one d-simplex are called *boundary* simplexes).

(c) Any two d-simplexes can be connected by a sequence $\sigma_1, \ldots, \sigma_k$ of d-simplexes, such that σ_i and σ_{i+1} share a $(d-1)$-simplex, for $i = 1, \ldots, k-1$.

Note that the simplicial complexes shown in Fig. 6.2 are not pseudo-manifolds.

We say that two simplicial complexes K and L are isomorphic if there exists a bijective mapping $f : K \mapsto L$ such that σ' is a face of $\sigma \in K$ if and only if $f(\sigma')$ is a face of $f(\sigma) \in L$. *An (abstract) triangulation may then be defined as an isomorphism class of simplicial complexes.*

Triangulations obtained from arbitrary connected simplicial complexes K are the most general triangulations allowed by the above definition. A larger class of triangulations is obtained if we relax condition (ii) in the definition of a simplicial complex and allow the intersection of two simplexes to consist of more than one face. Such triangulations may be called *unrestricted triangulations* in the terminology of Chapter 4. More precisely, the unrestricted triangulations that occur in the power series representation of the loop functions $w(g, z_1, ..., z_b)$ in two-dimensional gravity are connected and satisfy in addition condition (b) for pseudo-manifolds. Thus the second complex shown in Fig. 6.2 is not an unrestricted triangulation in the terminology of Chapter 4. For the partition function, on the other hand, only triangulations corresponding to pseudo-manifolds appear. Note that in two dimensions pseudo-manifolds are automatically manifolds in the usual sense, as we discuss below.

We expect in higher dimensions, as in two dimensions, that any scaling limit of the model we construct below is, to a large degree, insensitive to the class of triangulations used since the wild behaviour of unrestricted triangulations can be imitated on a slightly larger length-scale by more regular triangulations.

The most restrictive (smallest) class of triangulations that we consider corresponds in a given dimension d to simplicial complexes K such that $|K|$ is a d-dimensional *combinatorial manifold*. In order to define this concept it is useful to introduce a few standard concepts from piecewise linear (PL) topology.

A *subdivision* of a simplicial complex K is another simplicial complex L such that $|K| = |L|$, with the further requirement that any simplex in L is contained in some simplex in K. An example of a subdivision in two dimensions is obtained by placing a new vertex in the interior of a triangle and connecting it by 1-simplexes (links) to the three vertices of the triangle. Two simplicial complexes K and L are *combinatorially equivalent* if there exist isomorphic subdivisions K' and L' of K and L, respectively. Given a simplicial complex K and a vertex v in K we define the *star* of v in K, denoted by $st(v, K)$, as the simplicial complex consisting of all simplexes in K containing v, together with their faces. The *link* of v in K, denoted by $lk(v, K)$, is the subcomplex of $st(v, K)$ consisting of the simplexes that do not contain v.

We are now in a position to define a d-dimensional *combinatorial manifold* as a d-dimensional simplicial complex K with the property that $lk(v, K)$ is combinatorially equivalent to the boundary of a d-simplex for each interior vertex in $|K|$ and to a $(d-1)$-simplex if v is on the boundary of $|K|$. In addition we assume that $|K|$ is connected. It follows that any combinatorial manifold is also a pseudo-manifold and $|K|$ is a continuous manifold in the usual sense. Triangulations corresponding to combinatorial manifolds will be called *regular triangulations*. This notion coincides with the one introduced in Chapter 4 for $d = 2$.

Now let M be a continuous manifold. A *triangulation* of M is defined to be a pair (K, h), where K is a combinatorial manifold and $h : |K| \mapsto M$ is a homeomorphism. Letting T denote the triangulation corresponding to

K we also say that T is a triangulation of M. Two triangulations (K_1, h_1) and (K_2, h_2) of M are said to be *compatible* if $h_2^{-1} \circ h_1 : |K_1| \mapsto |K_2|$ is a picewise linear mapping. This means that there exist subdivisions L_1 and L_2 of K_1 and K_2, respectively, such that $h_2^{-1} \circ h_1$ maps the vertices of L_1 onto those of L_2 and is linear on each simplex in L_1. It is clear that compatibility is an equivalence relation on the set of triangulations of M. The corresponding equivalence classes are called *piecewise linear (or PL) structures* on M. A *PL-manifold* is a continuous manifold M together with a PL-structure on M.

There is a natural notion of PL-maps between PL-manifolds. Two PL-structures on a continuous manifold M are said to be equivalent if they are related by a PL-homeomorphism. It is known [184] that not all continuous manifolds admit a PL-structure and some admit several inequivalent ones. In dimensions ≤ 3, however, any topological manifold has a unique PL-structure up to equivalence [303]. Moreover, any smooth manifold admits at least one PL-structure [108]. This implies that any smooth manifold can be triangulated. In dimensions ≤ 6 this PL-structure is unique up to equivalence. This last fact indicates that the setting of PL-manifolds is the appropriate one for discussing the summation over topologies (i.e. diffeomorphism classes of manifolds) in the path integral (6.6) if the manifolds can be assumed to be smooth.

6.2.2   The metric structure

Let us now turn to the metric aspects of the theory. We can use Regge's prescription [331] to assign a metric to a d-dimensional piecewise linear manifold M for any triangulation (T, h) of M. The curvature is concentrated on the $(d-2)$-subsimplexes. To a given $(d-2)$-subsimplex σ^{d-2} we associate a volume

$$V_{\sigma^{d-2}} = \frac{2}{d(d+1)} \sum_{\sigma^d \supset \sigma^{d-2}} V_{\sigma^d}, \qquad (6.9)$$

where the sum runs over the d-simplexes that contain σ^{d-2} and V_{σ^d} is the volume of σ^d regarded as a subset of $\mathbb{R}^d$. The factor in front of the sum implies that the volume of a given simplex is shared equally among its $(d-2)$-subsimplexes. Likewise we can associate a deficit angle to σ^{d-2} as follows. Let σ^{d-2} be a face of the simplex σ^d. Then σ^{d-2} is the common face of precisely two $(d-1)$-subsimplexes of σ^d, and these two subsimplexes intersect in σ^{d-2} at an angle $\alpha(\sigma^{d-2}, \sigma^d)$. In two dimensions it is the angle between two links intersecting at a vertex, in three dimensions it is the angle between two triangles intersecting along a link, etc. For a triangulation of flat d-dimensional space the sum of these angles around a given $(d-2)$-simplex equals 2π. The *deficit angle* at σ^{d-2} is defined by

$$\varepsilon_{\sigma^{d-2}} = 2\pi - \sum_{\sigma^d \ni \sigma^{d-2}} \alpha(\sigma^{d-2}, \sigma^d). \tag{6.10}$$

The curvature associated with σ^{d-2} is then given by

$$2\varepsilon_{\sigma^{d-2}} V_{\sigma^{d-2}}. \tag{6.11}$$

The total volume and the total curvature of the manifold equal

$$\sum_{\sigma^{d-2} \in T_{d-2}} V_{\sigma^{d-2}} \tag{6.12}$$

and

$$\sum_{\sigma^{d-2} \in T_{d-2}} 2\varepsilon_{\sigma^{d-2}} V_{\sigma^{d-2}}, \tag{6.13}$$

respectively, where T is a triangulation of the piecewise linear manifold M and T_k is the corresponding collection of k-simplexes. One can think of Eqs. (6.12) and (6.13) as the integrals over the manifold M of a volume density and a curvature density, i.e. as

$$\int_M d^d\xi \sqrt{g} \quad \text{and} \quad \int_M d^d\xi \sqrt{g}\, R,$$

respectively.

In quantum gravity we propose to sum or integrate over all the Riemannian structures of a given manifold and over all diffeomorphism classes of manifolds. In order to imitate the construction in two dimensions, we take the lengths of the links in the simplexes to be all equal to a fixed number ε which can be thought of as the cutoff in the theory. We take $\varepsilon = 1$ unless explicitly stated otherwise.

From Eqs. (6.12) and (6.13) it is then clear how to express the Einstein–Hilbert action in geometric terms for a given piecewise linear manifold. When the lengths of all links are equal to ε the angles α in Eq. (6.10) are all equal ($\alpha = \arccos(1/d)$) and the volumes V_{σ^d} and $V_{\sigma^{d-2}}$ are likewise independent of the d-simplex, or $(d-2)$-simplex, that we are considering.

Let $o(\sigma^{d-2})$ denote the *order* of the subsimplex σ^{d-2}, i.e. the number of d-simplexes which contain σ^{d-2}, and let V_d and V_{d-2} be the volumes of the d- and the $(d-2)$-simplexes which are now all equal by assumption. The expressions (6.12) and (6.13) can now be written as

$$\frac{2V_d}{d(d+1)} \sum_{\sigma^{d-2}} o(\sigma^{d-2}) = N_d(T)V_d, \tag{6.14}$$

and

$$\sum_{\sigma^{d-2}} 2[2\pi - \alpha o(\sigma^{d-2})] V_{d-2} = (4\pi N_{d-2}(T) - \alpha d(d+1)N_d(T)) V_{d-2}, \tag{6.15}$$

where $N_d(T)$ and $N_{d-2}(T)$ denote the number of d- and $(d-2)$-simplexes in the triangulation T and we have used

$$\sum_{\sigma^{d-2}} o(\sigma^{d-2}) = \frac{d(d+1)}{2} N_d(T). \tag{6.16}$$

Absorbing the dimensionfull quantities V_{d-2} and V_d in the gravitational and cosmological coupling constants we can finally write the following expression for the Einstein–Hilbert action (6.1) of a given piecewise linear manifold with a triangulation T:

$$S_T(k_{d-2}, k_d) = k_d N_d(T) - k_{d-2} N_{d-2}(T), \tag{6.17}$$

where k_d and k_{d-2} are the the coupling constants. Clearly, $S_T(k_{d-2}, k_d)$ depends only on the isomorphism class of T. The regularized partition function which replaces Eq. (6.6) for quantum gravity is defined in analogy with the two-dimensional case as

$$Z(k_{d-2}, k_d) = \sum_T \frac{1}{C_T} e^{-S_T(k_{d-2}, k_d)}, \tag{6.18}$$

where the summation is over all closed (abstract) triangulations T, and C_T, as before, is the order of the automorphism group of T. However, *the sum in Eq. (6.18) is not convergent* for similar reasons as in the two-dimensional case; see [34] for an explicit demonstration. It is therefore natural, as a first step, to restrict the sums to manifolds with a fixed topology. The simplest case is $M = S^d$. We shall consider the case $d = 4$ and occasionally $d = 3$, in which cases we obtain

$$Z(k_2, k_4) = \sum_{T \in \mathcal{T}_4} \frac{1}{C_T} e^{-k_4 N_4(T) + k_2 N_2(T)} \tag{6.19}$$

and

$$Z(k_1, k_3) = \sum_{T \in \mathcal{T}_3} \frac{1}{C_T} e^{-k_3 N_3(T) + k_1 N_1(T)}, \tag{6.20}$$

respectively, and $\mathcal{T}_d$ denotes the class of triangulations of S^d under consideration. One might suspect that this action is much too simple to describe a theory of gravity. We take the opposite point of view: it is an encouraging feature, which hopefully reflects an underlying simplicity of the theory.

Including higher-derivative terms in the action certainly spoils the simplicity of Eqs. (6.19) and (6.20). The higher-derivative terms will contain explicitly the order of the subsimplexes which carry the curvature. We shall not discuss the lattice versions of such theories.

A priori it is certainly not clear that the series (6.19) and (6.20) are convergent for any values of the coupling constants. As in two dimensions, we need an exponential bound on the number of triangulations of S^d, $d = 3, 4$, as a function of the number of d-simplexes. Unfortunately no such bound is known at the present time. Extensive computer simulations, which will be discussed below, do however support the existence of such a bound and some partial results towards proving the bound in three dimensions were obtained in [160].

It is interesting to note that *the number of different curvature assignments to triangulations* with a fixed number of d-simplexes can be exponentially bounded in any dimension as a function of the volume [69]. In two dimensions one finds in this way a bound whose exponential part is sharp in the sense that it agrees with the known exact exponential growth of the number of two-dimensional triangulations with a fixed topology. Let us explain this in some detail; see also [69]. First consider the two-dimensional case and recall that the curvature associated to a vertex i in a closed triangulation T is equal to $\pi(6 - o(i))/3$, where $o(i)$ is the order of i. Bounding the number of curvature assignments is then equivalent to bounding the number of different sequences $(o(1), o(2), \ldots, o(N_0))$, $N_0 = N_0(T)$, that can arise when one glues together $N_2 = N_2(T)$ different triangles by identifying their sides pairwise. In the gluing process the vertices are also identified so we end up with N_0 vertices. Note that the identification of vertices is completely determined by how the sides of the triangles are identified and we have assumed that the vertices have been ordered by some convention. For regular triangulations we have $o(i) \geq 3$ for all i and it follows that the total number of sequences $(o(1), o(2), \ldots, o(N_0))$ for given N_2 and N_0 is bounded from above by

$$\binom{3N_2 - 2N_0 - 1}{3N_2 - 3N_0} = \binom{4N_0 - 1 - 6\chi}{3N_0 - 6\chi}, \tag{6.21}$$

where we have used the fact that the Euler characteristic $\chi = N_0 - \frac{1}{2}N_2$. If we keep χ fixed we obtain as $N_0 \to \infty$ an asymptotic upper bound (from Stirling's formula) of the form

$$c_\chi N_0^{-\frac{1}{2}} \left(\frac{4^4}{3^3}\right)^{N_0} \tag{6.22}$$

for the number of different curvature assignments for closed regular triangulations. A refinement of the above argument even gives the correct subleading power $N_0^{-7/2}$ for the two-sphere [69]. *A priori* we cannot expect that the expression (6.22) is an upper bound on the number of triangulations with N_0 vertices and a fixed Euler characteristic since, in general, different triangulations may have identical curvature assignments, as can be seen by considering, for example, triangulations of the torus with all vertices of order 6. Nevertheless, the exponential part of the bound (6.22) actually gives the correct exponential behaviour of the number of triangulations with N_0 vertices and fixed χ first found by Tutte [363] for $\chi = 2$. Thus, the number of triangulations with a fixed curvature assignment only contributes a subleading power correction to (6.22), which depends of course on χ. It should be noted that the number (6.21) can be trivially bounded from above by 8^{N_2} so the number of curvature assignments is exponentially bounded without regard to topology. Since the number of arbitrary triangulations increases factorially with N_2 we conclude that, generally, there are many triangulations with a prescribed curvature assignment, and (6.21) is therefore neither an upper bound nor a lower bound on the number of triangulations, but, surprisingly, it has the correct exponential growth for fixed topology.

For d-dimensional triangulations we obtain, by the same reasoning as above, the upper bound

$$\begin{pmatrix} \frac{1}{2}d(d+1)N_d - 2N_{d-2} - 1 \\ N_{d-2} - 1 \end{pmatrix} \tag{6.23}$$

on the number of curvature assignments with N_d and N_{d-2} fixed. Here we have also used the fact that the order of each $(d-2)$-simplex is at least three. It is natural to speculate whether the quantity (6.23) gives the exponential behaviour of the number of triangulations in d dimensions when we fix the topology. In the case of S^4 it seems that the exponential growth of (6.23) as a function of N_d, summed over all allowed values of N_2, is faster than the exponential growth determined by computer simulations [42, 104].

6.2.3 *Generalized matrix models*

It is natural to look for a generalized matrix model that generates (unrestricted) triangulations of dimension d since the corresponding strategy was very useful in two dimensions. Let us concentrate on the case $d = 3$ since generalization to higher dimensions is straightforward. Consider Fig. 6.3, which shows a building block for three-dimensional triangulations, i.e. a tetrahedron, and assume that the edges of the tetrahedron carry labels $\alpha, \beta, \gamma, \delta, \varepsilon, \rho$ which take values $1, ..., N$. To each oriented triangle in the boundary of the tetrahedron with labels α, β, γ, say, we assign a complex variable $\phi_{\alpha\beta\gamma}$ whose value is invariant under even permutations of the labels but is conjugated under odd permutations (which correspond to a reversal of orientation). To the oriented tetrahedron we associate a

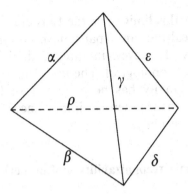

Fig. 6.3. A labelled tetrahedron.

generalization of Tr ϕ^3 in the case $d = 2$, i.e. we set

$$V(\phi) = \sum_{\alpha\beta\gamma\delta\varepsilon\rho} \phi_{\alpha\beta\gamma}\phi_{\gamma\delta\varepsilon}\phi_{\varepsilon\rho\alpha}\phi_{\beta\rho\delta}. \tag{6.24}$$

Consider the following integral:

$$Z(g) = \int d\phi \, \exp\left(-\frac{1}{6}\sum_{\alpha\beta\gamma} |\phi_{\alpha\beta\gamma}|^2 + gV(\phi)\right), \tag{6.25}$$

where $d\phi$ is defined in analogy with the two-dimensional case as a product of Lebesgue measures on the real and imaginary parts of the elements of ϕ. If $\exp(gV(\phi))$ is expanded in a power series in g and the integral is evaluated by performing all possible Wick contractions on the powers of $V(\phi)$ then all possible gluings of the tetrahedra along triangles are generated and we obtain, with a proper normalization of the measure, the representation

$$\log Z(g) = \sum_T \frac{1}{C_T} N^{N_1(T)} g^{N_3(T)}, \tag{6.26}$$

where the sum runs over all closed unrestricted triangulations representing pseudo-manifolds.

If we compare Eq. (6.20) with Eq. (6.26) derived from Eq. (6.25) we obtain the identification

$$k_3 = -\log g, \qquad k_1 = \log N, \tag{6.27}$$

which is similar to the one we encountered in two dimensions and generalizes to matrix models in any dimension as

$$k_d = -\log g, \qquad k_{d-2} = \log N. \tag{6.28}$$

From a formal point of view the limit $N \to \infty$ corresponds to taking the bare gravitational coupling constant G to 0. In two dimensions it was natural to consider this limit since the Einstein–Hilbert action in two dimensions is topological and an expansion in G automatically becomes an expansion in genus. However, in three and higher dimensions, the large N expansion is not topological. The use of matrix models in higher-dimensional quantum gravity has been limited so far but we mention them here because of their potential importance for non-perturbative approaches to quantum gravity.

6.3 Algorithmic recognizability and numerical methods

In the absence of analytic results one can turn to computer simulations which work well in the two-dimensional case, as we saw in the previous chapter. The action (6.17) is well suited for Monte Carlo simulations and the results to be discussed below for $d = 3$ and $d = 4$ have been obtained by such simulations. We first discuss some of the principles involved in such simulations since there are a number of interesting new aspects that arise in dimensions greater than two.

As in two-dimensional gravity, we need a set of ergodic moves in the set of triangulations. Such moves have been known for a long time as the Alexander moves [12]. We use a simplified version of these moves [320, 213]. In dimension d there are $d + 1$ of them in the case of regular triangulations, one corresponding to each n-simplex σ, $n = 0, 1, \ldots, d$, of the lowest allowed order, i.e. $(d - n + 1)$. We say that a simplex in T has order k if it is shared by exactly k different d-simplexes in T. The moves can be described as follows. Given an n-simplex of order $(d-n+1)$ the $(d-n+1)$ different d-simplexes sharing σ make up, together with their faces, a simplicial complex $B(\sigma)$ which is a combinatorial ball, i.e. a combinatorial manifold homeomorphic to a ball. We then remove all simplexes in T that contain σ, which amounts to removing the interior of $B(\sigma)$. Instead we insert the "dual" simplexes, i.e. all simplexes whose vertices are in $\partial B(\sigma)$ but not in σ. In this way $B(\sigma)$ is replaced by another combinatorial ball $B'(\sigma)$ with the same boundary as $B(\sigma)$ and a new triangulation T' has been obtained. In particular, σ is replaced by a $(d-n)$-dimensional simplex. The moves have already been illustrated in Fig. 5.1 for the case where $d = 2$ and they are shown for $d = 3$ in Fig. 6.4. Note that Fig. 5.1 really illustrates three different moves since the inverse of a vertex insertion is a vertex removal which counts as a separate move, but the inverse of a link flip is just another link flip. Likewise, Fig. 6.4 illustrates four moves rather than two. In four and higher dimensions it is convenient to exhibit the moves algorithmically, which is, anyway, what is needed in a computer program. The five moves in four dimensions can be described as follows, where the vertices of the simplexes are numbered

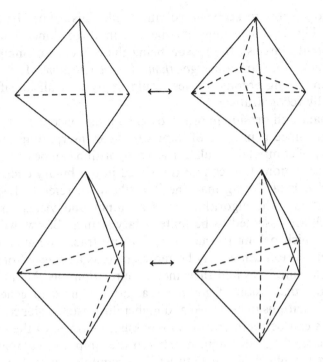

Fig. 6.4. The elementary moves in $d = 3$. Note that the moves are different from their inverses so there really are four different moves.

from 1 to 6:

$$12345 \longleftrightarrow 1234\underline{6} + 1235\underline{6} + 1245\underline{6} + 1345\underline{6} + 2345\underline{6}, \quad (6.29)$$

$$1\underline{3456} + 2\underline{3456} \longleftrightarrow \underline{12}345 + \underline{12}346 + \underline{12}456, \quad (6.30)$$

$$12\underline{456} + 13\underline{456} + 23\underline{456} \longleftrightarrow \underline{123}45 + \underline{123}46 + \underline{123}56. \quad (6.31)$$

Here the numbers underlined and in italics indicate subsimplexes which are common to two or more d-simplexes, so Eq. (6.29) describes the insertion of a vertex (labelled "6") in the interior of a 4-simplex (labelled (12345)) together with five new 4-simplexes. The inverse of this operation is of course another move. Eq. (6.30) likewise describes the replacement of two 4-simplexes with a common 3-simplex by three new 4-simplexes and its inverse and, finally, Eq. (6.31) is a "self-inverse" move (analogous to link flip in two dimensions) which preserves the number of 4-simplexes.

It is known [320, 213] that the set of moves given above is ergodic for a given PL-manifold M, i.e. for any two triangulations of M one can be deformed into the other by applying a finite sequence of moves. For $d = 2$ the link flip alone is ergodic in the set of triangulations with a fixed topology and fixed volume N_2. From a computational point of view it is convenient to be able to keep the volume fixed. This cannot be achieved in four and higher dimensions (maybe not in three dimensions either, but for different reasons), the reason being the existence of manifolds in these dimensions that are not *algorithmically recognizable*. The existence of volume preserving ergodic moves would imply that all manifolds are algorithmically recognizable.

We now make a digression in order to explain the concept of algorithmic recognizability since it may be of importance for interpreting the results and feasibility of computer simulations. An algorithm can be characterized as a set of instructions that can be translated into a binary code of finite length installed in a Turing machine, for which we refer to [196] for a precise definition. The algorithm takes an input and yields an output, both of which are assumed to be finite binary strings. Below, we have in mind inputs which belong to the set $\mathscr{A}\mathscr{T}$ of abstract triangulations. An abstract triangulation can clearly be represented as a finite set of integers which label the simplexes and their incidence relations, as should be clear from the above discussion of computer algorithms used to generate the abstract triangulations in both two and higher dimensions. Hence, abstract triangulations can be represented as finite binary strings. In the following discussion we shall not distinguish between the simplicial complexes and their abstract triangulations or between PL-manifolds and their abstract triangulations.

We say that a subset A of objects, which can be taken as inputs in the sense given above, is *algorithmically decidable* (or *recursive*) if there is a program which, when fed with an arbitrary input, is able to decide in a finite number of steps whether the input fits one of the objects in A. Given an equivalence relation on A we say that A is *algorithmically classifiable* by the relation if there exists an algorithm which can determine in a finite number of steps whether two arbitrary objects in A are equivalent or not. Finally, an object $x \in A$ is *algorithmically recognizable* in A if there exists an algorithm which can decide in a finite number of steps whether an arbitrary object in A is equivalent to x or not. Clearly, if A is algorithmically classifiable, then any element of A is algorithmically recognizable in A.

It is not difficult to check that the set of pseudo-manifolds of any fixed dimension d is recursive in $\mathscr{A}\mathscr{T}$. Whether the set of combinatorial manifolds of dimension d is recursive in $\mathscr{A}\mathscr{T}$ depends on the value of d. For $d = 2$ the notions of a manifold and pseudo-manifold coincide so two-dimensional manifolds are recursive in $\mathscr{A}\mathscr{T}$. For $d = 3$ it is known that a pseudo-manifold is a manifold exactly when the Euler characteristic χ vanishes, which implies immediately that three-dimensional manifolds are recursive. In general, it can be seen from the definition of a combinatorial manifold that the set of closed combinatorial d-manifolds is recursive if and only if the $(d-1)$-dimensional sphere S^{d-1} is algorithmically recognizable in $\mathscr{AT}$. It was recently shown that S^3 is algorithmically recognizable [337, 359]. For S^4 the answer is not known, but S^d with $d \geq 5$ is known not to be algorithmically recognizable [369]. Algorithmically unrecognizable four-manifolds do exist [293], which implies that four-manifolds are not algorithmically classifiable by topological equivalence, as mentioned previously.

The existence of algorithmically unrecognizable four-manifolds is a consequence of the unsolvability of the so-called word problem for finitely presented groups. A finitely presented group G is a group generated by a finite number of elements $a_1, a_2, \ldots, a_m$, among which there exist only finitely many independent relations. More precisely, if $\langle a_1, \ldots, a_m \rangle$ denotes the free group with generators $a_1, \ldots, a_m$ consisting of all possible words $a_{i_1}^{\pm 1} \cdots a_{i_N}^{\pm 1}$, in which no a_i is adjacent to a_i^{-1}, and with obvious multiplication, then G can be written as a quotient

$$G = \langle a_1, \ldots, a_m \rangle / N(r_1, \ldots, r_k),$$

where $N(r_1, \ldots, r_k)$ is the normal subgroup of $\langle a_1, \ldots, a_m \rangle$ generated by finitely many words $r_1, \ldots, r_k$ on $a_1, \ldots, a_m$. We then write

$$G = \langle a_1, \ldots, a_m ; r_1, \ldots, r_k \rangle \qquad (6.32)$$

and call $r_1, \ldots, r_k$ relations in G. Note that isomorphic finitely presented groups may have different presentations. For example,

$$Z_2 = \langle a ; a^2 \rangle = \langle a, b ; a^2, a^4, b \rangle.$$

The isomorphism problem for finitely presented groups is the question of whether these can be classified algorithmically up to isomorphism. The word problem for the finitely presented group defined by Eq. (6.32) is the question of whether there exists an algorithm which decides whether or not a given word w in $\langle a_1, \ldots, a_m \rangle$ equals the identity element in $\langle a_1, \ldots, a_m ; r_1, \ldots, r_k \rangle$, i.e. whether the normal subgroup $N(r_1, \ldots, r_k)$ is recursive in $\langle a_1, \ldots, a_m \rangle$. The existence of finitely presented groups for which the word problem is not solvable was demonstrated by Novikov [312]. Examples exist of such groups with as few as two generators and 32 relations. However, some of these relations are very long. The existence of certain finitely presented groups with unsolvable word problem can be shown to imply that the isomorphism problem for finitely presented groups is unsolvable; see [353] for a review.

The point now is that any finitely presented group can occur as the fundamental group $\pi_1(M)$ of some four-dimensional manifold M, a fact that can be traced back to the vanishing of the linking number of two arbitrary closed curves in a four- or higher-dimensional manifold. In fact, starting with a manifold M_0 whose fundamental group is $\langle a_1, \ldots, a_m \rangle$, e.g.

the connected sum of m copies of $S^1 \times S^{n-1}$ (where $n \geq 4$) and some curve c in M_0 representing a word r in $\pi_1(M_0)$, one can construct a new manifold M_1 with fundamental group $\langle a_1, \ldots, a_m; r \rangle$ as follows. Remove a tube homeomorphic to $S^1 \times B^{n-1}$. Its boundary is $S^1 \times S^{n-2}$, which may also be considered as the boundary of $B^2 \times S^{n-2}$. Gluing back this manifold one obtains the desired M_1. Repeating this construction one may obtain a manifold M with any prescribed finitely generated group as fundamental group.

By constructing for each finitely presented group G a four-dimensional manifold having G as fundamental group in such a way that two such manifolds are homeomorphic if their fundamental groups are isomorphic, Markov [293] showed that four-dimensional manifolds are not algorithmically classifiable, since otherwise it would contradict the non-solvability of the isomorphism problem for finitely presented groups. Similarly, one can prove the unrecognizability of certain manifolds whose fundamental groups have unsolvable word problems. We refer to [5] and references given there for further account of these issues.

Let us return to the discussion of the moves introduced above for arbitrary d. If there existed a finite collection of volume preserving local moves which were ergodic in the set of simplicial complexes of volume N_d homeomorphic to some fixed manifold M_0, it would follow that there would be a computable function $f(N_d)$ and an algorithm which could list all combinatorial manifolds of volume N_d homeomorphic to M_0 in at most $f(N_d)$ steps. This would imply that M_0 was algorithmically recognizable since for a given combinatorial manifold with volume N_d one could simply check whether this manifold appeared in the list generated by the program for the given N_d. We therefore see that the existence of algorithmically unrecognizable manifolds in dimensions four and higher prevents the existence of finite sets of volume preserving ergodic moves in these dimensions.

The reader may wonder how the existence of a finite set of ergodic moves can be compatible with algorithmic unrecognizability. The explanation is that if the volume is not preserved and there is no *a priori bound* on the volume of manifolds that are intermediate between two of the same topolgy and volume, then the recursive function f mentioned above will in general not exist. If M_0 is not algorithmically recognizable, it follows that in deforming, e.g. in a computer simulation, one triangulation of M_0 with N d-simplexes into another one with the same number of d-simplexes, large detours to triangulations with $N' \gg N$ d-simplexes must generally occur, where N' cannot be bounded by any function of N computable in a finite number of steps by any algorithm. In such a computer simulation some triangulations will be separated from a fixed reference triangulation by large barriers. So far it has not been possible to detect such barriers,

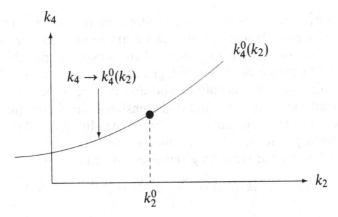

Fig. 6.5. A hypothetical phase diagram for four-dimensional gravity with the topology constrained to that of a sphere.

for either S^4 or for S^5, even though S^5 is known to be algorithmically unrecognizable.

There are two possible simple interpretations of the absence of barriers in computer simulations. Either the barriers are so huge that by the specific procedure chosen to move around in the space of triangulations, one never reaches outside the "valley" connected to the initial configuration, or the configurations which are separated from the rest by barriers form a small fraction of configuration space, a fraction which becomes increasingly unimportant as N grows. In the first case the calculation only explores an unknown fraction of configuration space, while in the other case crossing the bariers is irrelevant for the calculation of physical observables.

With these remarks in mind we now proceed to describe briefly the results of Monte Carlo simulations of simplicial quantum gravity for $d > 2$, noting that no problems with convergence of expectation values of relevant physical quantities have been observed up to now.

Let us first discuss the phase diagram for S^4. We assume an exponential upper bound on the number of triangulations as a function of the number of 4-simplexes. We have the inequality

$$N_2(T) \le \frac{10}{3} N_4(T),$$

which implies that for each k_2 there exists a value $k_4^0(k_2)$ such that the right-hand side of Eq. (6.19) is finite for $k_4 > k_4^0(k_2)$ and divergent for $k_4 < k_4^0(k_2)$; see Fig. 6.5. The model therefore makes sense above the critical curve $k_4 = k_4^0(k_2)$ and possible scaling limits will be obtained for fixed k_2 as $k_4 \to k_4^0(k_2)$. Depending on the nature of the critical point $(k_2, k_4^0(k_2))$ a sensible continuum theory may or may not be associated to this point. On Fig. 6.5 we have tentatively indicated a point, corresponding to $k_2 = k_2^0$, where a physical continuum limit might exist according to the discussion given below. Since here we are in an uncharted territory it is not at all obvious what to expect. There is even the possibility that some range of k_2-values yields a topological theory at the critical line, a theory where concepts such as volume and distance are irrelevant.

Critical exponents for higher-dimensional simplicial quantum gravity can be defined in a similar way as for the two-dimensional theories we have considered previously. First let us consider the entropy or susceptibility exponent γ. By the remarks given above we can write

$$\mathcal{N}(k_2, N_4) = \sum_{T \in \mathcal{T}_4} e^{k_2 N_2} = e^{k_4^0(k_2)N_4} f_{k_2}(N_4), \qquad (6.33)$$

where $f_{k_2}(N_4)$ denotes the subexponential contribution to the quantity $\mathcal{N}(k_2, N_4)$. If this contribution is asymptotically a power of N_4 for large N_4, we define $\gamma(k_2)$ by

$$\mathcal{N}(k_2, N_4) \sim N_4^{\gamma(k_2)-3} e^{k_4^0 N_4}. \qquad (6.34)$$

It is of course possible to imagine different asymptotic behaviour such as

$$\mathcal{N}(k_2, N_4) \sim e^{k_4^0 N_4 - c(k_2)N_4^{\varepsilon(k_2)}}, \qquad (6.35)$$

where $0 < \varepsilon < 1$. In fact there are indications from numerical simulations that there are several regions of k_2 with different asymptotic behaviour. This will be described in the following section.

Similarly it is possible to introduce the analogues of the critical exponents ν and η which we have defined in the context of random walks, strings and two-dimensional quantum gravity where they were determined from the properties of the two-point function. The same remark applies to the Hausdorff dimension. Since the generalization to higher dimensions is straightforward we refrain from giving details.

6.4   Numerical results

Here we summarize the results of computer simulations performed with the action (6.17) and regular triangulations of the manifold S^4. There seem to be two different phases depending on the value of the bare inverse gravitational coupling constant k_2. For small or negative values of k_2 a typical quantum universe is highly crumpled, with almost no extension and a large, if not infinite, Hausdorff dimension, while for large values of k_2 the typical triangulations are elongated with a Hausdorff dimension as small as two. Fig. 6.6 shows the average radius of universes with 9000, 16 000, 32 000 and 64 000 4-simplexes as a function of k_2. The two phases are separated by a transition which could be of order two or

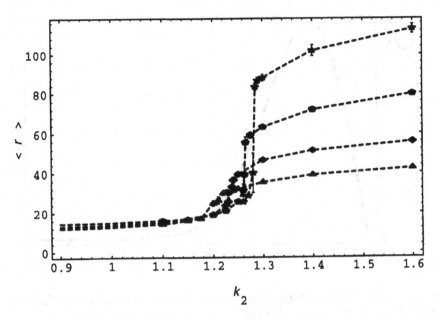

Fig. 6.6. The average radius of universes of sizes 9000 (triangles), 16 000 (squares), 32 000 (pentagons) and 64 000 (stars) versus k_2.

higher, although the possibility of a first-order transition cannot be ruled out. At the transition point, k_2^0, the Hausdorff dimension is possibly finite. The precise value is, however, not well determined.

Similar results are valid for three-dimensional simplicial quantum gravity with the topology restricted to S^3, except that the transition seems to be of first order rather than of higher order.

In the elongated phase there exists a very efficient computer algorithm which supplements the ordinary moves described above and which we refer to as *baby universe surgery*. The principle of this algorithm for $d = 4$ is to search for so-called bottlenecks of the triangulation where it is possible to cut it into two parts along a three-dimensional subcomplex isomorphic to the boundary of a 4-simplex. By cutting the triangulation along such bottlenecks and gluing the parts together again at different places (preserving the topology of S^4) it is possible to make *global moves* which affect many 4-simplexes at the same time. In this way the configuration space is explored orders of magnitude faster than with the standard local moves and it is possible to perform high statistics measurements of the quantity $\langle n(r) \rangle_{N_4}$, where $n(r)$ is the number of 4-simplexes at a distance r from a fixed marked 4-simplex. The distance r between two 4-simplexes σ_1 and σ_2 is defined as the minimal length of the paths via neighbouring 4-simplexes between σ_1 and σ_2. The average is taken over all closed

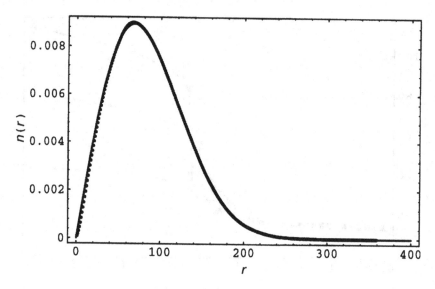

Fig. 6.7. The measured distribution (dots, the error bars smaller than the dots), and the fit (the solid curve) using the functional form $r\exp(-cr^2/N_4)$.

regular spherical triangulations. This quantity depends on the coupling constant k_2 and is related to the two-point function precisely as in two dimensions, i.e.

$$G(r;N_4) \sim N_4^{\gamma(k_2)-2} e^{k_4^0(k_2)N_4} \langle n(r) \rangle_{N_4} \qquad (6.36)$$

for large r, where G is the two-point function defined as in two dimensions by summing over all triangulations with two marked four-simplexes separated by a distance r. In Fig. 6.7 a measurement of $\langle n(r)/N_4 \rangle_{N_4}$ deep in the elongated phase is shown and a fit to the function $r\exp(-cr^2/N_4)$ which is seen to be excellent. Now recall the general scaling relations derived in the context of two-dimensional quantum gravity, but valid by the same arguments in any dimension:

$$\langle n(r) \rangle_{N_4} \sim r^{d_h-1} \qquad \text{for } 1 \ll r \ll N_4^{1/d_h},$$

$$\langle n(r) \rangle_{N_4} \sim r^\alpha \exp\left(-c\left(r^{d_h}/N_4\right)^{1/(d_h-1)}\right) \qquad \text{for } N_4^{1/d_h} < r < N_4.$$

We conclude that $d_h = 2$ (and $\alpha = 1$).

The critical exponent γ can likewise be effectively measured by baby universe counting, as described in Chapter 5 for two-dimensional gravity, since the baby universe surgery algorithm used in the simulations localizes the baby universes anyway. The result is shown in Fig. 6.8 and it suggests that $\gamma = 1/2$ in the elongated phase. From Eq. (6.36) and the measurement of $\langle n(r) \rangle_{N_4}$ we obtain $G(r;N_4)$ and in turn the two-point function $G_{k_4}(r;k_2)$

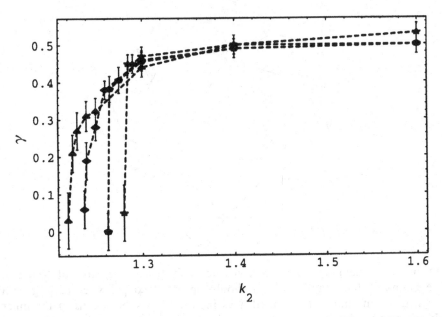

Fig. 6.8. Measurement of γ in the elongated phase for various size lattices (N_4 = 9000, 16 000, 32 000 and 64 000).

by Laplace transformation, i.e.

$$G_{k_4}(r;k_2) = \sum_{N_4=1}^{\infty} N_4^{-3/2} e^{-\Delta(k_4)N_4} \, r e^{-cr^2/N_4} \sim e^{-\bar{c}r\sqrt{\Delta k_4}}, \qquad (6.37)$$

where $\Delta(k_4) = k_4 - k_4^0(k_2)$ is assumed to be small. This function equals the two-point function of branched polymers, see Chapter 2, which are known to have intrinsic Hausdorff dimension $d_h = 2$. We conclude that the numerical simulations provide convincing evidence that the elongated phase of simplicial quantum gravity corresponds to a model of branched polymers due to the abundancy of baby universes in the generic triangulations.

At values of k_2 below k_2^0 the fractal structure of the ensemble of triangulations changes drastically. There are only few baby universes and they are small. At the same time the Hausdorff dimension seems to be infinite or at least large and at small values of k_2 the average curvature is negative. This might lead to a picture of a phase where triangulations corresponding to manifolds of negative curvature dominate. For such manifolds one expects that the volume of geodesic balls of radius r grows exponentially with r, which is what is actually observed. A closer analysis of typical triangulations reveals that they cannot be regarded as approximations of smooth manifolds. They have a few vertices of very

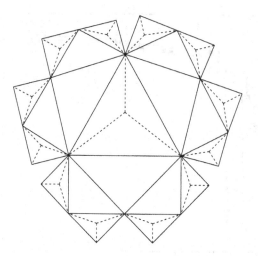

Fig. 6.9. A triangulation of S^2 constructed from a Cayley tree of triangles. The boundary links (and vertices) should be identified pairwise, starting with two neigbouring links. The dotted lines indicate that subdivisions of the inner triangle and the boundary triangles are inserted to ensure we obtain a type III triangulation after closing the boundary. The triangulation is constructed in such a way that the "link" diameter as well as the "triangle" diameter only grow logarithmically with the number of triangles.

high order which are neighbours to almost any other vertex such that the linear extension will be small. In Fig. 6.9 we illustrate an example of an analogous triangulation of S^2 where the geodesic distance between any pair of vertices or triangles grows only logarithmically with the number of triangles, irrespective of whether the distance is measured as a shortest-link distance between vertices or the distance is measured by the length of paths of neighbouring triangles connecting the chosen pair of triangles. In pure two-dimensional gravity we have seen that the Hausdorff dimension is four. This means that configurations of this type cannot be dominant in two-dimensional gravity. However, by coupling two-dimensional gravity to a large number of Gaussian fields it is possible to achieve the dominance of such triangulations. It is easy to obtain analogous four-dimensional triangulations and there is evidence that such configurations dominate for k_2 small or negative.

The numerical results we have discussed above suggest the following picture. If the gravitational coupling k_2 vanishes, the typical quantum universe has (almost) no extension. Its Hausdorff dimension might be infinite and internal distances between points are always at the Planck scale. By this, we simply mean that no consistent scaling can be found which is compatible with a finite continuum volume and a finite Hausdorff dimension. For a finite value of the bare gravitational coupling constant there is a phase transition (maybe second or higher order) to a phase with a completely different geometry and pronounced fractal characteristics. It is tempting to view the transition between the two types of geometry as being due to the liberation of excitations related to the conformal mode since large values of k_2 formally correspond to small values of the gravitational coupling constant. Right at the transition, genuine extended triangulations with a finite Hausdorff dimension may be dominant. If the transition at k_2^0 is of second (or higher) order it may provide a natural starting point for a non-perturbative definition of quantum gravity.

6.5 Notes

For a general discussion of the Euclidean path integral in quantum gravity we refer to [226, 199]. In these works the unboundedness of the action is dealt with by an additional rotation of the conformal mode. An alternative treatment of the conformal mode, inspired in part by the treatment in two dimensions, is advocated in [297]. In this approach flat Euclidean space becomes stable and the conformal mode results in a non-local term in the action. For a review of these ideas we refer to [305]. The idea of "asymptotic safety" as a tool to define a non-perturbative theory of quantum gravity goes back to Weinberg [373]. Attempts to regularize quantum gravity by adding R^2 terms to the action are found in [351, 181].

For a general introduction to piecewise linear topology we refer to [336]. The seminal work of Regge [331] *General relativity without coordinates* uses the concept of PL-manifolds. For further discussion see [121]. The intention of this work lies entirely within classical relativity, but later numerous attempts have been made to use "Regge calculus" as a starting point for a quantum theory of gravity [216, 217, 171, 185]; see [387, 386] for further references. So far, a major problem has been to determine the measure of integration. In Regge calculus the dynamical variables are the link-lengths, while the neighbour assignment is fixed. However, the lengths of the links are not in one-to-one correspondence with the assignment of metrics to the PL-manifolds by the Regge prescription, as becomes clear if we consider a compact region of the two-dimensional plane. Any triangulation of such a region by non-overlapping triangles corresponds to the same manifold and the same (flat) metric. We should divide out this degeneracy in the functional integration. In addition, the length assignments of the links have to fulfil the triangle inequalities and the action is unbounded from below if $d > 2$. Due to these problems little analytical progress has been made, even in two dimensions.

In this book we have followed a different path, first advocated in the pioneering work of Weingarten [375]. He used an embedding of the piecewise linear manifolds in a D-dimensional hyper-cubic lattice and he also discussed how to make the realization independent of the embedding. The version of simplicial gravity presented here arose as a generalization of the highly successful dynamically triangulated two-dimensional gravity models described in Chapter 4. Dynamically triangulated models in three dimensions were presented in [34, 6, 95, 54, 18]. Related matrix models were studied in [34, 205, 341]. The first four-dimensional randomly triangulated models were discussed in [41, 8, 368]. Later, a considerable number of numerical studies of four-dimensional simplicial quantum gravity were performed [44, 7, 39, 103, 66, 117]. In the description of the numerical results we have followed [43] closely, where the geodesic two-point function was used to extract the fractal dimension.

The remarkable problems with computer algorithms in four and higher dimensions were first noted in [307].

There is still no asymptotic estimate of the number of non-isomorphic triangulations with a given number of simplexes for $d > 2$. In [160] it is proved that the number of triangulations of S^3 is exponentially bounded if a plausible technical assumption holds. Progress has been made on related problems of counting so-called ball coverings as well as the counting of possible curvature assignments to a given manifold [69]. Computer simulations support these analytic results [42, 104] and indicate that the number of non-isomorphic triangulations of S^4 is exponentially bounded as a function of the number of d-dimensional simplexes.

7

Topological quantum field theories

7.1 Introduction

When quantizing gravity in the functional integral formalism, the metric
on the base manifold M is a dynamical variable that is integrated out in
the partition function. This was discussed at length in Chapters 4 and 6
in particular cases. Because of general covariance one actually integrates
over equivalence classes of metrics related by diffeomorphisms of M, and
expectation values of physical observables are therefore diffeomorphism
invariants of M. The central problem of quantum gravity in this formalism is to attribute a mathematical meaning to, as well as a physical
interpretation of, such expectation values. It is fair to say that substantial
results have been obtained in only two dimensions so far.

This chapter is devoted to a discussion of certain examples of theories
in which general covariance is realized in a simpler way than by averaging
over metrics. For example, a theory is generally covariant if its action is
a functional of a set of fields on the manifold M which does not involve
a metric at all and for which the functional integration measure over the
fields is also metric independent. Alternatively, it may occur that although
a metric enters the expression defining the action, the energy–momentum
tensor nevertheless vanishes on the physical state space. Perhaps the
best-known example of the former is the three-dimensional Chern–Simons
gauge theory [391]. Examples of the latter are cohomological field theories
[393], which include the two-dimensional twisted $N = 2$ superconformal
models [167]. It has become customary to call quantum field theories of
this type topological. We give a formal definition of topological quantum
field theories in the next section. It should be mentioned that, apart from
being of great mathematical interest, in particular in relation to geometry and topology, one of the most exciting speculations concerning the
relevance of topological quantum field theory in physics is the possibility

297

that realistic theories incorporating gravity may have topological phases corresponding to unbroken general covariance [390].

In the general spirit of this book we shall consider discretized versions of topological quantum field theories obtained by triangulating the manifold of interest M. The triangulation will not be considered dynamical any more. The topological character of the theory, corresponding to metric-independence, will be achieved by looking for statistical weights defined in terms of variables attached to subsimplexes of triangulations with the property that the resulting weight of a given triangulation depends only on its topological class. Theories so obtained are exact discretizations and may be viewed as defining underlying continuum topological quantum field theories. This is, of course, an uncommon feature of quantum field theories and makes topological quantum field theories particularly tractable for rigorous treatment.

The main problem we face is to produce suitable local weights associated with triangulations. It follows from the discussion given in Chapters 5 and 6 that it is sufficient to require invariance under a certain finite set of elementary changes of a triangulation (moves). The strategy will then be to express these requirements as algebraic equations, solutions of which can be obtained in a reasonably simple form in two and three dimensions. The achievements of this approach in dimensions ≥ 4 have been more modest and will not be discussed here; see however the notes to this chapter. The contents of the present chapter are as follows. In Section 7.2 we introduce axiomatically the notion of a topological quantum field theory. In Section 7.3 we discuss two-dimensional theories and Section 7.4 deals with three-dimensional models of the Chern–Simons type.

7.2 Generalities

7.2.1 The axioms

It is convenient for the following discussion to use the axiomatic approach to topological quantum field theory (TQFT) introduced in [57]. The following notion of a TQFT is not the most general one possible but it is adequate for our present purposes.

Let $d \geq 1$ be an integer and let $\Sigma, \Sigma', \Sigma_1, \Sigma_2, \ldots$ denote generic, compact, closed, smooth oriented $(d-1)$-dimensional manifolds, and similarly let $M, M', M_1, M_2, \ldots$ denote generic, compact, smooth, oriented d-manifolds. A d-dimensional TQFT associates* to each Σ a finite-dimensional complex

---

* Mathematicians find it convenient to describe these associations as *functors*.

vector space $\mathcal{H}_\Sigma$ such that

$$\mathcal{H}_\emptyset = \mathbb{C}, \tag{7.1}$$

and to each M a vector $Z(M) \in \mathcal{H}_{\partial M}$, where ∂M denotes the boundary of M. Furthermore, a TQFT associates to each orientation preserving diffeomorphism $f : \Sigma_1 \to \Sigma_2$ an isomorphism

$$U_f : \mathcal{H}_{\Sigma_1} \to \mathcal{H}_{\Sigma_2}.$$

These quantities fulfil the following conditions:

(i) (Diffeomorphism invariance) If $f : \Sigma_1 \to \Sigma_2$ and $f_2 : \Sigma_2 \to \Sigma_3$ are orientation preserving diffeomorphisms then

$$U_{f_2 \circ f_1} = U_{f_2} U_{f_1}, \tag{7.2}$$

and if $f : \partial M \to \partial M'$ is the restriction of an orientation preserving diffeomorphism $F : M \to M'$, then

$$Z(M') = U_f Z(M). \tag{7.3}$$

(ii) (Duality) If Σ^* denotes Σ with reversed orientation then

$$\mathcal{H}_{\Sigma^*} = \mathcal{H}_\Sigma^*, \tag{7.4}$$

where $\mathcal{H}_\Sigma^*$ denotes the dual space to $\mathcal{H}_\Sigma$. Equivalently, there exists a non-degenerate bilinear form $(\cdot, \cdot)_\Sigma : \mathcal{H}_\Sigma \times \mathcal{H}_{\Sigma^*} \to \mathbb{C}$ such that

$$(x, y)_\Sigma = (y, x)_{\Sigma^*}, \quad x \in \mathcal{H}_\Sigma, \ y \in \mathcal{H}_{\Sigma^*}, \tag{7.5}$$

and this form is preserved by diffeomorphisms, i.e.

$$(x, y)_\Sigma = (U_f x, U_{f^*} y)_{\Sigma'} \tag{7.6}$$

for $x \in \mathcal{H}_\Sigma$, $y \in \mathcal{H}_{\Sigma^*}$ and any orientation preserving diffeomorphism $f : \Sigma \to \Sigma'$, where $f^* : \Sigma^* \to \Sigma'^*$ denotes f regarded as an orientation preserving diffeomorphism from Σ^* onto Σ'^*.

(iii) (Factorization) If $\Sigma = \Sigma_1 \cup \Sigma_2$ is a disjoint union of $(d-1)$-manifolds then

$$\mathcal{H}_\Sigma = \mathcal{H}_{\Sigma_1} \otimes \mathcal{H}_{\Sigma_2} \tag{7.7}$$

and, correspondingly, the bilinear forms $(\cdot, \cdot)_\Sigma$ as well as the mappings U_f factorize. Moreover, if M_1 and M_2 are two d-manifolds such that $\partial M_1 = \Sigma_1 \cup \Sigma_3$, $\partial M_2 = \Sigma_2 \cup \Sigma_3^*$ and

$$M = M_1 \cup_{\Sigma_3} M_2 \tag{7.8}$$

denotes the manifold obtained by gluing M_1 and M_2 along Σ_3, then

$$Z(M_1 \cup_{\Sigma_3} M_2) = (Z(M_1), Z(M_2))_{\Sigma_3}, \tag{7.9}$$

where, by a slight abuse of notation, $(\cdot, \cdot)_{\Sigma_3}$ denotes the mapping

$$\mathscr{H}_{\Sigma_1} \otimes \mathscr{H}_{\Sigma_3} \otimes \mathscr{H}_{\Sigma_2} \otimes \mathscr{H}_{\Sigma_3^*} \to \mathscr{H}_{\Sigma_1} \otimes \mathscr{H}_{\Sigma_2}$$

defined by contracting with respect to $(\cdot, \cdot)_{\Sigma_3}$.

We note that (i) allows us to identify the spaces $\mathscr{H}_\Sigma$ and $\mathscr{H}_{\Sigma'}$ by the mapping U_f once a canonical identification $f : \Sigma \to \Sigma'$ is given. We shall tacitly do so in the following. Moreover, given M such that $\partial M = \Sigma^* \cup \Sigma'$, the requirement expressed by Eq. (7.4) allows us to identify

$$Z(M) \in \mathscr{H}_{\Sigma^*} \otimes \mathscr{H}_{\Sigma'} = \mathscr{H}_\Sigma^* \otimes \mathscr{H}_{\Sigma'}$$

with a linear mapping

$$Z(M) : \mathscr{H}_\Sigma \to \mathscr{H}_{\Sigma'}$$

in the standard way, i.e. if

$$Z(M) = \sum_i x_i \otimes y_i,$$

where $x_i \in \mathscr{H}_{\Sigma^*}$ and $y_i \in \mathscr{H}_{\Sigma'}$, then

$$Z(M)(x) = \sum_i (x, x_i)_\Sigma \, y_i, \quad x \in \mathscr{H}_\Sigma.$$

In terms of mappings Eq. (7.9) means that the gluing of d-manifolds corresponds to the composition of operators.

With these conventions the fourth and final requirement is:

(iv) (Non-degeneracy) For any Σ we have

$$Z(\Sigma \times [0,1]) = 1_\Sigma, \qquad\qquad (7.10)$$

where 1_Σ denotes the identity operator on $\mathscr{H}_\Sigma$.

Note that condition (iii) implies in general that $Z(\Sigma \times [0,1])$ is a projection so condition (iv) is more of a convenience than an important requirement.

The vector $Z(M)$ is called the partition function of M and may be thought of, in the Schrödinger representation, as a functional of the boundary values φ of the set of fields Φ appearing in the model. In terms of a functional integral $Z(M)$ is given as[*]

$$Z(M)(\varphi) = \int_{\Phi|\partial M = \varphi} D\Phi \, e^{-S(\Phi)}. \qquad\qquad (7.11)$$

The factorization property is then simply an expression of the fact that one can split the integration over Φ on $M_1 \cup_{\Sigma_3} M_2$ into integrations over $\Phi|_{M_1}$ and $\Phi|_{M_2}$ separately, followed by an integration over the common values of

---

[*] It is of course not at all clear whether every TQFT has such a representation.

the two on Σ_3, such that the last integration corresponds to a contraction with respect to the bilinear form $(\cdot\,,\,\cdot)_{\Sigma_3}$, or, in the language of quantum mechanics, to summing over a complete set of intermediate states. It is worth while emphasizing that this formal interpretation of $Z(M)$ makes it clear that the factorization property is peculiar to TQFTs. In topological-gravity-type theories with a dynamical metric, diffeomorphisms of $M_1 \cup_{\Sigma_3} M_2$ do not in general preserve Σ_3 and therefore diffeomorphism classes of metrics on $M_1 \cup_{\Sigma_3} M_2$ cannot be sensibly restricted to Σ_3.

Another point to note about the axioms is the finite dimensionality of the state spaces $\mathcal{H}_\Sigma$, which is a generic feature of currently known TQFTs.

We shall mostly be interested in the subclass of TQFTs which are *unitary*. These are theories whose spaces $\mathcal{H}_\Sigma$ are Hilbert spaces, with an inner product denoted by $\langle\,\cdot\,,\,\cdot\,\rangle_\Sigma$. In this case the isomorphisms U_f are assumed to be unitary and, moreover, one requires

$$Z(M^*) = Z(M)^*, \tag{7.12}$$

where the $*$-operation from $\mathcal{H}_\Sigma$ to $\mathcal{H}_{\Sigma^*}$ is defined by

$$(x, y^*)_\Sigma = \langle x, y \rangle_\Sigma, \quad x, y \in \mathcal{H}_\Sigma. \tag{7.13}$$

7.2.2 Some properties of TQFTs

Now let us note a few basic consequences of the axioms. First, a straight-forward application of (iii) and (iv) implies that the isomorphism U_f in (i) only depends on the homotopy class of f. In particular, given two oriented $(d-1)$-spheres S and S' there exists exactly one homotopy class of orientation preserving diffeomorphisms from S onto S' so the spaces $\mathcal{H}_S$ and $\mathcal{H}_{S'}$ can be canonically identified with a fixed vector space that we shall denote by V. In the case where $S' = S^*$ we obtain a canonical identification $I : V \to V^*$ and hence a symmetric bilinear form $(\cdot\,,\,\cdot)$ on V given by

$$(x, y) = (x, Iy)_S, \quad x, y \in V. \tag{7.14}$$

It is important to observe that V automatically has the structure of a commutative algebra, where the product is given by the mapping

$$Z(B_{2,1}) : V \otimes V \to V,$$

where $B_{2,1}$ denotes the manifold obtained by removing two disjoint balls from the interior of a ball. Associativity of the product is a simple consequence of the gluing property (7.9), and commutativity follows from the existence of an orientation preserving diffeomorphism of $B_{2,1}$ which permutes the two interior boundary spheres. Moreover, the element

$Z(B) \in V$, where B is a suitably oriented ball, is the unit element of V by Eqs. (7.9) and (7.10).

Letting

$$\mu = Z(B^*) \in V^*,$$

it follows from Eq. (7.9) and the definition of multiplication in V that

$$(x, y) = \mu(xy), \quad x, y \in V.$$

This implies that the pair (V, μ) is a commutative *Frobenius algebra*, i.e. V is a commutative algebra with a non-degenerate bilinear form which in the present case is $(x, y) \rightarrow \mu(xy)$. If, in addition, the TQFT is unitary, then V is a *-algebra and μ is positive, i.e. $\mu(x^*x) > 0$ for $x \neq 0$, where the *-operation is given by Eq. (7.13). The Frobenius algebra (V, μ) does not in general carry all information about the TQFT from which it emerges, except for unitary theories in two dimensions, as we shall see in the next section.

There is a natural notion of *direct sum* of TQFTs defined as follows. Let $\mathcal{T}^1$ and $\mathcal{T}^2$ be two d-dimensional TQFTs whose vector spaces and partition functions are denoted by $\mathcal{H}_\Sigma^1$, $Z^1(M)$ and $\mathcal{H}_\Sigma^2$, $Z^2(M)$ respectively. The direct sum of $\mathcal{T}^1$ and $\mathcal{T}^2$ is defined by setting

$$\mathcal{H}_\Sigma = \mathcal{H}_\Sigma^1 \oplus \mathcal{H}_\Sigma^2$$

for Σ non-empty and connected,

$$\mathcal{H}_{\Sigma_1 \cup \ldots \cup \Sigma_n} = \mathcal{H}_{\Sigma_1} \otimes \ldots \otimes \mathcal{H}_{\Sigma_n}$$

for $\Sigma_1, \ldots, \Sigma_n$ connected and disjoint, and

$$\begin{aligned} Z(M) &= Z^1(M) \oplus Z^2(M) \\ &\in \mathcal{H}_{\partial M}^1 \oplus \mathcal{H}_{\partial M}^2 \subseteq \mathcal{H}_{\partial M} \end{aligned}$$

for $\partial M \neq \emptyset$. For $\partial M = \emptyset$ one defines

$$Z(M) = Z^1(M) + Z^2(M).$$

Similarly, the bilinear forms $(\cdot, \cdot)_\Sigma$ and the isomorphisms U_f are defined in terms of direct sums of the corresponding data for $\mathcal{T}^1$ and $\mathcal{T}^2$. It is not difficult to verify that the objects defined in this fashion give rise to a TQFT which we denote by $\mathcal{T}^1 \oplus \mathcal{T}^2$. If $\mathcal{T}^1$ and $\mathcal{T}^2$ are both unitary then, requiring the direct sums to be orthogonal, it follows that $\mathcal{T}^1 \oplus \mathcal{T}^2$ is unitary as well.

We say that a TQFT is trivial if and only if $Z(M) = 0$ for all M. Furthermore, a TQFT is *indecomposable* if and only if it cannot be written as a direct sum of non-trivial TQFTs. By exploiting the decomposition properties of Frobenius algebras it can be shown [343] that any TQFT can be written as a direct sum of indecomposable TQFTs. Moreover, for any indecomposable unitary TQFT, the Frobenius algebra V is one-dimensional and hence can be identified with $\mathbb{C}$ and the linear form μ is given by multiplication by a positive scalar λ.

7.3 Two-dimensional TQFT

In this section we specialize to the case $d = 2$ and use the same notation as in the previous section. Here we shall give a simple explicit construction of all unitary two-dimensional TQFTs before discussing the richer and more interesting three-dimensional theories in Section 7.4.

7.3.1 TQFT on triangulations

We assume that the surfaces under discussion come equipped with a triangulation and we shall at times use the words "surface" and "triangulation" interchangably. Given a surface M we propose to construct a vector space $\mathscr{H}_{\partial M}$ and a vector $Z(M) \in \mathscr{H}_{\partial M}$ as data of a TQFT by starting at the "microscopic" level with a vector space V_t attached to the boundary of an oriented triangle t, as well as a partition function $Z_t \in V_t$ associated with t, and using these as building blocks. The space V_t will be given in terms of vector spaces associated to the oriented links in the boundary of t, as we now describe. Let n be a fixed positive integer and suppose that for each pair (a, b) of labels (colours) belonging to the set $J = \{1, \ldots, n\}$ a finite-dimensional complex vector space V_{ab} is given such that

$$V_{ab} = V_{ba}^*, \quad a, b \in J. \tag{7.15}$$

Given an oriented link ℓ whose endpoints are coloured by a and b, respectively, ordered according to the orientation of ℓ, we associate V_{ab} to this coloured oriented link, and define

$$V_t = \bigoplus_{a,b,c} V_{ab} \otimes V_{bc} \otimes V_{ca}, \tag{7.16}$$

where a, b, c denote the colours of the corners of the oriented triangle t in positive cyclic order; see Fig. 7.1. More generally, for a given oriented polygonal loop L with m vertices, we define the vector space

$$V_L = \bigoplus_{a_1, \ldots, a_m \in J} V_{a_1 a_2} \otimes \ldots \otimes V_{a_{m-1} a_m} \otimes V_{a_m a_1}, \tag{7.17}$$

where the $a_1, \ldots, a_m$ denote the colours of the m vertices arranged in a positive cyclic order along L. If L is a disjoint union of oriented polygonal loops $L_1, \ldots, L_b$ we set

$$V_L = V_{L_1} \otimes \ldots \otimes V_{L_b}. \tag{7.18}$$

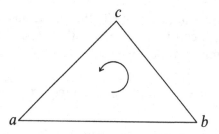

Fig. 7.1. An oriented triangle with cyclically ordered vertices with colours a, b and c.

Eq. (7.15) yields a canonical identification

$$V_{L^*} = V_L^*. \tag{7.19}$$

Clearly, if L is connected, V_L only depends on the length m of L and therefore we shall occasionally use the notation V_m instead of V_L. In particular,

$$V_t = V_3, \tag{7.20}$$

and we note that Eq. (7.19) gives an identification

$$V_m = V_m^*, \tag{7.21}$$

and an associated non-degenerate bilinear form on V_m.

Given a triangulation T of the surface M and a vector $Z_t \in V_t$ for each triangle t in T we can produce a vector in $V_{\partial T}$ by simply taking the tensor product of Z_t over all triangles t in T and using Eq. (7.15) to contract tensors with respect to components associated to the interior links in T. In addition we need a numerical factor w_a associated to each interior vertex of T with colour a and a factor $\sqrt{w_a}$ for each boundary vertex with colour a in order to conform to the gluing property (7.9). Thus, we define

$$Z_T(M) = \bigoplus_{\Phi:V(T)\to J} \left(\bigotimes_{t\in T} Z_t\right)_{\text{int } T} \prod_{i\in V(T\setminus\partial T)} w_{\Phi(i)} \prod_{i\in V(\partial T)} w_{\Phi(i)}^{\frac{1}{2}}, \tag{7.22}$$

where the first direct sum is over all colourings $\Phi : V(T) \to J$ of the vertices of T and $(\,\cdot\,)_{\text{int } T}$ indicates the aforementioned contraction with respect to the variables on the interior links.

Both $V_{\partial T}$ and $Z_T(M)$ will in general depend on T. In order to eliminate the dependence of $Z_T(M)$ on the interior of the triangulation T we use the fact, discussed in Sections 5.2 and 6.3, that the two moves depicted in Fig. 5.1 are ergodic in the space of triangulations of M with a fixed boundary. Denoting by Z_{abc} the component of Z_t in $V_{ab} \otimes V_{bc} \otimes V_{ca}$, it is

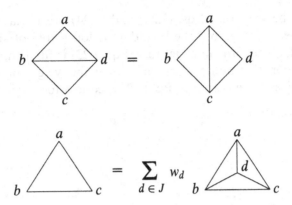

Fig. 7.2. A diagrammatic representation of Eqs. (7.23) and (7.24).

therefore sufficient to require that these fulfil the equations

$$(Z_{abd} \otimes Z_{cdb})_{bd} = (Z_{abc} \otimes Z_{acd})_{ac} \qquad (7.23)$$

and

$$Z_{abc} = \sum_{d \in J} w_d (Z_{abd} \otimes Z_{bcd} \otimes Z_{cad})_{ad,bd,cd} \qquad (7.24)$$

for any choice of colours $a, b, c, d \in J$, where $(\cdot)_{ab}$ denotes contraction with respect to the bilinear form $(\cdot, \cdot)_{ab}$ on $V_{ab} \times V_{ba}$ given by Eq. (7.15) and $(\cdot)_{ad,bd,cd}$ correspondingly denotes the contraction with respect to the appropriate tensor product of three such forms. A diagrammatic form of these equations is shown in Fig. 7.2. In order to exhibit solutions Z_t, w_a to these equations we make the simple choice

$$V_{ab} = \begin{cases} \mathbb{C} & \text{for } a = b \\ 0 & \text{otherwise,} \end{cases} \qquad (7.25)$$

which has the virtue that

$$V_L = \mathbb{C}^n, \qquad (7.26)$$

for any loop L, where n is the number of colours. Henceforth, we denote $V_{\partial T}$ by $\mathcal{H}_{\partial M}$.

It is understood that $\mathbb{C}$ is identified with its dual $\mathbb{C}^*$ in the standard way, i.e. the corresponding bilinear form on $\mathbb{C}$ is simply multiplication and, analogously, $\mathbb{C}^n$ is identified with $(\mathbb{C}^n)^*$ via the bilinear form

$$((x_1, \ldots, x_n), (y_1, \ldots, y_n)) = x_1 y_1 + \ldots + x_n y_n. \qquad (7.27)$$

With the choice defined by Eq. (7.25) the partition function Z_t is a vector in $\mathbb{C}^n$,

$$Z_t = (\mu_1, \ldots, \mu_n), \qquad (7.28)$$

and it can be seen that Eqs. (7.23) and (7.24) are trivially satisfied with both sides equal to zero if the boundary colours are not all equal. On the other hand, if they are all equal to a, Eq. (7.23) is fulfilled, with both sides equal to a complex number μ_a^2, where μ_a is the weight of a triangle all of whose vertices have colour a. Eq. (7.24) reduces to

$$\mu_a^3 w_a = \mu_a,$$

i.e.

$$w_a = \mu_a^{-2} \tag{7.29}$$

provided $\mu_a \neq 0$, which we shall assume to be the case.

We can now insert Eqs. (7.28) and (7.29) into Eq. (7.22). If M is connected, it follows from Eq. (7.25) that only terms for which all vertices have identical colour give a non-vanishing contribution to the partition function and this contribution equals $\mu_a^{|T|-2|N_v(T)|+|\partial T|}$, where we have used the square root $\sqrt{w_a} = \mu_a^{-1}$. Letting $e_1, \ldots, e_n$ denote the natural basis for $\mathbb{C}^n$ and using

$$\begin{aligned}
\chi(M) &= |T| - N_l(T) + N_v(T) \\
&= -\frac{1}{2}|T| + N_v(T) - \frac{1}{2}|\partial T|,
\end{aligned} \tag{7.30}$$

we conclude that

$$Z_T(M) = \sum_{a=1}^{n} \mu_a^{-2\chi(M)} e_a^{\otimes b}, \tag{7.31}$$

where b denotes the number of boundary components in ∂M. This expression shows explicitly that $Z_T(M)$ is actually independent of T and a topological invariant of M that depends only on the squares of the numbers μ_a. We shall denote this invariant by $Z(M)$.

By construction, the partition functions $Z(M)$ satisfy the gluing property (7.9). With the isomorphisms U_f given as the appropriate identity maps it is easy to check that the vector spaces and partition functions we have defined constitute a TQFT.

We note that since in Eq. (7.31) we have $\chi(M) = 2(1 - h) - b$, where h is the genus of M, it follows that replacing e_a by $-e_a$ for some $a \in J$ can be compensated for by replacing μ_a^{-2} by $-\mu_a^{-2}$ and similarly a permutation of the basis vectors $e_1, \ldots, e_n$, can be compensated for by the same permutation of the numbers $\mu_1^{-2}, \ldots, \mu_n^{-2}$. This shows that the sets $(\pm \mu_{\sigma(1)}^{-2}, \ldots, \pm \mu_{\sigma(n)}^{-2})$, where σ is any permutation of $1, \ldots, n$, yield equivalent TQFTs.

7.3.2 The unitary case

Let us now discuss the unitary TQFTs obtained by the construction given above. Regarding V_{ab} given by Eq. (7.25) as a Hilbert space with the standard inner product, then each V_L given by Eq. (7.26) becomes a Hilbert space and the *-operation is just complex conjugation of coordinates. Since $\chi(M)$ can take any integer value ≤ 2 it follows from Eq. (7.31) that the unitarity requirement (7.12) is satisfied if and only if μ_a^{-2} is a real number for all a. The unitary TQFTs obtained by the above construction can hence be labelled (up to equivalence) by finite non-decreasing sequences of positive numbers $0 < \lambda_1 \leq \lambda_2 \leq \ldots \leq \lambda_n$, where $\lambda_a = \mu_a^{-2}$, $a \in J$. In this subsection we show that these theories encompass all unitary two-dimensional TQFTs up to equivalence.

Given a two-dimensional unitary TQFT we may, as explained in the previous subsection, assume that a fixed Hilbert space V is associated to any oriented circle. It follows that $V^{\otimes b}$ is associated to the disjoint union of b oriented circles. Moreover, V is a commutative *-algebra such that the inner product is given by

$$\langle x, y \rangle_V = \mu(xy^*)$$

for some $\mu \in V^*$. We also note that since any surface M has an orientation reversing diffeomorphism it follows that $Z(M) = Z(M^*)$ and hence, by Eq. (7.14),

$$Z(M) = Z(M)^* \tag{7.32}$$

for all M.

Let us choose an orthonormal basis $e_1, \ldots, e_n$ for V such that

$$e_i^* = e_i, \tag{7.33}$$

$i = 1, \ldots, n$. This basis is also an orthonormal basis for the *real* Hilbert space $V_R = \{x \in V : x = x^*\}$, whose inner product is obtained by restricting the inner product on V to V_R. With the notation of Section 7.2 we can then write

$$Z(B) = \sum_{a=1}^{n} d_a e_a, \tag{7.34}$$

$$Z(S^1 \times [0,1]) = \sum_{a,b=1}^{n} q_{ab} e_a \otimes e_b, \tag{7.35}$$

$$Z(B_{2,1}) = \sum_{a,b,c=1}^{n} c_{abc} e_a \otimes e_b \otimes e_c. \tag{7.36}$$

Since any surface M can be constructed by gluing together a number of copies of the three surfaces B, $S^1 \times [0,1]$ and $B_{2,1}$, it follows from the gluing property (7.9) that it is sufficient to show that the three quantities given by Eqs. (7.34), (7.35) and (7.36) coincide with the corresponding ones of a representative from the list of TQFTs which we constructed explicitly in the previous subsection. Thus, according to Eq. (7.31), all we need to show is that the basis $e_1, \ldots, e_n$ can be chosen such that

$$d_a = \lambda_a, \tag{7.37}$$

$$q_{ab} = \delta_{ab}, \tag{7.38}$$

$$c_{abc} = \lambda_a^{-1} \delta_{ab} \delta_{bc}, \tag{7.39}$$

for some positive numbers $\lambda_1, \ldots, \lambda_n$.

Eq. (7.38) follows immediately from Eq. (7.10). In order to check Eqs. (7.37) and (7.39) we first note that d_a and c_{abc} are real numbers as a consequence of Eqs. (7.32) and (7.33), and c_{abc} is completely symmetric in the indices, since the boundary components of $B_{2,1}$ can be arbitrarily permuted by diffeomorphisms of $B_{2,1}$. According to the definition of multiplication in V the operator C_a on V_R defined by

$$C_a = \sum_{b,c=1}^{n} c_{abc} \, e_b \otimes e_c$$

represents multiplication by e_a for each $a = 1, \ldots, n$. Since V is commutative, these n operators commute and since they are in addition symmetric they can be simultaneously diagonalized on V_R, i.e. we can choose the basis $e_1, \ldots, e_n$ such that $c_{abc} = \mu_{ab} \delta_{bc}$ for suitable numbers $\mu_{ab} \in \mathbb{R}$. Due to the symmetry of c_{abc} it follows that $\mu_{ab} = \mu_a^2 \delta_{ab}$, so

$$c_{abc} = \mu_a^2 \delta_{ab} \delta_{bc}$$

for suitable real numbers $\mu_1^2, \ldots, \mu_n^2$. By an appropriate choice of sign of $e_1, \ldots, e_n$ we can, in addition, assume that these numbers are non-negative. Next, using the fact that $S^1 \times [0, 1]$ can be obtained by gluing B onto $B_{2,1}$ along one of the boundary components of the latter, it follows from Eqs. (7.9) and (7.38) that

$$\delta_{ab} = \sum_{c=1}^{n} c_{abc} d_c = \mu_a^2 d_a \delta_{ab},$$

from which we obtain that $\mu_a^2 \neq 0$ and

$$d_a = \mu_a^{-2}.$$

Setting $\lambda_a = \mu_a^{-2}$ we have thus verified Eqs. (7.37), (7.38) and (7.39).

It follows that all the information in a two-dimensional unitary TQFT is contained in the numbers λ_i in the sense that two such theories with the same λ_is are equivalent. Let us make it perfectly clear what we mean by the equivalence of two unitary TQFTs. Let $\mathscr{T}$ be a TQFT satisfying the axioms of Section 7.2 with vector spaces $\mathcal{H}_\Sigma$ and partition functions $Z(M)$, etc. Let $\mathcal{T}'$ be another one whose objects are distinguished from those of $\mathcal{T}$ by a prime. We say that $\mathcal{T}$ and $\mathcal{T}'$ are *equivalent* if for any $(d-1)$-manifold Σ there exists a unitary mapping

$$A_\Sigma : \mathcal{H}_\Sigma \mapsto \mathcal{H}'_\Sigma \tag{7.40}$$

such that the following conditions hold:

1. For any orientation preserving diffeomorphism $f : \Sigma_1 \mapsto \Sigma_2$ between $(d-1)$-manifolds

$$U'_f = A_{\Sigma_2} U_f A_{\Sigma_1}^*. \tag{7.41}$$

2. For any oriented d-manifold M

$$Z'(M) = A_{\partial M}(Z(M)), \tag{7.42}$$

with the understanding that $Z'(M) = Z(M)$ if M is closed.

3. For any $(d-1)$-manifold Σ we have

$$(A_\Sigma x, A_\Sigma \cdot y)'_\Sigma = (x, y)_\Sigma. \tag{7.43}$$

To sum up we have proved the following theorem.

Theorem 7.1 *Up to equivalence the partition functions $Z(M)$ of a two-dimensional unitary TQFT are given by Eq. (7.31), where $e_1, \ldots, e_n$ form an orthonormal basis consisting of self-adjoint elements of the Hilbert space V associated to the circle, considered as a $*$-algebra, and $\lambda_1 \leq \ldots \leq \lambda_n$ are positive numbers.*

We conclude this subsection with some remarks. The unitary theories we have classified above are uniquely determined by the values of the partition functions for closed surfaces. We conjecture that this is also true for higher-dimensional unitary TQFTs. In fact, one can give a simple proof of this in the special case where the partition functions $Z(M)$ with $\partial M = \Sigma$ span the Hilbert space $\mathcal{H}_\Sigma$ for all $(d-1)$-manifolds Σ. We say that a TQFT has the *spanning property* if it fulfils this condition. It can easily be seen that two-dimensional unitary theories have the spanning property exactly when the positive numbers λ_i are all different.

Let $\mathcal{T}^i$ be two unitary TQFTs with the spanning property whose partition functions and Hilbert spaces are $Z^i(M)$ and $\mathcal{H}^i_\Sigma$, $i = 1, 2$. Suppose that

$$Z^1(M) = Z^2(M)$$

for all closed d-manifolds M. In order to to conclude that $\mathcal{T}^1$ and $\mathcal{T}^2$ are equivalent it suffices to exhibit unitary isomorphisms $A_\Sigma : \mathcal{H}^1_\Sigma \rightarrow \mathcal{H}^2_\Sigma$

which take $Z^1(M)$ to $Z^2(M)$ for all M with $\partial M = \Sigma$. Let $x \in \mathcal{H}^1_\Sigma$. Then, by hypothesis, there are manifolds M_i with boundary Σ and complex numbers a_i such that

$$x = \sum_i a_i Z^1(M_i).$$

We define $y = A_\Sigma x \in \mathcal{H}^2_\Sigma$ by

$$y = \sum_i a_i Z^2(M_i).$$

In order to verify that this is a well-defined unitary isomorphism it suffices to check that if

$$x = \sum_i a_i Z^1(M_i) = \sum_j b_j Z^1(M'_j)$$

then

$$\sum_i a_i Z^2(M_i) = \sum_j b_j Z^2(M'_j). \tag{7.44}$$

One can show by a straightforward calculation, using the gluing axiom and the fact that partition functions for the two theories agree on closed manifolds, that the Hilbert space norm of the difference of the left-hand side and the right-hand side in Eq. (7.44) is zero and the desired result follows.

The classification problem for non-unitary two-dimensional TQFT boils down to classifying finite-dimensional indecomposable commutative Frobenius algebras; see [343]. In this case it is easy to see that the partition functions for closed surfaces do not in general determine the theory and the Frobenius algebra may have nilpotent elements.

7.4 Three-dimensional unitary TQFT

In this section we extend the construction of the previous section to three dimensions using similar methods as in the two-dimensional case, starting from triangulations of three-dimensional manifolds. The requirement of topological invariance will lead to equations for the weights assigned to simplexes and we shall exhibit explicitly the simplest solutions to these equations which are intimately related to the representation theory of $su(2)$. Before turning to the mathematical development we review in the following subsection some formal considerations linking three-dimensional gravity to TQFT.

7.4.1 TQFT and three-dimensional gravity

Two-dimensional gravity without any matter fields is almost a topological theory. The curvature part of the Einstein–Hilbert action is a topological invariant, and even in the presence of matter and a non-zero cosmological constant we saw explicitly in Chapter 4 that the Hartle–Hawking wave functionals of the higher genus surfaces have a topological interpretation as intersection indices on the moduli space of punctured Riemann surfaces. In three dimensions the Einstein–Hilbert action is not a topological invariant. Nevertheless, three-dimensional gravity exhibits topological features. If we formulate the theory in terms of the metric, the number of independent field theoretical degrees of freedom (obtained by analysing the classical initial value problem) is equal to zero. This is the statement that there is no dynamical graviton in three dimensions. This suggests that a corresponding quantum field theory will only have a finite number of degrees of freedom. We encountered the same phenomenon in two-dimensional gravity, where the theory could be solved almost explicitly.

It becomes more transparent how a topological theory can emerge from the Einstein–Hilbert action if we change variables from the metric $g_{\mu\nu}$ to a *drei-bein* e_μ^a and a *spin connection* ω_μ^{ab} [389]. It is convenient to start the discussion by assuming that the space-time manifold M has a Lorentzian signature $(++-)$ rather than the Euclidean signature $(+++)$ which has been used up to now. The drei-bein can be regarded as a vector-valued one-form e and in the same way the spin connection can be viewed as an $so(2,1)$-valued one-form ω on M. In this formalism the curvature tensor can be expressed as

$$R(\omega) = d\omega + \omega \wedge \omega \qquad (7.45)$$

and it is a two-form on M so that

$$S = \int_M d^3\xi \, \sqrt{-g}\, R(g) = \int_M e \wedge R(\omega), \qquad (7.46)$$

where $R(g)$ is the scalar curvature. The "Lorentz" indices $a, b, \ldots$ are lowered and raised with a (constant) metric tensor η_{ab} of signature $(++-)$, while the indices $\mu, \nu, \ldots$ are lowered and raised by the metric tensor $g_{\mu\nu}$. The relation between the metric and the drei-bein is given by

$$g_{\mu\nu} = e_\mu^a e_\nu^b \eta_{ab}. \qquad (7.47)$$

Let J_{ab} and P_a be the generators of Lorentz transformations and translations in the inhomogeneous Lorentz group $ISO(2,1)$. These generators span the Lie algebra of $ISO(2,1)$ and with $J_a = \varepsilon_{abc}J^{bc}$ they satisfy the commutation relations

$$[J_a, J_b] = \varepsilon_{abc}J^c, \quad [J_a, P_b] = \varepsilon_{abc}P^c, \quad [P_a, P_b] = 0, \qquad (7.48)$$

and there is an invariant, non-degenerate quadratic form on the Lie algebra given by

$$\langle J_a, P_b \rangle = \delta_{ab}, \quad \langle P_a, P_b \rangle = \langle J_a, J_b \rangle = 0. \tag{7.49}$$

Setting $\omega_\mu^a = -\frac{1}{2}\varepsilon_{abc}\omega_\mu^{bc}$ we can now introduce a Lie algebra valued gauge field A_μ by

$$A_\mu = e_\mu^a P_a + \omega_\mu^a J_a, \tag{7.50}$$

in terms of which the action (7.46) can be written as

$$S_{CS}(A) = \int_M \left\langle A, dA + \frac{2}{3}A \wedge A \right\rangle, \tag{7.51}$$

which takes the form of the action of a *Chern–Simons* theory with gauge group $ISO(2,1)$. However, it is important to realize that the first-order formalism using e_μ^a and ω_μ^a is not equivalent to Einstein's gravity formulated in terms of the metric $g_{\mu\nu}$ unless we impose the condition expressed by Eq. (7.47) by hand. This condition is not present if we view the theory defined by Eq. (7.51) as a genuine gauge theory where the A_μ field can take arbitrary values. If we ignore the condition (7.47), i.e. allow configurations which are singular from a metric point of view, three-dimensional gravity can be viewed as a Chern–Simons theory, which is a topological theory in the sense that there is no reference to a metric on the underlying manifold.

The above considerations can be generalized to include a positive cosmological constant. The Einstein–Hilbert action then becomes

$$S = \int_M d^3\xi \, \sqrt{-g}\,(R + \lambda) = \int_M \left(e \wedge R(\omega) + \frac{\lambda}{3}e \wedge e \wedge e \right). \tag{7.52}$$

The invariance group is no longer $ISO(2,1)$, but $SO(3,1)$. The commutation relations between the P_a generators are changed to

$$[P_a, P_b] = \lambda \varepsilon_{abc} J^c, \tag{7.53}$$

while the other commutation relations remain unchanged and the quadratic form (7.49) is still invariant. With the definition (7.50) the Einstein–Hilbert action (7.52) can still be written as in Eq. (7.51) but the appropriate gauge group is $SO(3,1)$. This is seen explicitly by introducing

$$J_a^\pm = \frac{1}{2}\left(J_a \pm \frac{1}{\sqrt{\lambda}}P_a \right). \tag{7.54}$$

The Lie algebra for $P_a, J_a^\pm$ becomes

$$[J_a^\pm, J_b^\pm] = \varepsilon_{abc} J_c^\pm, \quad [J_a^+, J_b^-] = 0, \tag{7.55}$$

i.e. the Lie algebra of $SL(2,\mathbb{R}) \times SL(2,\mathbb{R})$ which is isomorphic to that of $SO(3,1)$. The corresponding gauge fields are

$$A_{\mu\pm}^a = \omega_\mu^a \pm \sqrt{\lambda}e_\mu^a, \tag{7.56}$$

and the action becomes

$$S(A_+, A_-) = \frac{k(\lambda)}{4\pi}(S_{CS}(A_+) - S_{CS}(A_-)), \quad k(\lambda) = \frac{4\pi}{\sqrt{\lambda}}, \tag{7.57}$$

where the Chern–Simons action $S_{CS}(A)$ for the group $SL(2, \mathbb{R})$ is given by Eq. (7.51).

Let us now attempt a path integral quantization of this system. As usual we rotate to Euclidean space in order to define the path integral. In this case an interesting possibility arises since the Chern–Simons theory only contains first-order derivatives. In ordinary field theory this results in a purely imaginary Euclidean Chern–Simons action. If we insist that the quantity we call action is real, this means that we have to include it in the path integral as a phase factor, precisely as is done in ordinary field theory in Minkowskian space-time. In this way the path integral for Euclidean three-dimensional quantum gravity with a positive cosmological constant is defined by the partition function

$$
\begin{aligned}
Z(\lambda) &= \int_M \mathcal{D}e\mathcal{D}\omega \, \exp\left(i \int \left(e \wedge R + \frac{\lambda}{3} e \wedge e \wedge e\right)\right) \\
&= \int_M \mathcal{D}A_+ \mathcal{D}A_- \, \exp\left(\frac{ik(\lambda)}{4\pi} S_{CS}(A_+) - \frac{ik(\lambda)}{4\pi} S_{CS}(A_-)\right) \\
&= |Z_{CS}(M; k(\lambda))|^2, \tag{7.58}
\end{aligned}
$$

where $Z_{CS}(M; k)$ denotes the partition function for a Chern–Simons theory on the manifold M with gauge group $SU(2)$ and coupling constant k.

A number of remarks are in order. First it should be made clear that we have chosen a rather special way of analytically continuing (rotating) to Euclidean space-time offered by the first-order formalism and the analogy with Chern–Simons theory in flat space-time. It is natural in this rotation to also include the continuation of the Lorentz group $SO(2, 1)$ to the rotation group $SO(3)$ (or $SU(2)$ which has the same Lie algebra). In this way we can maintain the interpretation of e^a_μ as a drei-bein on the manifold M with *Euclidean signature*. If the cosmological constant λ vanishes we obtain an $ISO(3)$ Chern–Simons theory, while in the case of a positive cosmological constant we obtain an $SO(4)$ Chern–Simons theory and the last equality sign in Eq. (7.58) reflects the isomorphism between $SO(4)$ and $SU(2) \times SU(2)/\mathbb{Z}_2$.

It follows from this way of carrying out the rotation that the value of the cosmological constant is quantized:

$$\lambda = \left(\frac{4\pi}{k}\right)^2, \tag{7.59}$$

since the quantum Chern–Simons theory of $SU(2)$ is only gauge invariant for k an integer. Under a topologically non-trivial gauge transformation

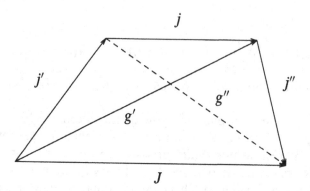

Fig. 7.3. The six vectors that enter in the addition of three angular momenta form a tetrahedron.

$A \rightarrow A'$ with a winding number n, the Chern–Simons action changes as

$$S_{CS}(A') = S_{CS}(A) + 2\pi k n, \qquad (7.60)$$

so $e^{iS_{CS}(A)}$ is only invariant for integer k. Note that in this formalism the unboundedness of the Euclidean action from below is not a serious problem since the action appears as a phase. On the other hand, it is notoriously difficult to give a rigorous definition of the path integral of such oscillating factors. The purpose of the following sections is to show that there exists a rigorous definition of the Chern–Simons gravity theory in the framework of topological field theories.

The topological aspect of the $ISO(3)$ Chern–Simons gravity (corresponding to $\lambda = 0$) was "almost" discovered in a remarkable paper by Ponzano and Regge [329] more than 25 years ago in the study of $6j$-symbols used in quantum mechanics. Let us outline some of their results, which serve as an additional motivation for the formalism to be developed in the following. The $6j$-symbols were introduced in quantum mechanics to describe the coupling of three angular momenta. If j, j' and j'' are angular momenta and $J = j + j' + j''$, we can think of the addition in two ways:

$$j + j' = g', \quad J = g' + j'' \quad \text{and} \quad j + j'' = g'', \quad J = j' + g''. \qquad (7.61)$$

In Fig. 7.3 the two ways of adding the angular momenta are illustrated by a *tetrahedron* whose edges have length equal to those of the six angular momenta which enter the discussion. The two different ways of adding the angular momenta give rise to two different sets of basis vectors in the angular momentum space

$$\{|(j'j)g', j''; JM\rangle\}, \quad \text{and} \quad \{|j', (jj'')g''; JM\rangle\}, \qquad (7.62)$$

using standard notation. We can of course expand one set of basis vectors in terms of the other using the inner product

$$\langle j', (jj'')g''; JM | (j'j)g', j''; J'M' \rangle \tag{7.63}$$

$$= \delta_{JJ'} \delta_{MM'} \sqrt{(2g'+1)(2g''+1)} (-1)^{j+j'+j''+J} \left\{ \begin{matrix} j' & j & g' \\ j'' & J & g'' \end{matrix} \right\}.$$

The coefficients

$$\left\{ \begin{matrix} j' & j & g' \\ j'' & J & g'' \end{matrix} \right\}$$

are known as the Wigner $6j$-symbols. They are symmetric under a permutation of the columns and they have the following asymptotic behaviour for $j_1, \ldots, j_6$ uniformly large (and fulfilling the triangle inequalities):

$$\left\{ \begin{matrix} j_1 & j_2 & j_3 \\ j_4 & j_5 & j_6 \end{matrix} \right\} \sim \frac{1}{\sqrt{12\pi V}} \cos \left(\sum_{l=1}^{6} j_l \theta_l + \frac{\pi}{4} \right), \tag{7.64}$$

where θ_l is the angle between the outer normals to the two faces sharing j_l in the tetrahedron associated with the $6j$-symbol in Eq. (7.64) and V denotes the volume of the tetrahedron. For an exhaustive discussion of $6j$-symbols we refer to [84].

Let us consider the partition function for Euclidean Chern–Simons gravity with cosmolocial constant $\lambda = 0$ at a formal level. In the functional integral we integrate over all possible values of $\det e_i^a(\xi)$, positive and negative, so formally we can write

$$Z(M) = \int_M \mathscr{D}e \mathscr{D}\omega \, e^{i \int e \wedge R} = \sum_{\det e > 0} \prod_{\xi \in M} 2 \cos (e(\xi) \wedge R(\xi)) \, de(\xi) d\omega(\xi). \tag{7.65}$$

Take a triangulation T of M. As usual we denote the set of vertices by $V(T)$, the set of links by $L(T)$ and (with a slight abuse of notation) the set of tetrahedra by T. We assume that the lengths l_i of the links $i \in L$ only take integer or half-integer values corresponding to the representations of $SU(2)$. For a given length assignment to links, compatible with the triangle inequality, we can define a piecewise linear metric on M and, together with this metric, a connection and a drei-bein. According to Regge's prescription we have

$$\int e \wedge R = \int d^2 \xi \sqrt{g} \, R = \sum_{i \in L} \varepsilon_i l_i, \tag{7.66}$$

where ε_i is the deficit angle of link i. For the special values of the lengths l_i of the links we can write

$$e^{i \int e \wedge R} = \prod_{i \in L(T)} (-1)^{2l_i} \prod_{t \in T} \exp \left(i \sum_{i \in t} \theta_i^{(t)} l_i + i\pi \sum_{i \in t} l_i \right), \tag{7.67}$$

where the angles $\theta_i^{(t)}$ are defined as in Eq. (7.64) for a tetradron $t \in T$ and a link $i \in t$. Eq. (7.67) follows since the deficit angle is related to the angles θ_i by

$$\varepsilon_i = 2\pi - \sum_{t \ni i}(\pi - \theta_i^{(t)}). \tag{7.68}$$

In the path integral we integrate over all values of e corresponding to both positive and negative orientation in each space-time point (here each tetrahedron) and as in Eq. (7.65) these contributions combine to

$$\prod_{i \in L(T)}(-1)^{2l_i} \prod_{t \in T}(-1)^{\sum_{i \in t} l_i}2\cos\left(\sum_{i \in t}\theta_i^{(t)}l_i\right). \tag{7.69}$$

By Eq. (7.64) it is suggested to replace $\cos\left(\sum_{i \in t}\theta_i l_i\right)$ by the Wigner $6j$-symbols whenever the l_is are all uniformly large integers, and by summing over all (large) values of l_is consistent with the triangle inequalities we are summing over different length assignments of the links in the triangulation T. This can be viewed as a path integral over different metric assignments.* While this can at most be considered as a suggestive representation of the formal path integral (7.65) it led Ponzano and Regge to study the following partition function for a closed manifold M:

$$Z_T(M) = \sum_{\{l_i\}}\prod_{i \in L}(-1)^{2l_i}(2l_i + 1)\prod_{t \in T}(-1)^{\sum_{i \in t} l_i}\left\{\begin{matrix} l_1^{(t)} & l_2^{(t)} & l_3^{(t)} \\ l_4^{(t)} & l_5^{(t)} & l_6^{(t)} \end{matrix}\right\}. \tag{7.70}$$

The sum is ill defined since l_i can take any non-negative half-integer value. As presented here it seems a strange approximation to the formal continuum path integral, but Ponzano and Regge observed the following: *the formal expression $Z_T(M)$ is independent of the triangulation T, i.e. it is a topological invariant of the manifold M*[†]. The reason for this invariance can be traced to the so called Biedenharn–Elliott–Racah identities between $6j$-symbols. These arise from othogonality and completeness relations in the tensor decompositions of three angular momenta [84]. As an example we mention the following identity:

$$\sum_x(-1)^{a+\cdots+j+x}(2x + 1)\left\{\begin{matrix} a & b & x \\ c & d & g \end{matrix}\right\}\left\{\begin{matrix} c & d & x \\ e & f & h \end{matrix}\right\}\left\{\begin{matrix} e & f & x \\ b & a & j \end{matrix}\right\} \tag{7.71}$$

$$= \left\{\begin{matrix} g & h & j \\ e & a & d \end{matrix}\right\}\left\{\begin{matrix} g & h & j \\ f & b & c \end{matrix}\right\},$$

which has a graphical interpretation illustrated in Fig. 7.4. By this identity

---

* We will not discuss whether two different metric assignments defined this way may correspond to equivalent metrics.

[†] The idea of topological invariance was not emphasized by Ponzano and Regge, since that concept only became important in physics much later.

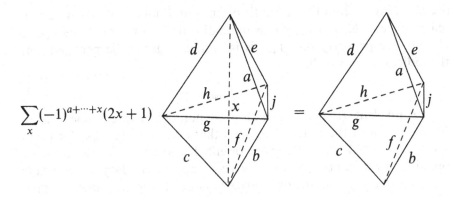

Fig. 7.4. A graphical illustration of Eq. (7.71).

we can replace any three tetrahedra in T which are glued together along a common link, which only belongs to the three tetrahedra, by the two new tetrahedra obtained by removing the common link and introducing a new triangle spanned by the three common links (g, h, j in Fig. 7.4). We recognize this as one of the *moves* discussed in Section 6.3 which change the triangulation T of M to a new triangulation T' of M and it can now be seen that the factor $\prod_{i \in L}(2l_i + 1)$ in the definition (7.70) of the partition function makes $Z_T(M)$ invariant under such a move. Some of the other identities between the $6j$-symbols can be given a similar interpretation and the set of moves obtained in this way is ergodic in the terminology of Chapters 5 and 6, i.e. we can obtain any triangulation of M by repeated application of such moves. This shows that the formal expression $Z_T(M)$ defined by Eq. (7.70) is independent of the triangulation T.

The above considerations suggest that the expression $Z_T(M)$ in some sense is the partition function of three-dimensional Euclidean Chern–Simons gravity in the limit $\lambda \to 0$ which is of course quite singular. We now show how the concepts of topological field theory lead in a natural way to analogues of $6j$-symbols and how it is possible to regularize the formal expression (7.70) while at the same time preserving the topological nature of $Z_T(M)$. This leads to a definition of the path integral for Chern–Simons theories.

7.4.2 The discrete framework

From now on $d = 3$ and we consider triangulated, smooth, compact and oriented three-manifolds (M, T), where T denotes the triangulation, together with the induced triangulation ∂T of ∂M.

Given a finite set J of colours we colour the triangulation T by assigning an element from J to each link (one-simplex) in T. Thus, a colouring of

T is a mapping Φ from the set $L(T)$ of links in T into J. We assume that for each triple (a, b, c) of colours a finite-dimensional vector space V_{abc} is given with the property that V_{abc} is invariant under cyclic permutations of the indices and such that

$$V_{abc} = V_{cba}^*. \tag{7.72}$$

This is equivalent to the existence of a non-degenerate bilinear form $(\cdot, \cdot)_{abc}$ on each of the spaces V_{abc} such that $(x, y)_{abc} = (y, x)_{cba}$ for every $x \in V_{abc}$ and $y \in V_{cba}$. To a given oriented triangle $\triangle$ whose links are coloured by a, b, c in positive cyclic order along its boundary, we associate the vector space V_{abc} and the dual space is associated to the same coloured triangle with the opposite orientation.

Given a surface Σ with a triangulation S we define the vector space $V_{(\Sigma, S)}$ as the direct sum, over all colourings Φ_0 of the links in S, of the tensor product of the vector spaces associated to oriented coloured triangles in S, i.e.

$$V_{(\Sigma, S)} = \bigoplus_{\Phi_0 : L(S) \to J} \bigotimes_{\triangle \in S} V_{\Phi_0(\triangle)}, \tag{7.73}$$

where $\Phi_0(\triangle)$ is the colouring of the triangle $\triangle$ in S induced by Φ_0. As a consequence of Eq. (7.72) we have a canonical identification

$$V_{(\Sigma^*, S)} = V_{(\Sigma, S)}^*. \tag{7.74}$$

By Eq. (7.73) a vector space V_t is associated to the boundary of each oriented tetrahedron t. Since the same summands enter the definition (7.73) for any oriented tetrahedron, the spaces V_t may be viewed as being identical to a single vector space V_t, and Eq. (7.72) yields an identification

$$V_t = V_t^*, \tag{7.75}$$

given by a symmetric non-degenerate bilinear form $(\cdot, \cdot)_t$ on V_t.

Let us now choose a vector $Z_t \in V_t$ and scalars w_a, $a \in J$. These quantities should be thought of as "initial values" of the TQFT we are in the process of defining. We now define the partition function of a triangulated manifold (M, T) in analogy with Eq. (7.22) by

$$Z(M, T) = w^{-2(N_v(T) - \frac{1}{2}N_v(\partial T))} \bigoplus_{\Phi : L(T) \to J} \prod_{\ell \in L(T \setminus \partial T)} w_{\Phi(\ell)}^2 \left(\bigotimes_{t \in T} Z_{\Phi(t)} \right)_{int\ T}, \tag{7.76}$$

where the tensor product is over all tetrahedra t in T and the colouring of t is denoted by $\Phi(t)$. The vector space V_t is given by the direct sum over all colourings of a tetrahedron and we denote the component of the vector Z_t along the colouring $\Phi(t)$ by $Z_{\Phi(t)}$. In the definition (7.76) we have also anticipated the need for a factor w^{-2} associated to each interior vertex and $w^{-1} > 0$ to each boundary vertex. In a similar way as in Eq. (7.22) the notation $(\cdot)_{int\ T}$ indicates contraction over the components of the vector

$$\bigotimes_{t \in T} Z_{\Phi(t)}$$

corresponding to interior triangles, using the duality relation (7.72), so

$$Z(M, T) \in V_{(\partial M, \partial T)}. \qquad (7.77)$$

It should be noted that we have not included any w_a-factors associated to boundary links with colour a, as might seem necessary in order for the gluing property to hold. We take care of this problem by including a factor w_a^2 for each link with colour a in the bilinear form behind the duality relation (7.74), i.e. we define the bilinear form by

$$\left(\bigotimes_{\Delta \in S} \alpha_{\Phi_0(\Delta)}, \bigotimes_{\Delta' \in S} \beta_{\Phi_0(\Delta')} \right)_{(\Sigma, S)} = \prod_{\ell \in L(S)} w_{\Phi_0(\ell)}^2 \prod_{\Delta \in S} \left(\alpha_{\Phi_0(\Delta)}, \beta_{\Phi_0(\Delta)} \right)_{\Phi_0(\Delta)} \qquad (7.78)$$

for any colouring Φ_0 of the links in S and vectors $\alpha_{\Phi_0(\Delta)} \in V_{\Phi_0(\Delta)}$, $\beta_{\Phi_0(\Delta)} \in V_{\Phi_0(\Delta)}^*$. With these definitions one can check that $Z(M, T)$ fulfils the analogue of the gluing property Eq. (7.9), which is important for the following discussion of the removal of the dependence of $V_{(\Sigma, S)}$ and $Z(M, T)$ on the triangulations. For $Z(M, T)$ this will be quite analogous to the considerations of the previous section, whereas for $V_{(\Sigma, S)}$ it will turn out to be less trivial since it is not possible to construct non-trivial TQFTs by just making sufficiently simple choices of the spaces V_{abc}.

In order to ensure that $Z(M, T)$ is independent of the interior of the triangulation T we appeal again to the results of Section 6.3 which for $d = 3$ show that it is sufficient that $Z(M, T)$ be invariant under the moves depicted in Fig. 6.4. In view of the definition (7.76) we thus have to require that Z_t, w_a and w fulfil the following two equations written in symbolic form:

$$\sum_x \omega_x^2 \quad \text{} \quad = \quad \text{} \qquad (7.79)$$

and

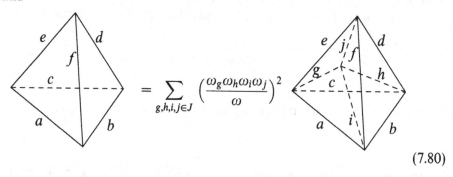

$$(7.80)$$

We note that by summing first over j in Eq. (7.80) it follows that Eq. (7.80) is a consequence of Eq. (7.79) and the following identity:

$$\sum_{g,h,i \in J} \left(\omega_g \omega_h \omega_i \right)^2 \quad = \omega^2 1_{V_{abc}} \qquad (7.81)$$

where, on the left-hand side, we have the partition function of two tetrahedra glued together along three common triangles and with labels as indicated. This partition function is an element of $V_{abc} \otimes V^*_{abc}$, which, when interpreted as a linear mapping on V_{abc}, is required to be the identity up to the factor w^2.

Below we shall derive Eq. (7.81) as a consequence of the identity

$$\sum_g \omega_g^2 \quad = \delta_{ab} \, \omega_a^{-2} \, 1_{V_{acd} \otimes V_{afe}} \qquad (7.82)$$

where on the left-hand side we have the partition function of two tetrahedra glued together along two common triangles. This partition function can be interpreted as an operator in a similar way as the left-hand side of Eq. (7.81). By summing first over i in Eq. (7.81), using Eq. (7.82) with $a = b$ and contracting over one of the spaces on its right-hand side it follows that Eq. (7.81) holds provided

$$\sum_{g,h \in J} w_g^2 w_h^2 N_{ghc} = w_c^2 w^2 \qquad (7.83)$$

for each $c \in J$, where

$$N_{ghc} = \dim V_{ghc}. \tag{7.84}$$

Let us for the moment assume that we are given quantities Z_t, w_a and w satisfying Eqs. (7.79) and (7.80). It follows that $Z(M, T)$ only depends on T through ∂T. This dependence, as well as the dependence of $V_{(\Sigma,S)}$ on S, can then be eliminated by restricting to a suitable subspace of each $V_{(\Sigma,S)}$. We now proceed to explain this restriction. Let Σ be a surface and let S_1 and S_2 be two triangulations of Σ. Picking an arbitrary triangulation T_{12} of $\Sigma \times [0, 1]$ which coincides with S_1 on $\Sigma \times \{0\}$ and with S_2 on $\Sigma \times \{1\}$ we obtain a linear mapping $P^\Sigma_{S_2,S_1} : V_{(\Sigma,S_1)} \rightarrow V_{(\Sigma,S_2)}$ by setting

$$P^\Sigma_{S_2,S_1} = Z(\Sigma \times [0, 1], T_{12}), \tag{7.85}$$

where Σ^* is identified with $\Sigma \times \{0\}$ and Σ with $\Sigma \times \{1\}$ in the obvious way. We note that $P^\Sigma_{S_2,S_1}$ is independent of the choice of T_{12}. As a consequence of the gluing property for $Z(M, T)$ it follows that if S_3 is a third triangulation of Σ then

$$P^\Sigma_{S_3,S_2} P^\Sigma_{S_2,S_1} = P^\Sigma_{S_3,S_1}. \tag{7.86}$$

In particular, the operator $P^\Sigma_S = P^\Sigma_{S,S}$ is a projection for each triangulation S of Σ. We now define a vector space

$$\mathscr{H}_\Sigma = P^\Sigma_S V_{(\Sigma,S)}, \tag{7.87}$$

which is independent of S, since Eq. (7.86) implies that $P^\Sigma_{S_2,S_1}$ maps $P^\Sigma_{S_1} V_{(\Sigma,S_1)}$ isomorphically onto $P^\Sigma_{S_2} V_{(\Sigma,S_2)}$. These maps give rise to a consistent identification of the vector spaces $P^\Sigma_S V_{(\Sigma,S)}$ which we use in the remainder of this chapter.

With this identification we note that the partition function $Z(\Sigma \times [0, 1], T)$ is independent of the triangulation T of $\Sigma \times [0, 1]$ and equals the identity operator on $\mathscr{H}_\Sigma$. More generally, it follows from the gluing property that if T_1 and T_2 are two triangulations of M then

$$Z(M, T_1) = P^{\partial M}_{\partial T_1, \partial T_2} Z(M, T_2).$$

Hence, the partition functions $Z(M, T)$, where T is an arbitrary triangulation of M, define a unique vector $Z(M) \in \mathscr{H}_{\partial M}$.

Next, we observe that the bilinear forms $(\cdot, \cdot)_{(\Sigma,S)}$ defined in Eq. (7.78) induce a well-defined non-degenerate bilinear form $(\cdot, \cdot)_\Sigma$ on $\mathscr{H}_\Sigma \times \mathscr{H}_{\Sigma^*}$ by restriction. The gluing property (7.9) for the partition function $Z(M)$ and the vector spaces $\mathscr{H}_\Sigma$ follows immediately from the gluing property for triangulated manifolds.

It remains to define the isomorphism $U_f : \mathscr{H}_\Sigma \rightarrow \mathscr{H}_{\Sigma'}$ for a given orientation preserving diffeomorphism $f : \Sigma \rightarrow \Sigma'$. This is done by picking an arbitrary triangulation S of Σ and considering the triangulation $f(S)$

of Σ' obtained as the image of S under f. Given a colouring Φ_0 of the links in S and vectors $\alpha_{\Phi_0(\triangle)} \in V_{\Phi_0(\triangle)}$ we define the linear mapping $\bar{U}_f : V_{(\Sigma,S)} \to V_{(\Sigma',f(S))}$ by

$$\bar{U}_f \left(\bigotimes_{\triangle \in S} \alpha_{\Phi_o(\triangle)} \right) = \bigotimes_{\triangle \in S} \alpha_{f(\Phi_0(\triangle))}, \qquad (7.88)$$

where $f(\Phi_0(\triangle))$ is the image under f of the coloured triangle $\Phi_0(\triangle)$. Clearly, $\bar{U}_f$ is an isomorphism. From the definition of P_S^Σ it follows that

$$\bar{U}_f P_S^\Sigma = P_{f(S)}^{\Sigma'} \bar{U}_f$$

and hence the restriction of $\bar{U}_f$ to $\mathcal{H}_\Sigma \subseteq V_{(\Sigma,S)}$ induces an isomorphism $U_f : \mathcal{H}_\Sigma \to \mathcal{H}_{\Sigma'}$. It is clear from the definition (7.88) that the isomorphisms U_f fulfil Eq. (7.2) as well as Eqs. (7.3) and (7.6).

The factorization of vector spaces, bilinear forms and isomorphisms is obvious so we have completed our construction of a TQFT using the building blocks V_{abc}, Z_t, w_a and w fulfilling equations Eqs. (7.79) and (7.80). Next we face the problem of finding such building blocks explicitly. Before doing so we make two remarks. First, assume that each of the spaces V_{abc} is a Hilbert space with an inner product $\langle \cdot\,,\,\cdot \rangle_{abc}$. We can then define the conjugate linear isomorphism $\beta \to \bar{\beta}$ from V_{abc} onto V_{abc}^* by

$$(\alpha, \bar{\beta})_{abc} = \langle \alpha, \beta \rangle_{abc}, \qquad \alpha, \beta \in V_{abc}. \qquad (7.89)$$

The inner products $\langle \cdot\,,\,\cdot \rangle_{abc}$ induce in the standard way an inner product $\langle \cdot\,,\,\cdot \rangle_{(\Sigma,S)}$ on $V_{(\Sigma,S)}$ given by

$$\left\langle \bigotimes_{\triangle \in S} \alpha_{\Phi_0(\triangle)}, \bigotimes_{\triangle' \in S} \beta_{\Phi_0(\triangle')} \right\rangle_{(\Sigma,S)} = \prod_{\triangle \in S} \langle \alpha_{\Phi_0(\triangle)}, \beta_{\Phi_0(\triangle)} \rangle_{\Phi_0(\triangle)}.$$

We define the conjugate linear isomorphism $x \to x^*$ from $V_{(\Sigma,S)}$ onto $V_{(\Sigma^*,S)}$ by

$$(x, y^*)_{(\Sigma,S)} = \langle x, y \rangle_{(\Sigma,S)}, \qquad x, y \in V_{(\Sigma,S)}. \qquad (7.90)$$

According to the definition of $(\cdot\,,\,\cdot)_{(\Sigma,S)}$, including the aforementioned factors ω_a^2 for each a-coloured link, it follows that

$$\left(\bigotimes_{\triangle \in S} \alpha_{\Phi_0(\triangle)} \right)^* = \prod_{\ell \in L(S)} w_{\Phi_0(\ell)}^{-2} \bigotimes_{\triangle \in S} \bar{\alpha}_{\Phi_0(\triangle)}.$$

From this identity we see that $x = x^{**}$ for $x \in V_{(\Sigma,S)}$ if and only if

$$\bar{\bar{\alpha}} = w_a^2 w_b^2 w_c^2 \alpha, \qquad \alpha \in V_{abc}. \qquad (7.91)$$

This is the case we shall deal with below, and this is the main reason for introducing the factors $w_{\Phi_0(\triangle)}^2$ in the definition of $(\cdot\,,\,\cdot)_{(\Sigma,S)}$. By restricting

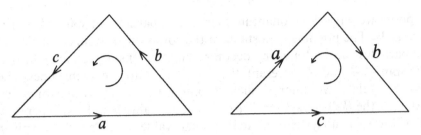

Fig. 7.5. Two coloured and arrow-decorated triangles corresponding to the vector spaces V_{abc} and $V_{cb^\vee a^\vee}$.

$\langle \cdot , \cdot \rangle_{(\Sigma,S)}$ and the mapping $x \to x^*$ to the subspace $\mathscr{H}_\Sigma \subseteq V_{(\Sigma,S)}$ it can then easily be checked that we obtain a unitary theory provided

$$Z_t = Z_t^* \tag{7.92}$$

considered as vectors in V_t.

Secondly, we shall need a slight generalization of the framework discussed above. It is convenient to equip the set of colours J with an involution $a \to a^\vee$, i.e. a bijective mapping fulfilling $a^{\vee\vee} = a$, $a \in J$, which we shall refer to as duality. In this case the assumption (7.72) is replaced by

$$V_{abc} = V^*_{c^\vee b^\vee a^\vee}. \tag{7.93}$$

We decorate the links in a given triangulation by arrows in some arbitrary way and associate to the oriented coloured triangle depicted on the left in Fig. 7.5 the vector space V_{abc} with the convention that switching the direction of an arrow is compensated for by replacing the colour of the link by its dual, so, for example, the space $V_{cb^\vee a^\vee}$ is associated to the triangle depicted at the right in Fig. 7.5. Moreover, reversing the orientation of the triangle means replacing the associated space with its dual by Eq. (7.93).

With these conventions it follows that the space $V_{(\Sigma,S)}$ defined by Eq. (7.73) as well as the partition function $Z(M, T)$ defined by Eq. (7.76) are independent of the chosen decoration by arrows and the whole construction of a TQFT given above can be carried through unaltered.

7.4.3 *Construction in terms of 6j-symbols*

We now proceed to discuss a rather general method for providing the necessary data underlying the construction of the previous subsection. The basic idea of this method has its origin in the "curious connection" [329] between three-dimensional Euclidean Einstein-gravity and angular momentum theory explained above. We now do this in a more precise fashion which allows for generalization and rigour.

Suppose we have three quantum mechanical particles of spin j, k and ℓ, respectively. The possible spin states of the composite system consisting of these particles can be found by decomposing the tensor product $j \otimes k \otimes \ell$ of the spin -j, -k and -ℓ representations of $SU(2)$ into irreducible ones. The 6j-symbols relate two different such decompositions, namely one for which the spin of the $(k\ell)$-subsystem is fixed to some value p and one where the spin of the (jk)-subsystem is fixed to some value q. To specify a spin p multiplet in the (jk)-subsystem means giving a map $\alpha : H_p \to H_j \otimes H_\ell$, where H_j, H_ℓ and H_p denote the respective state spaces, such that α commutes with rotations $g \in SU(2)$, i.e. such that

$$(j(g) \otimes \ell(g)) \circ \alpha = \alpha \circ p(g), \tag{7.94}$$

where $j(g)$ denotes the representation of the rotation g on H_j, etc. The mapping α is an *intertwiner* from the representation p to $j \otimes \ell$ and symbolically we write it as $\alpha : p \to j \otimes k$. In general, if ρ_1 and ρ_2 are two representations of a group G with representation spaces V_1 and V_2, respectively, an intertwiner from ρ_1 to ρ_2 is a mapping α such that

$$\alpha \circ \rho_1(g) = \rho_2(g) \circ \alpha$$

for all $g \in G$. The standard notion of Clebsch–Gordan coefficients is nothing but the matrix elements of an intertwiner with respect to the standard bases obtained by applying lowering operators to highest-spin states, and is written in standard notation as

$$|p, m\rangle \to |(j, \ell)p, m\rangle,$$

$m = -p, -p + 1, \ldots, p$. Note also that the adjoint α^* is an intertwiner from $j \otimes \ell$ to p, which follows by taking the adjoint of Eq. (7.94) and remembering that j, k and p are *-representations, i.e. that

$$j(g)^* = j(g^*). \tag{7.95}$$

Moreover, it is clear that composing intertwiners as well as taking tensor products of intertwiners yields new intertwiners.

We are now ready to define the 6j-symbols. Given representations j, k, ℓ, p, q as above, the two ways of obtaining a spin-i multiplet in the composite $(jk\ell)$-system mentioned above are given by

$$j \otimes (k \otimes \ell) \xrightarrow{1_j \otimes \beta} j \otimes p \xrightarrow{\alpha} i$$

and

$$(j \otimes k) \otimes \ell \xrightarrow{\gamma \otimes 1_\ell} q \otimes \ell \xrightarrow{\delta} i,$$

where α, β, γ and δ are intertwiners and 1_j is the identity mapping on H_j etc. By composition we thus obtain an intertwiner

$$\alpha \circ (1_j \otimes \beta) \circ (\gamma^* \otimes 1_\ell) \circ \delta^* : i \to i. \tag{7.96}$$

Let V_{jp}^i denote the space of all intertwiners from $j \otimes p$ to i. Since i is irreducible it follows from Schur's lemma that the intertwiner given by Eq. (7.96) is a number (multiple of the identity) which we shall interpret, up to a constant multiple, as the matrix element of an operator

$$\begin{vmatrix} i & p & j \\ k & q & \ell \end{vmatrix} \; : \; V_{jp}^i \otimes V_{k\ell}^p \longrightarrow V_{q\ell}^i \otimes V_{jk}^q.$$

In order to specify this precisely we define an inner product on V_{ab}^c by

$$\langle \alpha, \beta \rangle_{ab}^c \, 1_c = \alpha \circ \beta^* \tag{7.97}$$

for $\alpha, \beta \in V_{ab}^c$. Then we have

$$\alpha \circ (1_j \otimes \beta) \circ (\gamma^* \otimes 1_\ell) \circ \delta^* = w_p^2 \left\langle \begin{vmatrix} i & p & j \\ k & q & \ell \end{vmatrix} \, \alpha \otimes \beta, \, \gamma \otimes \delta \right\rangle 1_i, \tag{7.98}$$

where

$$w_p^2 = \eta_p \dim H_p \tag{7.99}$$

with

$$\eta_p = (-1)^{2p} = \begin{cases} +1 & \text{if } p \text{ is integer} \\ -1 & \text{if } p \text{ is half-integer}, \end{cases} \tag{7.100}$$

and the inner product $\langle \cdot, \cdot \rangle$ is simply the tensor product of the inner products on the intertwiner spaces. The $6j$-symbol defined by Eq. (7.98) is related to the Wigner $6j$-symbols introduced in Eq. (7.63) by a phase factor, see e.g. [84].

Note that definition (7.98) makes good sense, since the left-hand side is linear in α and β and conjugate linear in γ and δ. The operator

$$\begin{vmatrix} i & p & j \\ k & q & \ell \end{vmatrix}$$

is called a $6j$-symbol. It might seem artificial to define the $6j$-symbol as an operator since the dimensions of the intertwiner spaces are either 0 or 1. The $6j$-symbols are traditionally defined (up to certain factors) as the numbers given by Eq. (7.96) (or, rather, their complex conjugates) that one obtains by expressing the intertwiners in terms of Clebsch–Gordan coefficients. However, from an invariant point of view, the operator definition is more natural and, in addition, the latter is immediately generalizable to unitary representations of an arbitrary compact group. Furthermore, this point of view fits beautifully into the formalism for three-dimensional TQFTs developed above.

First we recall that for each spin a the tensor product representation $a \otimes a$ contains a unique singlet representation and the unique (up to a phase factor) invariant normalized vector can be written in the form

$$\Omega_a = d_a^{-\frac{1}{2}} \sum_i e_i^a \otimes e_i'^a, \qquad (7.101)$$

where d_a is the dimension of H_a and $(e_1^a, \ldots, e_{d_a}^a)$ and $(e_1'^a, \ldots, e_{d_a}'^a)$ are orthonormal bases for H_a. Moreover, Ω_a is a symmetric tensor if a is an integer and antisymmetric if a is a half-integer. Corresponding to Ω_a we fix once and for all an intertwiner $\psi_a^* : 0 \to a \otimes a$ by setting

$$\psi_a^*(\lambda) = \lambda \, \Omega_a, \ \lambda \in \mathbb{C}, \qquad (7.102)$$

where as usual we have chosen $H_0 = \mathbb{C}$. It follows that the adjoint intertwiner $\psi_a : a \otimes a \to 0$ is given by

$$\psi_a(x \otimes y) = d_a^{-\frac{1}{2}} \sum_i (x, e_i^a)(y, e_i'^a), \qquad (7.103)$$

where $(\cdot, \cdot)$ denotes the inner product on H_a. Since Ω_a is normalized it follows that ψ_a is a partial isometry, i.e.

$$\psi_a \psi_a^* = 1_0. \qquad (7.104)$$

From ψ_c we obtain a canonical isomorphism $\alpha \to \alpha'$ from V_{ab}^c onto the space V_{abc}^0 of intertwiners from $a \otimes b \otimes c$ to 0 by setting

$$\alpha' = \psi_c \circ (\alpha \otimes 1_c), \qquad (7.105)$$

i.e.

$$\alpha' : a \otimes b \otimes c \xrightarrow{\alpha \otimes 1_c} c \otimes c \xrightarrow{\psi_c} 0.$$

We say that α' is obtained from α by tensoring with c from the right. The inverse of the mapping $\alpha \to \alpha'$ is obtained, up to the factor w_c^{-2}, by tensoring once more with c from the right. More precisely, we define for $\beta \in V_{abc}^0$

$$\beta' : a \otimes b \to a \otimes b \otimes 0 \xrightarrow{1_{a \otimes b} \otimes \psi_c^*} a \otimes b \otimes c \otimes c \xrightarrow{\beta \otimes 1_c} 0 \otimes c \to c, \qquad (7.106)$$

where we have omitted indicating the canonical intertwiners $_b\varphi : b \otimes 0 \to b$ and $\varphi_c : 0 \otimes c \to 0$ which are given by

$$_b\varphi(x \otimes \lambda) = \lambda x = \varphi_b(\lambda \otimes x), \quad \lambda \in \mathbb{C}, \ x \in H_b. \qquad (7.107)$$

It is easy to see that for $\alpha \in V_{ab}^c$ we have

$$(\alpha')' : a \otimes b \xrightarrow{\alpha} c \to c \otimes 0 \xrightarrow{1_c \otimes \psi_c^*} c \otimes c \otimes c \xrightarrow{\psi_c \otimes 1_c} 0 \otimes c \to c.$$

Omitting the first step in this string we see that the remainder when applied to a vector $x \in H_c$ equals

$$\varphi_c \circ (\psi_c \otimes 1_c) \left(d_c^{-\frac{1}{2}} x \otimes \sum_i e_i^c \otimes e_i'^c \right) = d_c^{-1} \sum_{i,j} (x, e_j^c)(e_i^c, e_j'^c) e_i'^c$$

$$= \eta_c d_c^{-1} \sum_{i,j} (x, e_j'^c)(e_i^c, e_j^c) e_i'^c$$

$$= w_c^{-2} \sum_j (x, e_j'^c) e_j'^c$$

$$= w_c^{-2} x.$$

We thus obtain

$$(\alpha')' = w_c^{-2} \alpha, \tag{7.108}$$

as claimed.

In the following we shall identify the spaces V_{ab}^c and V_{abc}^0 by the mapping (7.105) and we propose to associate this space with labelled triangles. For this purpose we note that V_{abc}^0 can be canonically identified with V_{cab}^0 by the mapping $\alpha \to \tilde{\alpha}$ given by

$$\tilde{\alpha}(z \otimes x \otimes y) = \eta_c \, \alpha(x \otimes y \otimes z) \tag{7.109}$$

for $x \in H_a$, $y \in H_b$, $z \in H_c$ and $\alpha \in V_{abc}^0$. Here we have inserted the sign factor η_c for later convenience. Since

$$\eta_a \eta_b \eta_c = 1 \tag{7.110}$$

whenever $V_{abc}^0 \neq \{0\}$. we have $\tilde{\tilde{\alpha}} = \alpha$ and hence we may consistently invoke the identifications implied by the $\sim$-mappings, thus obtaining a space, which we denote by V_{abc}, that is invariant under cyclic permutations of the indices. We remark that $\tilde{\alpha}$ may also be defined from α by tensoring with c from both left and right, i.e. we have

$$\omega_c^{-2} \tilde{\alpha} : c \otimes a \otimes b \to c \otimes a \otimes b \otimes c \otimes c \xrightarrow{1_c \otimes \alpha \otimes 1_{c \otimes c}} c \otimes 0 \otimes c \to c \otimes c \to 0, \tag{7.111}$$

where we have omitted to indicate the fixed intertwiners ψ_c and $_c\varphi$ or φ_c, which we shall always do in the following. We leave the verification of Eq. (7.111) to the reader.

The additional requirement (7.72) on V_{abc} is satisfied if we define the bilinear form $(\cdot, \cdot)_{abc}$ by

$$(\alpha, \beta)_{abc} \, 1_b = w_c^{-2} (b \to a \otimes a \otimes b \xrightarrow{1_a \otimes \alpha} a \otimes c \xrightarrow{\beta} b), \tag{7.112}$$

where $\alpha \in V_{ab}^c$ and $\beta = V_{ac}^b$. Here the factor w_c^{-2} has been inserted in order to ensure symmetry of $(\cdot, \cdot)_{abc}$. Let us verify this. Applying both sides of Eq. (7.112) to $x \in H_b$ we obtain

$$
\begin{aligned}
(\alpha, \beta)_{abc} x &= w_c^{-2} \beta \circ (1_a \otimes \alpha)(\Omega_a \otimes x) \\
&= w_c^{-2} d_a^{-\frac{1}{2}} \sum_i \beta(e_i^a \otimes \alpha(e_i'^a \otimes x)) \\
&= w_c^{-2} d_a^{-\frac{1}{2}} \sum_{i,j} \beta(e_i^a \otimes e_j^c)(\alpha(e_i'^a \otimes x), e_j^c). \quad (7.113)
\end{aligned}
$$

Inserting $x = e_k^b$, taking the inner product with e_k^b on both sides and summing over k we obtain

$$
(\alpha, \beta)_{abc} = \eta_c (d_b d_c)^{-1} d_a^{-\frac{1}{2}} \sum_{i,j,k} (\alpha(e_i'^a \otimes e_k^b), e_j^c)(\beta(e_i^a \otimes e_j^c), e_k^b). \quad (7.114)
$$

From this expression and the symmetry properties of Ω_a it follows that

$$
(\alpha, \beta)_{abc} = \eta_a \eta_b \eta_c (\beta, \alpha)_{acb} = (\beta, \alpha)_{acb} \quad (7.115)
$$

as desired.

It is important to note that the definition of $(\cdot, \cdot)_{abc}$ is consistent with the cyclic symmetry. This can perhaps be seen most easily by expressing $(\cdot, \cdot)_{abc}$ in terms of intertwiners $\alpha : a \otimes b \otimes c \to 0$ and $\beta : a \otimes c \otimes b \to 0$, for which one finds that

$$
(\alpha, \beta)_{abc} = \eta_a (d_a d_b d_c)^{-\frac{1}{2}} \sum_{i,j,k} \alpha(e_i^a \otimes e_k^b \otimes e_j^c) \beta(e_i'^a \otimes e_j'^c \otimes e_k'^b) \quad (7.116)
$$

instead of Eq. (7.114). Combining this formula with Eq. (7.109) it follows that

$$
\begin{aligned}
(\alpha, \beta)_{abc} &= \eta_c (d_a d_b d_c)^{-\frac{1}{2}} \sum_{i,j,k} \tilde{\alpha}(e_j^c \otimes e_i^a \otimes e_k^b) \hat{\beta}(e_j'^c \otimes e_k'^b \otimes e_i'^a) \\
&= (\tilde{\alpha}, \hat{\beta})_{cab}, \quad (7.117)
\end{aligned}
$$

where $\beta \to \hat{\beta}$ denotes the inverse of the mapping $\alpha \to \tilde{\alpha}$. This proves our claim.

Below we show, in addition, that $(\cdot, \cdot)_{abc}$ is non-degenerate. We have thus defined the desired bilinear forms $(\cdot, \cdot)_{abc} : V_{abc} \times V_{acb} \to \mathbb{C}$ which allow us to make the identification Eq. (7.72). Let us now return to the $6j$-symbol (7.98) and note that in force of Eq. (7.72) we have

$$
\begin{vmatrix} i & p & j \\ k & q & \ell \end{vmatrix} \in V_{ipj} \otimes V_{p\ell k} \otimes V_{jkq} \otimes V_{iq\ell}. \quad (7.118)
$$

The vector spaces occurring in Eq. (7.118) are exactly those associated with the four coloured triangles in the boundary of the tetrahedron depicted in Fig. 7.6 and suitably oriented. Taking the direct sum over colourings in Eq. (7.118) we thus obtain our candidate partition function

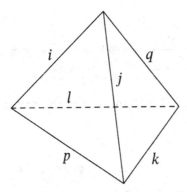

Fig. 7.6. A tetrahedron with coloured edges.

Z_t associated with the oriented tetrahedron. This definition is, however, potentially ambiguous, since the association of the $6j$-symbol in (7.98) with the coloured tetrahedron in Fig. 7.6 depends on the way the latter is drawn. In order for Z_t to be uniquely defined we need to show that the $6j$-symbol respects the symmetries of the tetrahedron. It is well known and easy to verify that the symmetry group of the tetrahedron is generated by two elements. This entails that it is sufficient to verify the following two identities:

$$\begin{vmatrix} i & p & j \\ k & q & \ell \end{vmatrix} = \begin{vmatrix} k & p & \ell \\ i & q & j \end{vmatrix} \tag{7.119}$$

and

$$\begin{vmatrix} i & p & j \\ k & q & \ell \end{vmatrix} = \begin{vmatrix} j & i & p \\ \ell & k & q \end{vmatrix}. \tag{7.120}$$

These identities have been known for a long time (and the same is true of the association of $6j$-symbols with coloured tetrahedra). They can be obtained rather simply by tensoring the definition (7.98) of $6j$-symbols from the left and/or right by appropriate representations. Thus, tensoring Eq. (7.98) by i from the left and by ℓ from the right yields the relation

$$w_p^{-2} \left\langle \begin{vmatrix} i & p & j \\ k & q & \ell \end{vmatrix} \alpha \otimes \beta, \delta \otimes \gamma \right\rangle = w_q^{-2} \left\langle \tilde{\beta} \otimes \tilde{\alpha}, \begin{vmatrix} \ell & q & i \\ j & p & k \end{vmatrix} \tilde{\delta} \otimes \gamma \right\rangle. \tag{7.121}$$

Applying this identity twice, the relation (7.119) results. Eq. (7.120) can be verified similarly, see e.g. [156]. For these relations to hold, the presence of the factor w_p^2 in Eq. (7.98) is essential.

We have thus constructed the desired building blocks V_{abc} and Z_t and we need to establish the relations (7.79) and (7.80). The first of these relations has been known for a long time as the Biedenharn–Elliott identity [170, 83]. It is a simple consequence of two ways of writing the intertwiner

$$u \xrightarrow{\eta^*} r \otimes \ell \xrightarrow{\delta^* \otimes 1_\ell} m \otimes k \otimes \ell \xrightarrow{\gamma^* \otimes 1_{k \otimes \ell}} n \otimes j \otimes k \otimes \ell$$
$$\xrightarrow{1_{n \otimes j} \otimes \beta} n \otimes j \otimes p \xrightarrow{1_n \otimes \alpha} n \otimes i \xrightarrow{\varepsilon} u$$

for given intertwiners $\varepsilon, \alpha, \beta, \gamma, \delta$ and η. On the one hand, we can express it as

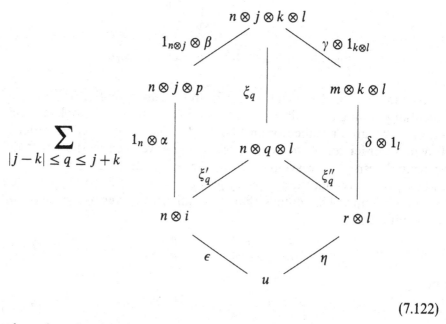

$$(7.122)$$

and on the other hand as

$$(7.123)$$

where the squares in each diagram are to be traversed in succession, each representing a 6j-symbol, and the intertwiners ξ_q, ξ_q', ξ_q'' and ξ are partial isometries. To obtain Eq. (7.122) we have used the fact that the tensor product $j \otimes k$ is completely reducible, i.e. that it is equivalent to the direct sum of the spin-q representations, which may also be stated as

$$\sum_q \xi_q^* \xi_q = 1_{j \otimes k}. \tag{7.124}$$

The reader may easily verify by inspection that the equality of the intertwiners (7.122) and (7.123) is equivalent to the identity (7.79) in which the factor w_j^2 originates from the factor w_p^{-2} in the definition (7.98) of the $6j$-symbol.

As mentioned previously, the validity of Eq. (7.80) is a consequence of Eq. (7.79) and the relation (7.82), provided that the factor w^2 defined by Eq. (7.83) is independent of c. We shall now see that Eq. (7.82) actually holds but w^2, unfortunately, is divergent due to the fact that there are infinitely many spin-representations. This is the origin of the divergence mentioned previously in connection with the Ponzano–Regge approach.

Before discussing how to avoid this problem let us derive Eq. (7.82). Given representations k, ℓ, p, q and j and $\beta \in V_{k\ell}^p$, $\gamma \in V_{k\ell}^q$ we have that

$$\delta_{pq} \langle \beta, \gamma \rangle 1_{j \otimes p} = (1_j \otimes \beta)(1_j \otimes \gamma^*),$$

since there exists no non-trivial intertwiner between p and q if $p \neq q$. Given an additional representation i and $\alpha \in V_{jp}^i$, $\delta \in V_{jq}^i$ we obtain

$$\delta_{pq} \langle \alpha \otimes \beta, \delta \otimes \gamma \rangle 1_i = \alpha \circ (1_j \otimes \beta) \circ (1_j \otimes \gamma^*) \circ \delta^*. \tag{7.125}$$

The right-hand side of this equation can be decomposed into two $6j$-symbols as

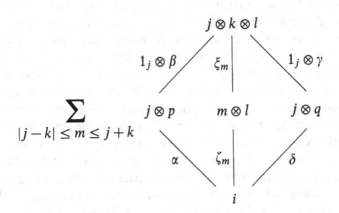

$$\tag{7.126}$$

where $\xi_m \in V_{jk}^m$ and $\zeta_m \in V_{m\ell}^i$ are partial isometries. Here we have again used the complete reducibility of the tensor product $j \otimes k$. By inspection one finds that the equality of the right-hand side of Eq. (7.125) and the expression (7.126) is equivalent to the operator identity

$$w_q^2 w_p^2 \sum_m \begin{vmatrix} i & q & j \\ k & m & \ell \end{vmatrix}^* \begin{vmatrix} i & p & j \\ k & m & \ell \end{vmatrix} = \delta_{pq} \, 1_{V_{jp}^i \otimes V_{k\ell}^p}.$$

Combining this with Eq. (7.121) one obtains Eq. (7.82).

As concerns Eq. (7.83), we note that N_{ghc} equals the multiplicity of g in $h \otimes c$ (which is 1 if $|h - c| \le g \le h + c$ and 0 otherwise). Thus,

$$\dim H_h \otimes H_c = \sum_g N_{ghc} \dim H_g$$

and hence, using Eq. (7.110),

$$w_h^2 w_c^2 = \sum_g N_{ghc} w_g^2. \tag{7.127}$$

We therefore see that the validity of Eq. (7.83) requires

$$w^2 = \sum_h w_h^4 = \sum_{m=0}^{\infty} (2m + 1)^2$$

which reveals the divergence we expected.

We conclude that the data V_{abc} and Z_t constructed on the basis of angular momentum theory satisfies all the necessary requirements for the construction of a three-dimensional TQFT except for the divergence of the vertex factor w^2 in Eq. (7.83). Introducing a cutoff in the form of an upper bound on the spin values clearly violates the relations (7.79) and (7.80) and hence, in a sense, breaks general covariance. An easy way out is simply to replace $SU(2)$ by some finite group. Indeed, the whole construction given above applies to an arbitrary compact group except that one has to take into account that representations are in general not self-dual, so the role played before by $\psi_a : a \otimes a \to 0$ is in this case taken over by a partial isometry $\psi_a : a \otimes a^\vee \to 0$, where $a^\vee$ denotes the dual representation to a which is characterized by the fact that a non-trivial ψ_a exists and is unique up to a phase factor. Moreover, one needs to ensure that the phase factors η_a reflecting the symmetry properties of the invariant vectors in $a \otimes a^\vee$ satisfy Eq. (7.110). Furthermore, in Eqs. (7.122), (7.123) and (7.126), one has to replace ξ_m, etc., by orthonormal bases $\{\xi_m^i\}$ for the appropriate intertwiner spaces, which are in general no longer one-dimensional. With these modifications the building blocks V_{abc} and Z_t can be used for the construction of a TQFT up to the possible divergence of w^2. For a finite group one finds that w^2 is finite and equals the number of elements in the group. We refer to [150] for a discussion of TQFT based on finite groups. For such theories the (formal) connection to quantum gravity is lost.

A beautiful solution to the divergence problem for $SU(2)$ (as well as for general classical Lie groups) was found by Turaev and Viro [362], who observed that one may replace $SU(2)$ by one of its *quantum deformations*. In this case the divergence problem disappears while the theory remains topological. We conclude this chapter by explaining the main features of the work of Turaev and Viro which necessitates a brief detour into the representation theory for quantum groups.

The deformed Lie algebra of $SU(2)$, where q is a complex number of modulus 1, is defined by the relations

$$KS_{\pm} = q^{\pm 1}S_{\pm}K, \qquad (7.128)$$

$$[S_+, S_-] = \frac{K - K^{-1}}{q - q^{-1}}, \qquad (7.129)$$

where $S_{\pm}$ are the standard raising and lowering operators and one should think of K as being given by the z-component S_z of the spin as

$$K = q^{2S_z}.$$

In the limit $q \to 1$ one then recovers the well-known relations defining $sl(2)$. We also note that the requirement $|q| = 1$ ensures that the relations (7.128) and (7.129) are invariant under the adjoint operation given by

$$S_z^* = S_z , \qquad S_+^* = S_-.$$

The algebra generated by $S_{\pm}$ and $K^{\pm}$ with the relations given by the Eqs. (7.128) and (7.129) is a deformation of the *universal enveloping algebra* of $sl(2)$. This algebra is denoted by $U_q sl(2)$ and is a *-algebra in the obvious way.

It is known that for generic values of q, i.e. for q not a root of unity, the representation theory of $U_q sl(2)$ is basically the same as for $sl(2)$, i.e. the irreducible representations are heighest weight representations labelled by a half-integer spin j and may be constructed in the standard fashion using the raising and lowering operators $S_{\pm}$, [335]. One finds that a basis $|jm>$, $m = -j, -j+1, \ldots, +j$, constructed in this way fulfils

$$|jm\rangle = \sqrt{(j \pm m)_q (j \pm m + 1)_q} \, |jm \pm 1\rangle, \qquad (7.130)$$

where we have introduced the notation

$$\alpha_q = \frac{q^\alpha - q^{-\alpha}}{q - q^{-1}}. \qquad (7.131)$$

If q is not a root of unity the zeros of the coefficients in Eq. (7.130) are identical to the ones in the classical case of $sl(2)$. This is basically the reason why there is a one-to-one correspondence between finite-dimensional irreducible representations of $U_q sl(2)$ and $sl(2)$, the dimensions being equal, while the matrices representing the generators are obtained by a suitable replacement of integers by their q-analogues defined by Eq. (7.131).

Assume now that q is a root of unity and let ℓ be the smallest positive integer such that

$$q^{2\ell} = 1. \tag{7.132}$$

For $j < \ell/2$ the coefficients in Eq. (7.130) are non-vanishing for $|m| < j/2$, which is sufficient to allow the standard construction of a spin-j representation of dimension $2j + 1$. For $j \geq \ell/2$, on the other hand, we can find m with $|m| < j$ such that $(j+m+1)_q = 0$ and hence, according to Eq. (7.130), $S_+|jm >= 0$, which shows that the standard procedure breaks down. It turns out that the spin representations with $j < \frac{\ell}{2}$ constructed in this way are the only irreducible finite-dimensional representations of $U_q sl(2)$, whereas there exist other finite-dimensional indecomposable but not irreducible representations, see e.g. [189] for a complete account. These are generated when decomposing tensor products of irreducible representations, as we explain below.

Let H_j be the representation space for the spin j representation of $U_q sl(2)$. In order to obtain a positive definite inner product on H_j by requiring the vectors $|jm>$ to have norm 1 it is necessary that the argument of the square root in Eq. (7.130) be positive for all m. From Eq. (7.131) it is easy to see that for $j < \frac{\ell}{2}$ this is only the case if ℓ is an *even integer*, which we assume to be the case from now on. Let us define the *q-dimension* $d_\pi(q)$ of a representation π as the trace of K^2, i.e.

$$d_\pi(q) = \mathrm{Tr}\,\pi(K^2). \tag{7.133}$$

For the irreducible representations j described above one finds that

$$d_j(q) = \sum_{m=-j}^{j} q^{2m} = (2j+1)_q, \tag{7.134}$$

from which it follows that these representations have a non-vanishing q-dimension except in the case $j = \frac{\ell-1}{2}$. It turns out that the indecomposable but non-irreducible representations mentioned all have q-dimension 0. It is useful to distinguish indecomposable representations according to whether their q-dimension vanishes or not. We call those with vanishing q-dimension type I representations and the others, i.e. those of spin

$$j \leq \frac{\ell - 2}{2}, \tag{7.135}$$

type II representations.

The definition of the tensor product of two representations of $U_q sl(2)$ depends on the existence of a so-called co-product

$$\Delta : U_q sl(2) \rightarrow U_q sl(2) \otimes U_q sl(2),$$

which is defined on the generators by

$$\Delta(S^{\pm}) = K \otimes S^{\pm} - S^{\pm} \otimes K^{-1}, \tag{7.136}$$

$$\Delta(K) = K \otimes K, \tag{7.137}$$

and is extended linearly to the entire algebra. One can check that Δ preserves the relations defining $U_q sl(2)$ and hence defines an algebra homomorphism, i.e.

$$\Delta(xy) = \Delta(x)\Delta(y). \tag{7.138}$$

Moreover, Δ fulfils the co-associativity relation

$$(1 \otimes \Delta)\Delta = (\Delta \otimes 1)\Delta. \tag{7.139}$$

Using Δ we define the tensor product $\pi \underline{\otimes} \rho$ of two representations π and ρ by

$$\pi \underline{\otimes} \rho(x) = \pi \otimes \rho(\Delta(x)). \tag{7.140}$$

If we write

$$\Delta(x) = \sum_i x_i^{(1)} \otimes x_i^{(2)},$$

the right-hand side of Eq. (7.140) equals

$$\sum_i \pi\left(x_i^{(1)}\right) \otimes \rho\left(x_i^{(2)}\right) \tag{7.141}$$

and the terms in the sum (7.141) are ordinary tensor products of operators $\pi(x_i^{(1)})$ and $\rho(x_i^{(2)})$ acting on the tensor product $H_\pi \otimes H_\rho$ of the two representation spaces H_π and H_ρ. It follows from Eq. (7.138) that $\pi \underline{\otimes} \rho$ defines a representation on $H_\pi \otimes H_\rho$.

Let us now consider two irreducible type II representations of spin j and k. It is known that the tensor product $j \underline{\otimes} k$ may be decomposed in a unique way in the form

$$H_j \otimes H_k = H_{j,k} \oplus Z_{j,k}, \tag{7.142}$$

where $H_{j,k}$ and $Z_{j,k}$ are direct sums of representation spaces of type II and type I, respectively [189]. In fact $H_{j,k}$ is obtained by the usual procedure used for $sl(2)$ by only retaining spins less than $\frac{\ell-1}{2}$. In other words, we have

$$H_{j,k} = \bigoplus_{|j-k| \le \ell \le \min\{\frac{1}{2}(\ell-2), j+k\}} H_\ell.$$

We now define the reduced tensor product of j and k as the restriction of $j \underline{\otimes} k$ to $H_{j,k}$, and denote this representation by $j \otimes k$ and the space $H_{j,k}$ by $H_{j \otimes k}$. This means that we simply disregard representations of type I. The important fact is that this may be done consistently such that the reduced tensor product possesses all the properties of the ordinary tensor product which are needed in order for the constructions described above to work. In particular, we note that the tensor product $\otimes$ is associative as a consequence of Eq. (7.139). Applying the decomposition (7.142) to a tensor product $j \underline{\otimes} k \underline{\otimes} \ell$ by either decomposing $(j \underline{\otimes} k)$ first or by decomposing $(k \underline{\otimes} \ell)$ first, one obtains two potentially different spaces $H_{(j,k),\ell}$ and $H_{j,(k,\ell)}$. It can be shown that these spaces are canonically isomorphic [56]. This fact implies that the definition of $6j$-symbols, Eq. (7.98), is still applicable, bearing in mind the fact that tensor products of intertwiners respect the decomposition (7.142). We thus obtain q-analogues of the $6j$-symbols which render the construction of Ponzano and Regge meaningful.

As a consequence of Eq. (7.137) the q-dimension of $j \underline{\otimes} k$ factorizes. Since the q-dimension of type II representations vanishes it follows from Eq. (7.142) that

$$d_{j \underline{\otimes} k}(q) = d_j(q) d_k(q). \tag{7.143}$$

Defining

$$w_j^2 = \eta_j \, d_j(q), \tag{7.144}$$

with η_j as in Eq. (7.100), we obtain the relation (7.127), from which we in turn conclude that Eq. (7.83) holds with

$$w^2 = \sum_h w_h^4 = \sum_{j=0}^{\ell-2} (j+1)_q^2, \tag{7.145}$$

which, of course, is finite. Eqs. (7.79) and (7.82) follow as in the classical case by using the complete reducibility of the reduced tensor product.

We have now shown that type II representations of $U_q sl(2)$ lead to a well-defined TQFT which can be regarded as a regularization of the Ponzano–Regge theory. The latter theory corresponds formally to the limit $q \to 1$. We emphasize that it is possible to define deformations of the universal enveloping algebra of any of the classical semisimple Lie algebras and that the arguments presented above for $sl(2)$ can be generalized in a straightforward manner to yield TQFTs associated with these algebras, when the deformation parameter is an even simple root of unity. We refer to [156, 361] for details.

It is now appropriate to return briefly to the question as to whether the models constructed in this way have any connection to quantum gravity. Euclidean gravity with a positive cosmological constant is formally related to a Chern–Simons theory with gauge group $SU(2) \times SU(2)$, as discussed earlier in this chapter. We saw that the coupling constant in front of the action in a three-dimensional Chern–Simons theory is quantized. It is reasonable to expect that this quantization of the coupling constant reflects the quantization of q in the Turaev–Viro theory. Since the continuum Chern–Simons theory given in terms of functional integrals is, strictly speaking, not well-defined, a proof of the equivalence of the two models is at best an agreement between calculations that can be carried out in both cases. Perhaps the strongest result of this type is the factorization of the partition functions of the Turaev–Viro model into a product of two topological invariants (corresponding to the two copies of $SU(2)$; see Eq. (7.58)), and that the latter equal the so-called Reshetikhin–Turaev invariants which were constructed on the basis of knot-invariants associated with quantum groups in [332]. By the well-established connection between knot theory [391] and Chern–Simons theory these invariants are known to be the ones obtained from Chern–Simons theory.

This area of research is still very much in a flux and, no doubt, many results will be discovered in the coming years, revealing a deep connection between TQFT and quantum gravity.

7.5 Notes

The concept of topological quantum field theory was introduced by Witten in [390], where, in particular, a quantum field theoretic formulation of the Donaldson theory [145] was proposed. Since then the topic has undergone a rapid development in tying together the subjects of quantum field theory and low-dimensional geometry. For reviews of continuum topological field theory see [393, 364, 90]. The most spectacular successes of this development are perhaps, on the one hand, the establishment of an inherently three-dimensional framework for understanding the Jones polynomial of knot theory in [391] and, on the other hand, the uncovering of the profound relation between two-dimensional gravity and intersection theory on the moduli spaces of Riemann surfaces discussed in Chapter 4. In the former three-dimensional Chern–Simons gauge theory is a key ingredient and the arguments are based on convincing, but formal, considerations of the Chern–Simons path integral; see [190, 298]. Hamiltonian quantization of Chern–Simons theory is discussed in great detail in [58].

A rigorous construction of the partition functions of three-dimensional Chern–Simons theory using link invariants derived from quantun groups combined with the fact that any 3-manifold can be obtained from S^3 by surgery along some link was initiated in [332]. To say that this construction yields the invariants of the Chern–Simons theory basically means that they transform as predicted under surgery. Subsequently, Turaev and Viro gave a statistical mechanical construction, based on representation theoretic data for $U_q sl(2)$, of a three-dimensional TQFT [362]. As mentioned earlier, many of the ideas of that paper are also present in [329] and form the basis of our approach to TQFT.

The axiomatic formulation of TQFT was first given by Atiyah in [57]. Our exposition in Section 7.2 follows the latter closely, albeit in a less general form. For an extensive discussion of the axiomatic aspects of TQFT see [330]. Decomposition properties of TQFTs are discussed from an axiomatic point of view in [343].

The presentation of two-dimensional TQFT in Section 7.3 relies heavily on [159] and

[248], see also [378]. As is evident, a main ingredient in the classification result given in Theorem 7.1 is the Frobenius algebra structure of the state space corresponding to the circle or, more generally, to S^{d-1}. This structure was first pointed out in [142], where the exponential form of the partition functions for closed surfaces is also given. Special constructions of discretized two-dimensional unitary models much in the spirit of the Turaev–Viro construction, but based on a finite-dimensional associative algebra instead of a quantum group, appear in [59, 192]. In [192] discrete realizations of some well-known two-dimensional continuum models, e.g. twisted $N = 2$ superconformal models [167], as well as gauge theories [394], are given.

The reformulation of three-dimensional gravity as a Chern–Simons theory is due to Witten [389], see also [334, 4]. Further references to work in this direction may be found in [90].

Our description of the Ponzano–Regge approach to three-dimensional gravity follows the original paper [329] closely. A related paper discussing a connection with gravity is [224]. The arguments relating this approach to Chern–Simons theory are partly borrowed from [315, 313]. The subsequent discussion aims at isolating the essential ingredients of the Ponzano–Regge construction and thus fitting it into a framework which makes it possible to discuss a more general class of three-dimensional TQFTs, although we have chosen to stick to the case of $SU(2)$ in the text. For details that we have left out in the derivation of the general version of the Biedenharn–Elliott identity, as well as the symmetry properties of the generalized $6j$-symbols, we refer to [156]. The case of an arbitrary simple Lie-algebra replacing $sl(2)$ is also discussed there. Alternative derivations can be found in [361].

Generalizations of the original theory by Turaev and Viro to include observables in the form of graphs embedded in the manifolds or on their boundary were given in [360] and [257]; see also [75]. Such generalizations are needed in order to prove the factorization property of the Turaev–Viro invariant, a proof of which can be found in [361]. In [361] one can also find a discussion of the relation between this approach to TQFT and the ribbon-graph construction [332] of invariants of 3-manifolds. Different proofs of the factorization property for the case of $U_q sl(2)$ appeared in [371] and [333] and, for general quantum Lie algebras, in [75, 74].

Other combinatorial approaches to three-dimensional TQFT or 3-manifold invariants, related to the ones discussed here, can be found in [259, 261, 258, 260, 285, 172, 124]. For work on TQFTs based on finite groups see [141, 183, 150, 14, 397]. Finally, some attempts to construct four-dimensional discretized TQFTs can be found in [314, 102, 126, 125].

References

[1] ABRAHAM, D. B., CHAYES, J. T., AND CHAYES, L. Random surface correlation functions. *Commun. Math. Phys.*, **96** (1984a), 439.

[2] ABRAHAM, D. B., CHAYES, J. T., AND CHAYES, L. Statistical mechanics of lattice tubes. *Phys. Rev. D*, **30** (1984b), 841.

[3] ABRAHAM, D. B., CHAYES, J. T., AND CHAYES, L. Nonperturbative analysis of a model of random surfaces. *Nucl. Phys. B*, **251** (1985), 553.

[4] ACHUCARRO, A. AND TOWNSEND, P. K. A Chern–Simons action for three-dimensional anti-de-Sitter supergravity theories. *Phys. Lett.*, **B180** (1986), 89.

[5] ACQUISTAPACE, F., BENEDETTI, R., AND BROGLIA, F. Effectiveness and non-effectiveness in semialgebraic and PL geometry. *Invent. Math.*, **102** (1990), 141.

[6] AGISHTEIN, M. E. AND MIGDAL, A. A. Numerical study of two point correlation function and Liouville field properties in two-dimensional quantum gravity. *Mod. Phys. Lett. A*, **6** (1991), 1863.

[7] AGISHTEIN, M. E. AND MIGDAL, A. A. Critical behavior of dynamically triangulated quantum gravity in four dimensions. *Nucl. Phys. B*, **385** (1992a), 395.

[8] AGISHTEIN, M. E. AND MIGDAL, A. A. Simulations of four-dimensional simplicial quantum gravity. *Mod. Phys. Lett. A*, **7** (1992b), 1039.

[9] AIDA, T., KITAZAWA, Y., KAWAI, H., AND NINOMIYA, M. Conformal invariance and renormalization group in quantum gravity near two-dimensions. *Nucl. Phys. B*, **427** (1994), 158.

[10] AIZENMAN, M. Geometric analysis of ϕ^4 fields and Ising models. *Commun. Math. Phys.*, **86** (1982), 1.

[11] AIZENMAN, M., CHAYES, J. T., CHAYES, L., FRÖHLICH, J., AND RUSSO, L. On a sharp transition from area law to perimeter law in a system of random surfaces. *Commun. Math. Phys.*, **92** (1983), 19.

[12] ALEXANDER, J. W. The combinatorial theory of complexes. *Ann. Mat.*, **31** (1930), 292.

[13] ALONSO, F. AND ESPRIU, D. On the fine structure of strings. *Nucl. Phys. B*, **283** (1987), 393.

[14] ALTSCHULER, D. AND COSTE, A. Quasi-quantum groups, knots, three-manifolds and topological field theory. *Commun. Math. Phys.*, **150** (1992), 83.

[15] ALVAREZ, O. Theory of strings with boundaries: Fluctuations, topology and quantum geometry. *Nucl. Phys. B*, **216** (1982), 125.

[16] ALVAREZ-GAUME, L., BARBON, J. L. F., AND CRNKOVIC, C. A proposal for strings at $d > 1$. *Nucl. Phys. B*, **394** (1993), 383.

[17] AMBARTZUMIAN, R. V., SAVVIDY, G. K., SAVVIDY, K. G., AND SUKIASIAN, G. Alternative model of random surfaces. *Phys. Lett. B*, **275** (1992), 99.

[18] AMBJØRN, J., BOULATOV, D. V., KRZYWICKI, A., AND VARSTED, S. The vacuum in three-dimensional simplicial quantum gravity. *Phys. Lett. B*, **276** (1992), 432.

[19] AMBJØRN, J., BURDA, Z., JURKIEWICZ, J., AND PETERSSON, B. A random surface theory with nontrivial γ_{string}. *Phys. Lett. B*, **341** (1995), 286.

[20] AMBJØRN, J., CHEKHOV, L., KRISTJANSEN, C. F., AND MAKEENKO, Y. Matrix model calculations beyond the spherical limit. *Nucl. Phys. B*, **404** (1993), 127.

[21] AMBJØRN, J., DEFORCRAND, P., KOUKIOU, F., AND PETRITIS, D. Monte Carlo simulations of regularized bosonic strings. *Phys. Lett. B*, **197** (1987), 548.

[22] AMBJØRN, J. AND DURHUUS, B. Regularized bosonic strings need extrinsic curvature. *Phys. Lett. B*, **118** (1987), 253.

[23] AMBJØRN, J., DURHUUS, B., AND FRÖHLICH, J. Diseases of triangulated random surface models, and possible cures. *Nucl. Phys. B*, **257** (1985), 433.

[24] AMBJØRN, J., DURHUUS, B., AND FRÖHLICH, J. The appearance of critical dimensions in string theories (II). *Nucl. Phys. B*, **275** (1986), 161.

[25] AMBJØRN, J., DURHUUS, B., FRÖHLICH, J., AND JONSSON, T. Regularized strings with extrinsic curvature. *Nucl. Phys. B*, **290** (1987), 480.

[26] AMBJØRN, J., DURHUUS, B., FRÖHLICH, J., AND JONSSON, T. A renormalization group analysis of lattice models of two-dimensional membranes. *J. Stat. Phys.*, **55** (1988), 29.

[27] AMBJØRN, J., DURHUUS, B., FRÖHLICH, J., AND ORLAND, P. The appearance of critical dimensions in string theories. *Nucl. Phys. B*, **270** (1986), 457.

[28] AMBJØRN, J., DURHUUS, B., AND JONSSON, T. Random paths with curvative dependent action. *Europhys. Lett.*, **3** (1987a), 1059.

[29] AMBJØRN, J., DURHUUS, B., AND JONSSON, T. Scaling of the string tension in a new class of regularized string theories. *Phys. Rev. Lett.*, **58** (1987b), 2619.

[30] AMBJØRN, J., DURHUUS, B., AND JONSSON, T. Statistical mechanics of paths with curvature dependent action. *J. Phys. A*, **21** (1988), 981.

[31] AMBJØRN, J., DURHUUS, B., AND JONSSON, T. Kinematical and numerical study of the crumpling transition in crystalline surfaces. *Nucl. Phys. B*, **316** (1989), 526.

[32] AMBJØRN, J., DURHUUS, B., AND JONSSON, T. A random walk representation of the Dirac propagator. *Nucl. Phys.*, **B330** (1990a), 509.

[33] AMBJØRN, J., DURHUUS, B., AND JONSSON, T. Summing over all genera for $d > 1$: a toy model. *Phys. Lett. B*, **244** (1990b), 403.

[34] AMBJØRN, J., DURHUUS, B., AND JONSSON, T. Three-dimensional simplicial quantum gravity and generalized matrix models. *Mod. Phys. Lett. A*, **6** (1991), 1133.

[35] AMBJØRN, J., DURHUUS, B., AND JONSSON, T. A solvable 2-d gravity model with $\gamma > 0$. *Mod. Phys. Lett. A*, **9** (1994), 1221.

[36] AMBJØRN, J., DURHUUS, B., JONSSON, T., AND THORLEIFSSON, G. Matter fields with $c > 1$ coupled to 2d gravity. *Nucl. Phys. B*, **398** (1993), 568.

[37] AMBJØRN, J. AND GREENSITE, J. Nonperturbative calculation of correlators in 2-D quantum gravity. *Phys. Lett. B*, **254** (1991), 66.

[38] AMBJØRN, J., IRBÄCK, A., JURKIEWICZ, J., AND PETERSSON, B. The theory of dynamical random surfaces with extrinsic curvature. *Nucl. Phys.*, **B 393** (1993), 571.

[39] AMBJØRN, J., JAIN, S., JURKIEWICZ, J., AND KRISTJANSEN, C. F. Observing 4d baby universes in quantum gravity. *Phys. Lett. B*, **305** (1993), 208.

[40] AMBJØRN, J., JAIN, S., AND THORLEIFSSON, G. Baby universes in 2d quantum gravity. *Phys. Lett. B*, **307** (1993), 34.

[41] AMBJØRN, J. AND JURKIEWICZ, J. Four-dimensional simplicial quantum gravity. *Phys. Lett. B*, **278** (1992), 50.

[42] AMBJØRN, J. AND JURKIEWICZ, J. On the exponential bound in four-dimensional simplicial gravity. *Phys. Lett. B*, **335** (1994), 355.

[43] AMBJØRN, J. AND JURKIEWICZ, J. Scaling in four-dimensional quantum gravity. *Nucl. Phys. B*, **451** (1995), 643.

[44] AMBJØRN, J., JURKIEWICZ, J., AND KRISTJANSEN, C. F. Quantum gravity, dynamical triangulations and higher-derivative regularization. *Nucl. Phys. B*, **393** (1993), 601.

[45] AMBJØRN, J., JURKIEWICZ, J., AND MAKEENKO, Y. M. Multiloop correlators for two-dimensional quantum gravity. *Phys. Lett. B*, **251** (1990), 517.

[46] AMBJØRN, J., JURKIEWICZ, J., AND WATABIKI, Y. On the fractal structure of two-dimensional quantum gravity. *Nucl. Phys. B*, **454** (1995), 313.

[47] AMBJØRN, J. AND KRISTJANSEN, C. F. From one-matrix model to Kontsevich model. *Mod. Phys. Lett. A*, **8** (1993a), 2875.

[48] AMBJØRN, J. AND KRISTJANSEN, C. F. Nonperturbative 2-d quantum gravity and hamiltonians unbounded from below. *Int. J. Mod. Phys. A*, **8** (1993b), 1259.

[49] AMBJØRN, J., KRISTJANSEN, C. F., AND MAKEENKO, Y. M. Higher genus correlations for the complex matrix model. *Mod. Phys. Lett. A*, **7** (1992), 3187.

[50] AMBJØRN, J. AND MAKEENKO, Y. M. Properties of loop equations for the hermitian matrix model and for two-dimensional quantum gravity. *Mod. Phys. Lett. A*, **5** (1990), 1753.

[51] AMBJØRN, J. AND THORLEIFSSON, G. A universal fractal structure of 2d quantum gravity for $c > 1$. *Phys. Lett. B*, **323** (1994), 7.

[52] AMBJØRN, J., THORLEIFSSON, G., AND WEXLER, M. New critical phenomena in 2-D quantum gravity. *Nucl. Phys. B*, **439** (1995), 187.

[53] AMBJØRN, J. AND VARSTED, S. Dynamical triangulated fermionic surfaces. *Phys. Lett. B*, **257** (1991), 305.

[54] AMBJØRN, J. AND VARSTED, S. Three-dimensional simplicial quantum gravity. *Nucl. Phys. B*, **373** (1992), 557.

[55] AMBJØRN, J. AND WATABIKI, Y. Scaling in quantum gravity. *Nucl. Phys. B*, **445** (1995), 129.

[56] ANDERSEN, H. H. Tensor products of quantized tilting modules. *Commun. Math. Phys.*, **149** (1992), 149.

[57] ATIYAH, M. Topological quantum field theories. *Pub. Math. IHES*, **68** (1989), 175.

[58] AXELROD, S., DELLAPIETRA, S., AND WITTEN, E. Geometric quantisation of Chern–Simons gauge theory. *J. Diff. Geom.*, **33** (1991), 787.

[59] BACHAS, C. AND PETROPOULOS, P. M. S. Topological models on the lattice and a remark on string theory cloning. *Commun. Math. Phys.*, **152** (1993), 191.

[60] BAIG, M., ESPRIU, D., AND TRAVESSET, A. Universality in the crumpling transition. *Nucl. Phys. B*, **426** (1994), 575.

[61] BAILLIE, C. AND JOHNSTON, D. A. Multiple Potts models coupled to two-dimensional quantum gravity. *Phys. Lett. B*, **286** (1992a), 44.

[62] BAILLIE, C. F., ESPRIU, D., AND JOHNSTON, D. A. Steiner variations on random surfaces. *Phys. Lett. B*, **305** (1993), 109.

[63] BAILLIE, C. F., IRBÄCK, A., JANKE, W., AND JOHNSTON, D. A. Scaling in Steiner random surfaces. *Phys. Lett. B*, **325** (1994), 45.

[64] BAILLIE, C. F. AND JOHNSTON, D. A. Modified Steiner functional string action. *Phys. Rev. D*, **45** (1992b), 3326.

[65] BAILLIE, C. F. AND JOHNSTON, D. A. A numerical test of KPZ scaling: Potts models coupled to two-dimensional quantum gravity. *Mod. Phys. Lett. A*, **7** (1992c), 1519.

[66] BAKKER, B. V. D. AND SMIT, J. Curvature and scaling in 4-d dynamical triangulation. *Nucl. Phys. B*, **439** (1995), 239.

[67] BARBER, M. N. Finite size scaling. In *Phase Transitions and Critical Phenomena, Vol. 8* (1983), C. Domb and J. L. Lebowitz, Eds., Academic Press, p. 145.

[68] BARS, I. AND GREEN, F. Complete integration of $U(n)$ lattice gauge theory in a large-N limit. *Phys. Rev. D*, **20** (1979), 3311.

[69] BARTOCCI, C., BRUZZO, U., CARFORA, M., AND MARZUOLI, A. Entropy of random coverings and 4-D quantum gravity. hep-th/9412097.

[70] BAUMANN, B. Noncanonical path and surface simulation. *Nucl. Phys. B*, **285** (1987), 391.

[71] BAUMANN, B. AND BERG, B. Non-trivial lattice random surfaces. *Phys. Lett. B*, **164** (1985), 131.

[72] BAUMANN, B., MUNSTER, G., AND BERG, B. A. Nontrivial critical behavior in a lattice model of random surfaces. *Nucl. Phys. B*, **305** (1988), 199.

[73] BELAVIN, A. A., POLYAKOV, A. M., AND ZAMOLODCHIKOV, A. B. Infinite conformal theory in two-dimensional quantum field theory. *Nucl. Phys. B*, **241** (1984), 333.

[74] BELIAKOVA, A. AND DURHUUS, B. On the relation between two quantum group invariants of 3-cobordisms. *J. Geom. Phys.*, to appear.

[75] BELIAKOVA, A. AND DURHUUS, B. Topological quantum field theory and invariants of graphs for quantum groups. *Commun. Math. Phys.*, **167** (1995), 395.

[76] BENDER, E. A. AND CANFIELD, E. R. The asymptotic number of rooted maps on a surface. *J. of Combinatorial Theory A*, **43** (1986), 244.

[77] BENDER, E. A., CANFIELD, E. R., AND ROBINSON, R. W. The enumeration of maps on the torus and the projective plane. *Canad. Math. Bull.*, **31** (1988), 257.

[78] BENDER, E. A. AND WORMALD, N. C. The asymptotic number of rooted nonseparable maps on a surface. *J. of Combinatorial Theory*, **49** (1988), 370.

[79] BERG, B. AND BILLOIRE, A. A Monte Carlo simulation of random surfaces. *Phys. Lett. B*, **139** (1984), 297.

[80] BERG, B., BILLOIRE, A., AND FOERSTER, D. Monte Carlo method for random surfaces. *Nucl. Phys. B*, **251** (1985), 665.

[81] BERG, B. AND FOERSTER, D. Random paths and random surfaces on a digital computer. *Phys. Lett.*, **106B** (1981), 323.

[82] BESSIS, D., ITZYKSON, C., AND ZUBER, J. B. Quantum field theory techniques in graphical enumeration. *Adv. Appl. Math.*, **1** (1980), 109.

[83] BIEDENHARN, L. C. An identity satisfied by Racah coefficients. *J. Math. Phys.*, **31** (1953), 287.

[84] BIEDENHARN, L. C. AND LOUCK, J. D. *The Racah–Wigner Algebra in Quantum Theory, Encyclopedia of Mathematics. vol. 9.* Addison-Wesley, 1981.

[85] BILLINGSLEY, P. *Convergence of Probability Measures.* John Wiley & Sons, New York, 1968.

[86] BILLOIRE, A. AND DAVID, F. Scaling properties of randomly triangulated planar random surfaces: a numerical study. *Nucl. Phys. B*, **275** (1986), 617.

[87] BINDER, K. Monte Carlo investigations of phase transitions and critical phenomena. In *Phase Transistion and Critical Phenomena (Vol. 5B)*, C. Domb and M. S. Green, Eds. Academic Press, London–New York–San Francisco, 1985.

[88] BINDER, K. Finite size effects at phase transitions. In *Schladming 1992 Proceedings, Computational methods in field theory* (1992a), p. 59.

[89] BINDER, K. *The Monte Carlo Method in Condensed Matter Physics.* Springer-Verlag, Berlin, 1992b.

[90] BIRMINGHAM, D., BLAU, M., RAKOWSKI, M., AND THOMPSON, G. Topological field theory. *Phys. Rep.*, **209** (1991), 129.

[91] BOERNER, H. *Representations of Groups.* North-Holland, Amsterdam, 1970.

[92] BOULATOV, D. V. AND KAZAKOV, V. A. The Ising model on a random planar lattice: the structure of the phase transition and the exact critical exponents. *Phys. Lett. B*, **186** (1987), 379.

[93] BOULATOV, D. V., KAZAKOV, V. A., KOSTOV, I. K., AND MIGDAL, A. A. Analytical and numerical study of a model of dynamically triangulated random surfaces. *Nucl. Phys. B*, **275** (1986a), 641.

[94] BOULATOV, D. V., KAZAKOV, V. A., KOSTOV, I. K., AND MIGDAL, A. A. Possible types of critical behaviour and the mean size of dynamically triangulated random surfaces. *Phys. Lett. B*, **174** (1986b), 87.

[95] BOULATOV, D. V. AND KRZYWICKI, A. On the phase diagram of three-dimensional simplicial quantum gravity. *Mod. Phys. Lett. A*, **6** (1991), 3005.

[96] BRÉZIN, E., ITZYKSON, C., PARISI, G., AND ZUBER, J. B. Planar diagrams. *Commun. Math. Phys.*, **59** (1978), 35.

[97] BREZIN, E. AND KAZAKOV, V. A. Exactly solvable field theories of closed strings. *Phys. Lett. B*, **236** (1990), 144.

[98] BREZIN, E., KAZAKOV, V. A., AND ZAMOLODCHIKOV, A. B. Scaling violation in a field theory of closed strings in one physical dimension. *Nucl. Phys. B*, **338** (1990), 673.

[99] BRINK, L., DIVECCHIA, P., AND HOWE, P. A locally supersymmetric and reparametrization invariant action for the spinning string. *Phys. Lett.*, **65B** (1976), 471.

[100] BRINK, L., DIVECCHIA, P., AND HOWE, P. A Lagrangian formulation of the classical and quantum dynamics of spinning particles. *Nucl. Phys.*, **B118** (1977), 76.

[101] BROCHARD, F. AND LENNON, J. F. Frequency spectrum of the flicker phenomenon in erythrocytes. *J. Physique*, **36** (1975), 1035.

[102] BRODA, B. Surgical invariants of four-manifolds. Preprint, 1993.

[103] BRUEGMANN, B. Nonuniform measure in four-dimensional simplicial quantum gravity. *Phys. Rev. D*, **47** (1993), 3330.

[104] BRUEGMANN, B. More on the exponential bound of four-dimensional simplicial quantum gravity. *Phys. Lett. B*, **349** (1995), 35.

[105] BRYDGES, D., FRÖHLICH, J., AND SPENCER, T. The random walk representation of classical spin systems and correlation inequalities. *Commun. Math. Phys.*, **83** (1982), 123.

[106] BRYDGES, D. AND SPENCER, T. Self-avoiding walk in 5 or more dimensions. *Commun. Math. Phys.*, **97** (1985), 125.

[107] BRYDGES, D. C., GIFFEN, C., DURHUUS, B., AND FRÖHLICH, J. Surface representation of Wilson loop expectations in lattice gauge theory. *Nucl. Phys. B*, **275** (1986), 459.

[108] CAIN, S. S. Triangulations of the manifold of class one. *Bull. Amer. Math. Soc.*, **41** (1935), 549.

[109] CARDY, J. AND GUTTMANN, A. J. Universal amplitude combinations for self-avoiding walks, polygons and trails. *J. Phys. A: Math. Gen.*, **26** (1993), 2485.

[110] CARDY, J. AND MUSSARDO, G. Universal properties of self-avoiding walks from two-dimensional field theory. *Nucl. Phys. B*, **410** (1993), 451.

[111] CARDY, J. L. Conformal invariance and the Yang–Lee edge singularity in two dimensions. *Phys. Rev. Lett.*, **54** (1985), 1354.

[112] CARDY, J. L. Conformal invariance. In *Phase, Transitions and Critical Phenoma (Vol. 11)*, C. Domb and J. L. Lebowitz, Eds. Academic Press, London–New York–San Francisco, 1987, p. 55.

[113] CARDY, J. L. Conformal invariance and statistical mechanics. In *Les Houches Summer School 1988* (1989), North Holland, p. 169.

[114] CASHER, A., FOERSTER, D., AND WINDEY, P. On the reformulation of the $d = 3$ Ising model in terms of random surfaces. *Nucl. Phys. B*, **251** (1985), 29.

[115] CATES, M. E. The Liouville field theory of random surfaces: When is the bosonic string a branched polymer? *Europhys. Lett.*, **8** (1988), 719.

[116] CATTERALL, S., KOGUT, J., AND RENKEN, R. Numerical study of $c > 1$ matter coupled to quantum gravity. *Phys. Lett. B*, **292** (1992), 44.

[117] CATTERALL, S., KOGUT, J., AND RENKEN, R. Phase structure of four-dimensional simplicial quantum gravity. *Phys. Lett. B*, **328** (1994), 277.

[118] CATTERALL, S., THORLEIFSSON, G., BOWICK, M., AND JOHN, V. Scaling and the fractal geometry of two-dimensional quantum gravity. *Phys. Lett. B*, **354** (1995), 58.

[119] CATTERALL, S. M., KOGUT, J. B., AND RENKEN, R. L. Scaling behavior of the Ising model coupled to two-dimensional quantum gravity. *Phys. Rev. D*, **45** (1992), 1519.

[120] CHANDRASEKHAR, S. Stochastic problems in physics and astronomy. *Rev. Mod. Phys.*, **15** (1943), 1.

[121] CHEEGER, J., MÜLLER, W., AND SCHRADER, R. On the curvature of piecewise flat spaces. *Commun. Math. Phys.*, **92** (1984), 405.

[122] CHEKHOV, L. Matrix models for moduli spaces. hep-th/9509001.

[123] CHEKHOV, L. Matrix model for discretized moduli space. *J. of Geometry and Physics*, **12** (1993), 153.

[124] CHUNG, S., FUKUMA, M., AND SHAPERE, A. Structure of topological lattice field theories in three dimensions. *Int. J. Mod. Phys. A*, **9** (1994), 1305.

[125] CRANE, L. AND FRENKEL, I. Four dimensional topological field theory, hopf categories and the canonical bases. *J. Math. Phys.*, **35** (1994), 5136.

[126] CRANE, L. AND YETTER, D. A categorical construction of 4D topological quantum field theories. preprint, 1994.

[127] DAS, S. R., DHAR, A., SENGUPTA, A. M., AND WADIA, S. R. New critical behaviour in $d = 0$ large-n matrix models. *Mod. Phys. Lett. A*, **5** (1990), 1041.

[128] DAVID, F. A model of random surfaces with non-trivial critical behaviour. *Nucl. Phys. B*, **257** (1985a), 543.

[129] DAVID, F. Planar diagrams, two-dimensional lattice gravity and surface models. *Nucl. Phys. B*, **257** (1985b), 45.

[130] DAVID, F. Randomly triangulated surfaces in two-dimensions. *Phys. Lett. B*, **159** (1985c), 303.

[131] DAVID, F. Conformal field theories coupled to 2-d gravity in the conformal gauge. *Mod. Phys. Lett. A*, **3** (1988), 1651.

[132] DAVID, F. Loop equations and non-perturbative effects in two-dimensional quantum gravity. *Mod. Phys. Lett. A*, **5** (1990), 1019.

[133] DAVID, F. Phases of the large-N matrix models and non-perturbative effects in 2D gravity. *Nucl. Phys. B*, **348** (1991), 507.

[134] DAVID, F. Self-avoiding random manifolds. hep-th/9511107, 1995.

[135] DAVID, F., DUPLANTIER, B., AND GUITTER, E. Renormalization and hyperscaling for self-avoiding manifold models. *Phys. Rev. Lett.*, **72** (1994), 311.

[136] DAVID, F., GUITTER, E., AND PELITI, L. Critical properties of fluid membranes with hexatic order. *J. Physique*, **48** (1987), 2059.

[137] DAVID, F., JURKIEWICZ, J., KRZYWICKI, A., AND PETERSSON, B. Critical exponents in a model of dynamically triangulated random surfaces. *Nucl. Phys. B*, **290** (1987), 218.

[138] DAWSON, K. A. Interfaces between phases in a lattice model of microemulsion. *Phys. Rev. A*, **35** (1987), 1766.

[139] DEULING, H. AND HELFRICH, W. Red blood cell shapes as explained on the basis of curvature elasticity. *Biophys. J.*, **16** (1976), 861.

[140] DIJKGRAAF, R., VERLINDE, E., AND VERLINDE, H. Loop equations and Virasoro contraints in non-perturbative 2-d quantum gravity. *Nucl. Phys. B*, **348** (1991), 435.

[141] DIJKGRAAF, R. AND WITTEN, E. Topological gauge theories and group cohomology. *Comm. Math. Phys.*, **129** (1990), 393.

[142] DIJKGRAAF, R. H. A geometrical approach to two-dimensional conformal field theory. PhD thesis, University of Utrecht, Utrecht, the Netherlands, 1989.

[143] DISTLER, J. AND KAWAI, H. Conformal field theory and 2-D quantum gravity. *Nucl. Phys. B*, **321** (1989), 509.

[144] DOMB, C. Self-avoiding random walk on lattices. In *Stochastic Processes in Chemical Physics*, K. E. Shuler, Ed. John Wiley and Sons, New York, 1969.

[145] DONALDSON, S. K. AND KRONHEIMER, P. B. *The Geometry of Four-manifolds*. Oxford University Press, Oxford, 1990.

[146] DOUGLAS, M. AND SHENKER, S. Strings in less than one-dimension. *Nucl. Phys. B*, **335** (1990), 635.

[147] DROUFFE, J. M., PARISI, G., AND SOURLAS, N. Strong coupling phase in lattice gauge theories at large dimension. *Nucl. Phys. B*, **161** (1980), 397.

[148] DURHUUS, B. Quantum theory of strings. Nordita preprint 82/36, 1982.

[149] DURHUUS, B. Critical properties of some discrete random surface models. In *Constructive Field Theory II*, A. Wightman, Ed. Springer-Verlag, Berlin–Heidelberg, 1989.

[150] DURHUUS, B. A discrete approach to topological quantum field theories. *J. Geometry and Phys.*, **11** (1993), 155.

[151] DURHUUS, B. Multi-spin systems on a randomly triangulated surface. *Nucl. Phys. B*, **426** (1994), 203.

[152] DURHUUS, B., FRÖHLICH, J., AND JONSSON, T. Critical properties of a model of planar random surfaces. *Phys. Lett.*, **137B** (1983a), 93.

[153] DURHUUS, B., FRÖHLICH, J., AND JONSSON, T. Self-avoiding and planar random surfaces on the lattice. *Nucl. Phys. B*, **225** (1983b), 185.

[154] DURHUUS, B., FRÖHLICH, J., AND JONSSON, T. Critical behaviour in a model of planar random surfaces. *Nucl. Phys. B*, **240** (1984), 453.

[155] DURHUUS, B., FRÖHLICH, J., AND JONSSON, T. Reflection positivity and tree inequalities in a theory of planar random surfaces. *Nucl. Phys. B*, **257** (1985), 779.

[156] DURHUUS, B., JAKOBSEN, H. P., AND NEST, R. Topological quantum field theories from generalized 6j-symbols. *Rev. Math. Phys.*, **5** (1993), 1.

[157] DURHUUS, B. AND JONSSON, T. Planar random surfaces with extrinsic action. *Phys. Lett. B*, **180**, 4 (1986), 385.

[158] DURHUUS, B. AND JONSSON, T. On subdivision invariant actions for random surfaces. *Phys. Lett. B*, **297** (1992), 271.

[159] DURHUUS, B. AND JONSSON, T. Classification and construction of unitary topological field theories in two dimensions. *J. Math. Phys.*, **35** (1994), 5306.

[160] DURHUUS, B. AND JONSSON, T. Remarks on the entropy of 3-manifolds. *Nucl. Phys. B*, **445** (1995), 182.

[161] DURHUUS, B., NIELSEN, H. B., OLESEN, P., AND PETERSEN, J. L. Dual models as saddle point approximations to Polyakov's quantized string. *Nucl. Phys. B*, **196** (1982a), 498.

[162] DURHUUS, B., OLESEN, P., AND PETERSEN, J. L. Polyakov's quantized string with boundary terms. *Nucl. Phys. B*, **198** (1982b), 157.

[163] DURHUUS, B., OLESEN, P., AND PETERSEN, J. L. Polyakov's quantized string with boundary terms II. *Nucl. Phys.*, **201** (1982c), 176.

[164] DVORETZKY, A., ERDÖS, P., AND KAKUTANI, S. Double points of paths of Brownian motion in n-space. *Acta Sci. Math. Szeged*, **12** (1950), 75.

[165] EGUCHI, T. AND KAWAI, H. The number of random surfaces on the lattice and the large-N limit. *Phys. Lett. B*, **110** (1980), 143.

[166] EGUCHI, T. AND KAWAI, H. Planar random surfaces on the lattice. *Phys. Lett.*, **114B** (1982), 247.

[167] EGUCHI, T. AND YANG, S. K. $N = 2$ superconformal models as topological field theories. *Mod. Phys. Lett. A*, **4** (1990), 1693.

[168] EGUCHI, T. AND YANG, S. K. On the genus expansion in topological string theory. *Rev. Math. Phys.*, **7** (1995), 279.

[169] EICHINGER, B. E. An approach to distribution functions for gaussian molecules. *Macromolecules*, **10** (1977), 671.

[170] ELLIOTT, J. P. Theoretical studies in nuclear structure V. The matrix elements of non-central forces with an application for the 2p-shell. *Proc. Royal Soc. (London)*, **A218** (1953), 370.

[171] FEINBERG, G., FRIEDBERG, R., LEE, T. D., AND REN, H. C. Lattice gravity near the continuum limit. *Nucl. Phys. B*, **245** (1984), 343.

[172] FELDER, G. AND GRANDJEAN, O. On combinatorial three-manifold invariants. ETH Zürich, preprint.

[173] FELLER, W. *An Introduction to Probability Theory and its Applications.* John Wiley & Sons, Inc., third edition, New York, 1968.

[174] FERGUSON, N. AND WHEATER, J. F. On the transition from crystalline to dynamically triangulated random surfaces. *Phys. Lett.*, **B 319** (1993), 104.

[175] FERGUSON, N. M. Continuous interpolations from crystalline to dynamically triangulated random surfaces. PhD thesis, Linacre College, University of Oxford, 1994.

[176] FERNANDEZ, R., FRÖHLICH, J., AND SOKAL, A. D. *Random Walks, Critical Phenomena and Triviality in Quantum Field Theory.* Springer-Verlag, Berlin–Heidelberg–New York, 1992.

[177] FISHER, M. E. Yang–Lee edge singularity and φ^3 field theory. *Phys. Rev. Lett.*, **40** (1978), 1610.

[178] FISHER, M. E., GUTTMANN, A. J., AND WHITTINGTON, S. G. Two-dimensional lattice vesicles and polygons. *J. Phys. A: Math. Gen.*, **24** (1991), 3095.

[179] FOERSTER, D. How to sum the planar diagrams: a reformulation of $U(N)$ lattice gauge theory, for $N \rightarrow \infty$, in terms of a statistical ensemble of non-interacting random surfaces. *Nucl. Phys. B*, **170** (1980), 107.

[180] FÖRSTER, D. On the scale dependence, due to thermal fluctuations, of the elastic properties of membranes. *Phys. Lett.*, **114A** (1986), 115.

[181] FRADKIN, E. S. AND TSEYTLIN, A. A. Renormalizable asymptotically free quantum theory of gravity. *Phys. Lett. B*, **104** (1981), 377.

[182] FRANCESCO, P. D., GINSPARG, P., AND ZINN-ZUSTIN, J. $2d$ gravity and random matrices. *Phys. Rep.*, **254** (1995), 1.

[183] FREED, D. S. AND QUINN, F. Chern–Simons theory with a finite gauge group. *Commun. Math. Phys.*, **156** (1993), 435.

[184] FREEDMAN, M. AND QUINN, F. *Topology of 4-manifolds*. Princeton University Press, Princeton, 1990.

[185] FRIEDBERG, R. AND LEE, T. D. Derivation of Regge's action from Einstein's theory of general relativity. *Nucl. Phys. B*, **242** (1984), 145.

[186] FRÖHLICH, J. On the triviality of $\lambda \phi^4$ theories and the approach to the critical point in $d \geq 4$ dimensions. *Nucl. Phys. B*, **200** (1982), 281.

[187] FRÖHLICH, J. Survey of random surface theory. In *Recent Developments in Quantum Field Theory*, J. Ambjørn, B. Durhuus, and J. L. Petersen, Eds. North Holland, Amsterdam, 1985.

[188] FRÖHLICH, J. The statistical mechanics of surfaces. Lectures Notes in Physics #216. In *Applications of Field Theory to Statistical Mechanics (Sitges, 1984)*, L. Garrido, Ed. Springer-Verlag, Berlin–Heidelberg–New York, 1986, p. 31.

[189] FRÖHLICH, J. AND KERLER, T. In *Quantum Groups, Quantum Categories and Quantum Field Theory. Lecture Notes in Physics* (Berlin–Heidelberg, New York, 1993), B. E. A. Dold and F. Takens, Eds., vol. 1542, Springer-Verlag.

[190] FRÖHLICH, J. AND KING, C. The Chern–Simons theory and knot polynimials. *Commun. Math. Phys.*, **126** (1989), 167.

[191] FRÖHLICH, J., PFISTER, C. E., AND SPENCER, T. On the statistical mechanics of surfaces. Lectures Notes in Physics #173. In *Stochastic Processes in Quantum Theory and Statistical Physics*. Springer-Verlag, Berlin–Heidelberg–New York, 1982, p. 169.

[192] FUKUMA, M., HOSONO, S., AND KAWAI, H. Lattice topological field theory in two dimensions. *Commun. Math. Phys.*, **161** (1994), 157.

[193] FUKUMA, M., KAWAI, H., AND NAKAYAMA, R. Continuum Schwinger–Dyson equations and universal structures in two-dimensional quantum gravity. *Int. J. Mod. Phys. A*, **6** (1991), 1385.

[194] FUKUMA, M., KAWAI, H., AND NAKAYAMA, R. Infinite dimensional grassmannian structure of two-dimensional quantum gravity. *Commun. Math. Phys.*, **143** (1992), 371.

[195] FUKUMA, N., ISHIBASHI, N., KAWAI, H., AND JOHN, V. Two-dimensional quantum gravity in temporal gauge. *Nucl. Phys. B*, **427** (1994), 139.

[196] GAREY, M. R. AND JOHNSON, D. S. *A Guide to the Theory of NP-completeness.* W.H. Freeman & Co., San Francisco, 1979.

[197] GERASIMOV, A., MAKEENKO, Y., MARSHAKOV, A., MIRONOV, A., MOROZOV, A., AND ORLOV, A. Matrix models as integrable systems: from universality to geometrodynamical principle of string theory. *Mod. Phys. Lett. A,* **6** (1991), 3079.

[198] GERVAIS, J.-L. AND NEVEU, A. Non-standard 2d critical statistical models from Liouville theory. *Nucl. Phys.,* **B257** (1985), 59.

[199] GIBBONS, G. W. AND HAWKING, S. W. Action integrals and partition functions in quantum gravity. *Phys. Rev. D,* **15** (1977), 2753.

[200] GIBBONS, G. W., HAWKING, S. W., AND PERRY, M. J. Path integrals and the indefiniteness of the gravitational action. *Nucl. Phys. B,* **138** (1978), 141.

[201] GINSPARG, P. AND ZINN-JUSTIN, J. 2-d gravity and 1-d matter. *Phys. Lett. B,* **240** (1990), 333.

[202] GLAUS, U. Monte Carlo simulation of self-avoiding surfaces in three dimensions. *Phys. Rev. Lett.,* **56** (1986), 1996.

[203] GLAUS, U. AND EINSTEIN, T. L. On the universality class of planar self-avoiding surfaces with fixed boundary. *J. Phys. A: Math. Gen.,* **20** (1987), L105.

[204] GLIMM, J. AND JAFFE, A. *Quantum Physics: A functional integral point of view.* Springer-Verlag, New York, 1981 (2nd edition 1987).

[205] GODFREY, N. AND GROSS, M. Simplicial quantum gravity in more than two dimensions. *Phys. Rev. D,* **43** (1991), R1749.

[206] GOTO, T. Relativistic quantum mechanics of one-dimensional mechanical continuum and subsidiary condition of dual resonance model. *Prog. Theor. Phys.,* **46** (1971), 1560.

[207] GREEN, M., SCHWARZ, J., AND WITTEN, E. *Superstring Theory.* Cambridge University Press, 1987.

[208] GREENSITE, J. AND HALPERN, M. Stabilizing bottomless action theories. *Nuclear Physics B,* **242** (1984), 167.

[209] GROSS, D. AND MIGDAL, A. A. Nonperturbative two-dimensional quantum gravity. *Phys. Rev. Lett.,* **64** (1990), 127.

[210] GROSS, D. AND NEWMAN, M. J. Unitary and hermitian matrices in an external field (II). The Kontsevich model and continuum Virasoro constraints. *Nucl. Phys. B,* **380** (1992), 168.

[211] GROSS, D. J. AND MILIKOVIĆ, N. A nonperturbative solution of D = 1 string theory. *Phys. Lett. B,* **238** (1990), 217.

[212] GROSS, D. J., PIRAN, T., AND WEINBERG, S., Eds. *Two-dimensional Quantum Gravity and Random Surfaces,* vol. 8. World Scientific, Singapore, 1992.

[213] GROSS, M. AND VARSTED, S. Elementary moves and ergodicity in d-dimensional simplicial quantum gravity. *Nucl. Phys. B,* **378** (1992), 367.

[214] GUBSER, S. S. AND KLEBANOV, I. R. Scaling functions for baby universes in two-dimensional quantum gravity. *Nucl. Phys. B,* **416** (1994), 827.

[215] GUITTER, E. AND KARDAR, M. Tethering, crumpling, and melting transitions in hexatic membranes. *Europhys. Lett.,* **13** (1990), 441.

[216]   HAMBER, H. W. Phases of four-dimensional simplicial quantum gravity. *Phys. Rev. D*, **45** (1992), 507.

[217]   HAMBER, H. W. Phases of simplicial quantum gravity in four-dimensions: estimates for the critical exponents. *Nucl. Phys. B*, **400** (1993), 347.

[218]   HAMMERSLEY, J. M. The number of polygons on a lattice. *Proc. Cambridge Phil. Soc.*, **57** (1961), 516.

[219]   HARARY, F. *Graph Theory*. Addison-Wesley, Reading, 1969.

[220]   HARNISH, R. G. AND WHEATER, J. F. The crumpling transition of crystalline random surfaces. *Nucl. Phys. B*, **350** (1991), 861.

[221]   HARRIS, M. G. AND WHEATER, J. F. Multiple Ising spins coupled to 2d quantum gravity. *Nucl. Phys. B*, **427** (1994), 111.

[222]   HARTLE, J. B. Simplicial minisuperspace I. General discussion. *J. Math. Phys.*, **26** (1985), 804.

[223]   HARTLE, J. B. AND HAWKING, S. W. Wave function of the Universe. *Phys. Rev. D*, **28** (1983), 2960.

[224]   HASSLACHER, B. AND PERRY, M. J. Spin networks are simplicial quantum gravity. *Phys. Lett.*, **103B** (1981), 21.

[225]   HAWKING, S. W. Spacetime foam. *Nucl. Phys. B*, **144** (1978), 349.

[226]   HAWKING, S. W. The path-integral approach to quantum gravity. In *General Relativity. An Einstein centenary survey* (1979), S. W. Hawking and W. Israel, Eds., Cambridge University Press, p. 746.

[227]   HELFRICH, W. Elastic properties of lipid bilayers: Theory and possible experiments. *Z. Naturforsch.*, **28C** (1973), 693.

[228]   HELFRICH, W. Effect of thermal modulations on the rigidity of fluid membranes and interfaces. *J. Physique*, **46** (1985), 1263.

[229]   HELFRICH, W. Size distribution of vesicles: The role of the effective rigidity of membranes. *J. Physique*, **47** (1986), 321.

[230]   HELFRICH, W. Measures of integration in calculating the effective rigidity of fluid membranes. *J. Physique*, **48** (1987), 285.

[231]   HELGASON, S. *Groups and Geometric Analysis*. Academic Press, New York, 1984.

[232]   HERMANNS, J. J. AND ULLMAN, R. The statistics of stiff chains with applications to light scattering. *Physica*, **18** (1952), 951.

[233]   HUSE, D. A. AND LEIBLER, S. Phase behaviour of an ensemble of nonintersecting random fluid films. *J. Physique*, **49** (1988), 605.

[234]   INCE, E. L. *Ordinary Differential Equations*. Dover Publications, New York, 1956.

[235]   ISAACSON, D. The continuum limit of a classical 3-component, one-dimensional Heisenberg model is Brownian motion on the surface of a sphere. *J. Math. Phys.*, **23** (1982), 138.

[236]   ISHIBASHI, N. AND KAWAI, H. String field theory of $c \leq 1$ noncritical strings. *Phys. Lett. B*, **322** (1994), 67.

[237]   ISHIBASHI, N. AND KAWAI, H. A background independent formulation of noncritical string theory. *Phys. Lett. B*, **352** (1995), 75–82.

[238]   ITZYKSON, C. Ising fermions. 2. three-dimensions. *Nucl. Phys. B*, **210** (1982), 477.

[239] ITZYKSON, C. AND DROUFFE, J.-M. *Statistical Field Theory*. Cambridge University Press, New York, 1989.

[240] ITZYKSON, C., KIRKPATRICK, S., PARISI, G., SOURLAS, N., AND VIRASORO, M. A., Eds. *Common Trends in Statistical Physics and Field Theory* (1989), vol. 184 of *Phys. Rep.*

[241] ITZYKSON, C. AND ZUBER, J.-B. The planar approximation. 2. *J. Math. Phys.*, **21** (1980), 411.

[242] ITZYKSON, C. AND ZUBER, J.-B. Combinatorics of the modular group. 2. the Kontsevich integrals. *Int. J. Mod. Phys. A*, **7** (1992), 5661.

[243] JAIN, S. AND MATHUR, S. World-sheet geometry and baby universes in 2D quantum gravity. *Phys. Lett. B*, **286** (1992), 239.

[244] JAROSZEWICZ, T. AND KURZEPA, P. S. Spin, statistics and geometry of random walks. *Annal of Physics*, **210** (1991), 255.

[245] JONSSON, T. Ornstein–Zernike theory for the planar random surface model. *Commun. Math. Phys.*, **106** (1986), 679.

[246] JONSSON, T. A lower bound on the size of crystalline surfaces. *Phys. Lett. B*, **221** (1989), 35.

[247] JONSSON, T. Random walk representation of propagators for particles with spin. In *Probabilistic Methods in Quantum Field Theory and Quantum Gravity*, P. H. Damgaard, H. Hüffel, and A. Rosenblum, Eds. Plenum Press, London–New York, 1990.

[248] JONSSON, T. Topological lattice theories in two dimensions. *Phys. Lett. B*, **265** (1991), 141.

[249] JONSSON, T. AND WHEATER, J. The phase diagram of an Ising model on a polymerized random surface. *Phys. Lett. B*, **345** (1995), 227.

[250] JURKIEWICZ, J., KRZYWICKI, A., AND PETERSSON, B. A numerical study of discrete Euclidean Polyakov surfaces. *Phys. Lett. B*, **168** (1986a), 273.

[251] JURKIEWICZ, J., KRZYWICKI, A., AND PETTERSSON, B. A grand-canonical ensemble of randomly triangulated surfaces. *Phys. Lett. B*, **177** (1986b), 89.

[252] KANTOR, Y., KARDAR, M., AND NELSON, D. R. Statistical mechanics of tethered surfaces. *Phys. Rev. Lett.*, **57** (1986), 791.

[253] KANTOR, Y., KARDAR, M., AND NELSON, D. R. Tethered surfaces: Statics and dynamics. *Phys. Rev. A*, **35** (1987), 3056.

[254] KANTOR, Y. AND NELSON, D. R. Crumpling transistion in polymerized membranes. *Phys. Rev. Lett.*, **58** (1987), 2774.

[255] KANTOR, Y. AND NELSON, D. R. Phase transition in flexible polymeric surfaces. *Phys. Rev. A*, **36** (1987), 4020.

[256] KARDAR, M. AND NELSON, D. R. ϵ-expansion for crumpled manifolds. *Phys. Rev. Lett.*, **58** (1987), 1289.

[257] KAROWSKI, M. AND SCHRADER, R. A combinatorial approach to topological quantum field theories and invariants of graphs. *Commun. Math. Phys.*, **151** (1993), 355.

[258] KAUFFMAN, L. Computing Turaev–Viro invariants for 3-manifolds. *Manuscripta Mathematica*, **72** (1991a), 81.

[259] KAUFFMAN, L. *Knots and Physics*. World Scientific, Singapore, 1991b.

[260] KAUFFMAN, L. q-Deformed spin-networks and the Turaev–Viro invariants for 3-manifolds. *Int. J. Mod. Phys. B*, **6** (1992), 1765.

[261] KAUFFMAN, L. AND LINS, S. L. *Temperley–Lieb Recoupling Theory and Invariants of 3-manifolds*. Princeton University Press, Princeton, N.J., 1994.

[262] KAVALOV, A. R. AND SEDRAKYAN, A. G. The sign factor of the three-dimensional Ising model and the quantum fermionic string. *Phys. Lett. A*, **123** (1986), 449.

[263] KAWAI, H., KAWAMOTO, N., MOGAMI, T., AND WATABIKI, Y. Transfer matrix formalism for two-dimensional quantum gravity and fractal structures of space-time. *Phys. Lett. B*, **306** (1993), 19.

[264] KAWAI, H., KITAZAWA, Y., AND NINOMIYA, M. Ultraviolet stable fixed point and scaling relations in $(2 + \varepsilon)$-dimensional quantum gravity. *Nucl. Phys. B*, **404** (1994), 684.

[265] KAWAI, H. AND NINOMIYA, M. Renormalization group and quantum gravity. *Nucl. Phys. B*, **336** (1990), 115.

[266] KAWAI, H. AND OKAMOTO, Y. Entropy of planar random surfaces on the lattice. *Phys. Lett.*, **130B** (1983), 415.

[267] KAWAMOTO, N., KAZAKOV, V., SAEKI, Y., AND WATABIKI, Y. Fractal structure of two-dimensional gravity coupled to d=-2 matter. *Phys. Rev. Lett.*, **68** (1992), 2113.

[268] KAZAKOV, V. A. Bilocal regularization of models of random surfaces. *Phys. Lett.*, **150B** (1985), 282.

[269] KAZAKOV, V. A. Ising model on a dynamical planar random lattice: Exact solution. *Phys. Lett. A*, **119** (1986), 140.

[270] KAZAKOV, V. A. The appearance of matter fields from quantum fluctuations of 2D-gravity. *Mod. Phys. Lett. A*, **4** (1989), 2125.

[271] KAZAKOV, V. A., KOSTOV, I. K., AND MIGDAL, A. A. Critical properties of randomly triangulated planar random surfaces. *Phys. Lett.*, **157B** (1985), 295.

[272] KAZAKOV, V. A. AND MIGDAL, A. A. Recent progress in the theory of noncritical strings. *Nucl. Phys. B*, **311** (1988), 171.

[273] KESTEN, H. On the number of self-avoiding walks. *J. Math. Phys.*, **4** (1963), 960.

[274] KHARCHEV, S., MARSHAKOV, A., MIRONOV, A., MOROZOV, A., AND ZABRODIN, A. Towards unified theory of 2d gravity. *Nucl. Phys. B*, **380** (1992), 181.

[275] KIRBY, R. AND SIEBENMANN, L. *Foundational Essays on Topological Manifolds, Smoothings and Triangulations*. Princeton University Press, Princeton M. J., 1977.

[276] KLEINERT, H. The membrane properties of condensing strings. *Phys. Lett.*, **174B** (1986), 335.

[277] KNIZHNIK, V. G., POLYAKOV, A. M., AND ZAMOLODCHIKOV, A. B. Fractal structure of 2d-quantum gravity. *Mod. Phys. Lett. A*, **3** (1988), 819.

[278] KONTSEVICH, M. L. Intersection theory on the moduli space of curves and the matrix Airy function. *Funkt. Anal. Pril.*, **25** (1991), 50.

[279] KOPLIK, J., NEVEU, A., AND NUSSINOV, S. Some aspects of the planar perturbation series. *Nucl. Phys. B*, **123** (1977), 109.

[280] KORCHEMSKY, G. P. Loops in the curvature matrix model. *Phys. Lett. B,* **296** (1992), 323.

[281] LABASTIDA, J., PERNICI, M., AND WITTEN, E. Topological gravity in two-dimensions. *Nucl. Phys. B,* **310** (1988), 611.

[282] LAWLER, G. F. A self-avoiding random walk. *Duke Math. J.,* **47** (1980), 655.

[283] LAWLER, G. F. The probability of intersection of independent random walks in four dimensions. *Commun. Math. Phys.,* **86** (1982), 539.

[284] LEIBLER, S., SINGH, R. R. P., AND FISHER, M. E. Thermodynamic behaviour of two-dimensional vesicles. *Phys. Rev. Lett.,* **59** (1987), 1989.

[285] LICKORISH, W. B. R. Three-manifold invariants and the Temperley–Lieb algebra. *Math. Ann.,* **290** (1991), 657.

[286] MAKEENKO, Y. Loop equations in matrix models and in 2-d gravity. *Mod. Phys. Lett. A,* **6** (1991), 1901.

[287] MAKEENKO, Y., MARSHAKOV, A., MIRONOV, A., AND MOROZOV, A. Continuum versus discrete Virasoro in one matrix models. *Nucl. Phys. B,* **356** (1991), 574.

[288] MAKEENKO, Y. AND MIGDAL, A. A. Quantum chromodynamics as dynamics of loops. *Nucl. Phys. B,* **188** (1981), 269.

[289] MARINARI, E. AND PARISI, G. The supersymmetric one-dimensional string. *Phys. Lett. B,* **240** (1990), 375.

[290] MARITAN, A. AND OMERO, C. The $N \to 0$ limit of a non-abelian gauge theory: A model for self-avoiding random surfaces. *Phys. Lett.,* **109B** (1982), 51.

[291] MARITAN, A. AND STELLA, A. Scaling behaviour of self-avoiding random surfaces. *Phys. Rev. Lett.,* **53** (1984), 123.

[292] MARITAN, A. AND STELLA, A. Some exact results for self-avoiding random surfaces. *Nucl. Phys. B,* **280** (1987), 561.

[293] MARKOV, A. A. Insolubility of the problem of homeomorphy. In *Proc. of the Int. Congr. of Math. 1958* (1960), Cambridge University Press, p. 300.

[294] MARSHAKOV, A., MIRONOV, A., AND MOROZOV, A. From Virasoro constraints in Kontsevich's model to W constraints in two matrix model. *Mod. Phys. Lett. A,* **7** (1992a), 1345.

[295] MARSHAKOV, A., MIRONOV, A., AND MOROZOV, A. On equivalence of topological and quantum 2-d gravity. *Phys. Lett. B,* **274** (1992b), 280.

[296] MARTELLINI, M., SPREAFICO, M., AND YOSHIDA, K. A generalized model for two-dimensional quantum gravity and dynamics of random surfaces for $c > 1$. *Mod. Phys. Lett. A,* **9** (1994), 2009.

[297] MAZUR, P. O. AND MOTTOLA, E. The gravitational measure, solution of the conformal factor problem and stability of the ground state of quantum gravity. *Nucl. Phys. B,* **341** (1990), 187.

[298] COTTA-RAMUSINO, P., GUADAGNINI, E., MARTELLINI, M., AND MINTCHEV, M. Quantum field theory and link invariants. *Nucl. Phys. B,* **330** (1990), 557.

[299] DE GENNES, P. AND TAUPIN, C. Microemulsions and the flexibility of oil/water interfaces. *J. Phys. Chem.,* **86** (1982), 2294.

[300] MERMIN, N. D. Absence of ordering in certain classical systems. *J. Math. Phys.,* **8** (1967), 1061.

[301] MICOVIC, A. AND SIEGEL, W. Random superstrings. *Phys. Lett. B*, **240** (1990), 363.

[302] MIGDAL, A. A. Loop equations and $1/N$ expansion. *Phys. Rep.*, **102** (1983), 199.

[303] MOISE, E. E. *Geometric Topology in Dimension 2 and 3*. Springer-Verlag, Berlin–Heidelberg–New York, 1977.

[304] MORRIS, T. Checkered surfaces and complex matrices. *Nucl. Phys. B*, **356** (1991), 703.

[305] MOTTOLA, E. Functional integration over geometries. *J. Math. Phys.*, **36** (1995), 697.

[306] MÜTTER, K. H. A random surface representation for correlation functions in Z(2) lattice gauge theory. *Nucl. Phys.B*, **251** (1985), 735.

[307] NABUTOVSKY, A. AND BEN-AV, R. Noncomputability arising in dynamical triangulation model of four-dimensional quantum gravity. *Commun. Math. Phys.*, **157** (1993), 93.

[308] NAMBU, Y. Quark model and the factorization of the Veneziano amplitude. In *Symmetries and Quark Models*, R. Chand, Ed. Gordon and Breach, New York, 1970, p. 269.

[309] NELSON, D., PIRAN, T., AND WEINBERG, S., Eds. *Statistical Mechanics of Membranes and Surfaces*. World Scientific, Singapore, 1989.

[310] NELSON, D. R. AND PELITI, L. Fluctuations in membranes with crystalline and hexatic order. *J. Physique*, **48** (1987), 1085.

[311] NIELSEN, H. B. AND NINOMIYA, M. Absence of neutrinos on a lattice. *Nucl. Phys. B*, **185** (1981), 20.

[312] NOVIKOV, P. S. On the algorithmic unsolvability of the word problem in group theory. *Trudy. Math. Inst. Steklov*, **44** (1955), 143.

[313] OOGURI, H. Partition functions and topology-changing amplitudes in the three-dimensional lattice gravity of Ponzano and Regge. *Nucl. Phys. B*, **382** (1992), 276.

[314] OOGURI, H. Topological lattice models in four dimensions. *Mod. Phys. Lett. A*, **7** (1992), 2799.

[315] OOGURI, H. AND SASAKURA, N. Discrete and continuum approaches to three-dimensional quantum gravity. *Mod. Phys. Lett. A*, **6** (1991), 3591.

[316] ORLAND, P. World sheet action for the three-dimensional Ising model. *Phys. Rev. Lett.*, **59** (1987), 2393.

[317] OSTERWALDER, K. AND SCHRADER, R. Axioms for Euclidean Green's functions. I. *Commun. Math. Phys.*, **31** (1973), 83.

[318] OSTERWALDER, K. AND SCHRADER, R. Axioms for Euclidean Green's functions. II. *Commun. Math. Phys.*, **42** (1975), 281.

[319] OTTO, H. J. AND WEIGT, G. Construction of exponential Liouville field operators for closed string models. *Zeit. Phys.*, **C31** (1986), 219.

[320] PACHNER, U. Bistellare Äquivalenz kombinatorischer Mannigfaltigkeiten. *Arch. Math.*, **30** (1978), 89.

[321] PELITI, L. Effective rigidity of membranes. *Physica A*, **140** (1986), 269.

[322] PELITI, L. AND LEIBLER, S. Effects of thermal fluctuations on systems with small surface tension. *Phys. Rev. Lett.*, **54** (1985), 1690.

[323] PISARSKI, R. Field theory of paths with a curvature dependent term. *Phys. Rev. D*, **34** (1986), 670.

[324] PLYUSHCHAY, M. S. Massive relativistic point particle with rigidity. *Int. J. Mod. Phys. A*, **4** (1989a), 3851.

[325] PLYUSHCHAY, M. S. Massless point particle with rigidity. *Mod. Phys. Lett. A*, **4** (1989b), 837.

[326] POLYAKOV, A. M. Quantum geometry of bosonic strings. *Phys. Lett.*, **103B** (1981), 207.

[327] POLYAKOV, A. M. Fine structure of strings. *Nucl. Phys. B*, **268** (1986), 406.

[328] POLYAKOV, A. M. *Gauge Fields and Strings*. Harwood Academic Publishers, Chur, 1987.

[329] PONZANO, G. AND REGGE, T. Semiclassical limit of Racah coefficients. In *Spectroscopic and Group Theoretical Methods in Physics*, F. Bloch, Ed. North-Holland, Amsterdam, 1968, p. 1.

[330] QUINN, F. Lectures on axiomatic topological quantum field theory. In *Geometry and Quantum Field Theory. (Vol. 1)*, D. S. Freed and K. K. Uhlenbeck, Eds. AMS, IMS, 1995, p. 323.

[331] REGGE, T. General relativity without coordinates. *Il Nuovo Cimento*, **19** (1961), 558.

[332] RESHETIKHIN, N. AND TURAEV, V. G. Invariants of 3-manifolds via link polynominals and quantum groups. *Invent. math.*, **103** (1991), 547.

[333] ROBERTS, J. *Skein theory and Viro–Turaev invariants*. Cambridge University, preprint, 1993.

[334] ROCEK, M. AND VAN NIEWENHUISEN, P. $N \geq 2$ supersymmetric Chern–Simons terms as $d = 3$ extended conformal supergravity. *Class. Quantum Grav.*, **3** (1986), 43.

[335] ROSSO, M. Finite dimensional representations of the quantum analog of the enveloping algebra of a complex simple Lie algebra. *Commun. Math. Phys.*, **117** (1988), 581.

[336] ROURKE, C. P. AND SANDERSON, B. J. *Introduction to Piecewise-linear Topology*. Springer-Verlag. Berlin–Heidelberg–New York, 1972.

[337] RUBINSTEIN, J. H. An algorithm to recognize the 3-sphere. Lecture at ICM, Zürich, 1994.

[338] RUELLE, D. *Statistical Mechanics*. W. A. Benjamin Inc., Reading, 1977.

[339] SAFRAN, S. A. *Statistical Thermodynamics of Surfaces, Interfaces, and Membranes*. Addison-Wesley, Reading, 1994.

[340] SAITO, N., TAKAHASHI, K., AND YUNOKI, Y. The statistical mechanical theory of stiff chains. *J. Phys. Soc. Japan*, **22** (1967), 219.

[341] SASAKURA, N. Tensor model for gravity and orientability of manifold. *Mod. Phys. Lett. A*, **6** (1991), 2613.

[342] SAVVIDY, G. K. AND SAVVIDY, K. G. Self-avoiding surfaces and spin systems. *Phys. Lett. B*, **324** (1994), 72.

[343] Sawin, S. Direct sum decompositions and indecomposable TQFTs. *J. Math. Phys.*, **36** (1995), 6673.

[344] Scherk, J. An introduction to the theory of dual models and strings. *Rev. Mod. Phys.*, **47** (1975), 123.

[345] Schrader, R. String tension and glueball mass in a lattice theory of disconnected, selfintersecting random surfaces. *Commun. Math. Phys.*, **102** (1985), 31.

[346] Seifert, U. Shape transformation of free, toroidal and bound vesicles. *Colloque de Physique, Colloque C7*, **23** (1990), 339.

[347] Silvestrov, P. G. 2-d gravity from a 1-d matrix model: double scaling limit. *Phys. Lett. B*, **276** (1992), 445.

[348] Simon, B. *Functional Integration and Quantum Physics*. Academic Press, New York, 1979.

[349] Spitzer, F. *Principles of Random Walk*. van Nostrand, New York, 1964.

[350] Staudacher, M. The Yang–Lee edge singularity on a dynamical planar random surface. *Nucl. Phys. B*, **336** (1990), 349.

[351] Stelle, K. S. Renormalization of higher-derivative quantum gravity. *Phys. Rev. D*, **16** (1977), 953.

[352] Sterling, T. and Greensite, J. Entropy of self-avoiding surfaces on the lattice. *Phys. Lett.*, **121B** (1983), 345.

[353] Stillwater, J. The word problem and the isomorphism problem for groups. *Bull. Amer. Math. Soc.*, **6** (1982), 23.

[354] Swensen, R. H. and Wang, J.-S. Nonuniversal critical dynamics in Monte Carlo simulations. *Phys. Rev. Lett.*, **58** (87), 86.

[355] Symanzik, K. Euclidean quantum field theory. In *Local Quantum Theory*, R. Jost, Ed. Academic Press, London–New York–San Francisco, 1969.

[356] 't Hooft, G. A planar diagram theory for the strong interactions. *Nucl. Phys. B*, **72** (1974), 461.

[357] Talmon, Y. and Prager, S. Statistical thermodynamics of phase equilibria in microemulsions. *J. Chem. Phys.*, **69** (1978), 2984.

[358] Talmon, Y. and Prager, S. Statistical thermodynamics of microemulsions. II. The interfacial region. *J. Chem. Phys.*, **76** (1982), 1535.

[359] Thompson, A. Thin position and the recognition problem for S^3. *Math. Res. Lett.*, **1** (1994), 613.

[360] Turaev, V. G. Quantum invariants of links of 3-manifolds. *Pub. Math. IHES*, **77** (1993), 121.

[361] Turaev, V. G. *Quantum Invariants of Knots and 3-manifolds*. Walter de Gruyter, 1994.

[362] Turaev, V. G. and Viro, O. State sum invariants of 3-manifolds and quantum $6j$-symbols. *Topology*, **31** (1992), 865.

[363] Tutte, W. T. A census of planar triangulations. *Can. J. Math.*, **14** (1962), 21.

[364] van Baal, P. An introduction to topological gauge theory. *Acta Phys. Polon. B*, **21** (1990), 73.

[365] van Rensburg, E. J. J. Surface in the hypercubic lattice. *J. Phys. A Math. Gen.*, **25** (1992), 3529.

[366] van Rensburg, E. J. J. Statistical mechanics and topology of surfaces in Z^d. *J. Knot Theory and its Ramifictions*, **3** (1994), 365.

[367] van Rensburg, E. J. J. and Whittington, S. G. Self-avoiding surfaces. *J. Phys. A: Math. Gen.*, **22** (1989), 4939.

[368] Varsted, S. Four-dimensional gravity by dynamical triangulations. *Nucl. Phys. B*, **412** (1994), 406.

[369] Volodin, I. A., Kuznetsov, M. E., and Fomensko, A. T. The problem of discriminating algorithmically the standard three-dimensional sphere. *Russ. Math. Surveys*, **29** (1974), 71.

[370] Wadia, S. Dyson–Schwinger equations approach to the large N limit : Model systems and string representation of Yang–Mills theory. *Phys. Rev. D*, **24** (1981), 970.

[371] Walker, K. On Witten's 3-manifold invariant. preprint, 1991.

[372] Watabiki, Y. Construction of noncritical string field theory by transfer matrix formalism in dynamical triangulation. *Nucl. Phys. B*, **441** (1995), 119.

[373] Weinberg, S. Ultraviolet divergences in quantum theories of gravitation. In *General Relativity. An Einstein centenary survey* (1979), S. W. Hawking and W. Israel, Eds., Cambridge University Press, p. 790.

[374] Weingarten, D. Pathological lattice field theory for interacting strings. *Phys. Lett. B*, **90** (1980), 280.

[375] Weingarten, D. Euclidean quantum gravity on a lattice. *Nucl. Phys. B*, **210** (1982), 229.

[376] Wexler, M. Critical behavior of target space mean field theory. *Mod. Phys. Lett. A*, **8** (1993a), 2703.

[377] Wexler, M. Matrix models on large graphs. *Nucl. Phys. B*, **410** (1993b), 377.

[378] Wheater, J. F. Topology and two-dimensional lattice gauge theories. *Phys. Lett.*, **B264** (1991), 161.

[379] Wheater, J. F. Random surfaces: From polymer membranes to strings. *J. Phys. A*, **27** (1994), 3323.

[380] Wheater, J. F. The critical exponents of crystalline random surfaces. *Nucl. Phys. B*, **458** (1995), 671.

[381] Widow, B. A model of microemulsion. *J. Chem. Phys.*, **81** (1984), 1030.

[382] Widow, B., Dawson, K. A., and Liplein, M. D. Hamiltonian and phenomenological models of microemulsion. *Physica*, **140A** (1986), 26.

[383] Wiegmann, P. B. Extrinsic geometry of superstrings. *Nucl. Phys. B*, **323** (1989), 330.

[384] Wigner, E. *Proc. Cambridge Philos. Soc.*, **47** (1951), 790.

[385] Wigner, E. *Ann. Math.*, **53** (1951), 36.

[386] Williams, R. M. Discrete quantum gravity: the Regge calculus approach. *Int. J. Mod. Phys. B*, **6** (1992a), 2097.

[387]   WILLIAMS, R. M. Regge calculus: a bibliography and brief review. *Class. Quant. Grav.*, **9** (1992b), 1409.

[388]   WILSON, K. Confinement of quarks. *Phys. Rev. D*, **10** (1974), 2445.

[389]   WITTEN, E. 2 + 1 dimensional gravity as exactly soluble system. *Nucl. Phys. B*, **311** (1988a), 46.

[390]   WITTEN, E. Topological quantum field theory. *Commun. Math. Phys.*, **117** (1988b), 353.

[391]   WITTEN, E. Quantum field theory and the Jones polynomial. *Commun. Math. Phys.*, **121** (1989), 351.

[392]   WITTEN, E. On the structure of the topological phase of two-dimensional quantum gravity. *Nucl. Phys. B*, **340** (1990), 281.

[393]   WITTEN, E. Introduction to cohomological field theories. *Int. J. Mod. Phys.*, **A6** (1991a), 2775.

[394]   WITTEN, E. On quantum gauge theories in two dimensions. *Commun. Math. Phys.*, **141** (1991b), 153.

[395]   WOLFF, U. Critical slowing down. *Nucl. Phys. B (Proc. Suppl.)*, **17** (1990), 93.

[396]   YANG, C. N. AND LEE, T. D. Statistical theory of equations of state and phase transitions II. Lattice gas and Ising model. *Phys. Rev.*, **87** (1952), 404.

[397]   YETTER, D. N. Topological quantum field theories associated to finite groups and crossed G-sets. *J. Knot Theory and its Ramifications*, **1** (1992), 1.

[398]   ZAMOLODCHIKOV, A. B. On the entropy of random surfaces. *Phys. Lett.*, **117B** (1982), 87.

Index